John Henry Gurney, Charles John Andersson

Notes on the Birds of Damara Land

And the Adjacent Countries of South-west Africa

John Henry Gurney, Charles John Andersson

Notes on the Birds of Damara Land
And the Adjacent Countries of South-west Africa

ISBN/EAN: 9783337125684

Printed in Europe, USA, Canada, Australia, Japan

Cover: Foto ©Andreas Hilbeck / pixelio.de

More available books at **www.hansebooks.com**

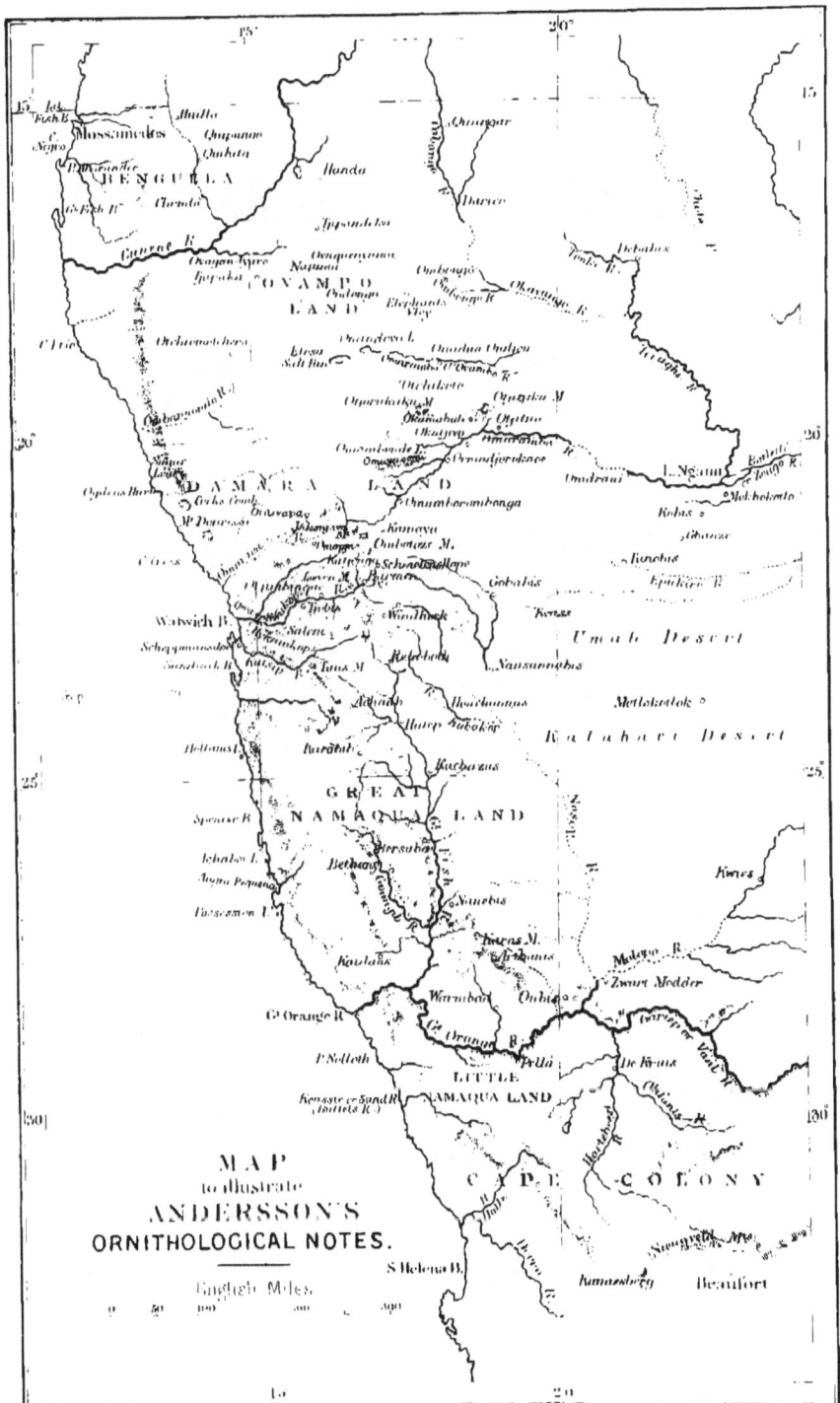

NOTES

ON

THE BIRDS OF DAMARA LAND

AND THE ADJACENT COUNTRIES OF

SOUTH-WEST AFRICA.

BY THE LATE
CHARLES JOHN ANDERSSON.
AUTHOR OF 'LAKE NGAMI' AND OF 'THE OKAVANGO RIVER.'

ARRANGED AND EDITED BY
JOHN HENRY GURNEY,

WITH SOME ADDITIONAL NOTES BY THE EDITOR, AND AN INTRODUCTORY
CHAPTER CONTAINING

A SKETCH OF THE AUTHOR'S LIFE,
ABRIDGED FROM THE ORIGINAL PUBLISHED IN SWEDEN.

LONDON:
JOHN VAN VOORST, PATERNOSTER ROW.
MDCCCLXXII.

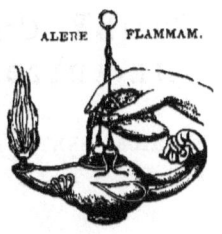

PRINTED BY TAYLOR AND FRANCIS,
RED LION COURT, FLEET STREET.

INTRODUCTORY CHAPTER,
BY THE EDITOR,

INCLUDING A BIOGRAPHICAL SKETCH OF THE AUTHOR
(TRANSLATED FROM THE SWEDISH).

In introducing to the public the following ornithological notes by my late friend Charles John Andersson, I purpose, in the first instance, to lay before my readers a short biographical sketch of Mr. Andersson's career which appeared in a Swedish periodical, the 'Svenska Jägarförbundets nya Tidskrift' for 1868, and for a translation of which I am indebted to my valued friend L. Lloyd, Esq., of Gothenburgh.

The details given in this sketch will enable the reader to judge of the circumstances under which Mr. Andersson made his ornithological observations, and will be followed by some remarks, in the form of a preface, relating chiefly to the mode which I have adopted in arranging the results of those observations as contained in his MS. notes, for the purpose of publication in the present volume.

BIOGRAPHICAL SKETCH.

Charles John Andersson was born in Wermeland † in 1827, was educated at the public high school in Wenersborg, and was afterwards a student at the University of Lund for a single term. In early life he evinced a great passion for the chase, natural history, and travel, * * * *; and, according to what he himself relates, his aspirations were at an early period turned towards the deserts of Africa.

† A province of Sweden.

In 1849 he may be said to have commenced his travels. Towards the end of that year he proceeded to England to dispose of many specimens (living and dead) of natural history that he had for some time past been collecting, purposing thereby to raise sufficient funds to enable him to visit distant countries. His most ardent wish, as mentioned above, was to proceed to Africa; but this project he, for the time, was obliged to forego, as the cost of such an expedition quite exceeded his means. Already, indeed, he had determined on visiting Iceland, when an altogether unexpected circumstance enabled him to carry into execution his original and favourite scheme.

David Livingstone had in 1849 commenced his wonderful discoveries in Africa. By means of the Kalahari desert he found his way to the " Lake Ngami," which had long previously been ascertained to exist, but to which no European had then penetrated, thereby making known to the world that to the north of South-western Africa lay a well-watered country, rich in vegetable and animal life, and especially in elephants. This intelligence, which promised so well for the naturalist, the man of business, and the hunter, created in England and elsewhere the greatest sensation. Amongst those seized with a desire to follow in the track of Livingstone was Francis Galton, who, as early as the first months of 1850, began to make preparations for a journey to " Ngami." Andersson met with this gentleman in London; and it was soon arranged between them that they should travel in company, Galton agreeing to bear all the expenses of the journey.

The needful outfit completed, Galton and Andersson sailed from England in the month of April 1850, and arrived at the Cape the following June. From thence, in the commencement of August, they proceeded by sea to Walvisch Bay, on the south-west coast, which they reached towards the end of the month. The starting-point of their projected journey was the missionary station of Scheppmansdorf, situated some twenty miles to the eastward of the bay in question. Galton was desirous that their little expedition should be thoroughly well equipped in every respect—more especially as regarded draft- and saddle-oxen, which were not to be had thereabouts, the Damaras in that part of the country being as poor in regard to cattle as in every thing else. The first journey of the travellers into the interior was, therefore, chiefly for the purpose of purchasing oxen trained as well to the saddle as to the yoke.

During this excursion they were exposed to all the hardships and inconveniences that usually follow a wandering life in South-western Africa—namely, burning heat, extreme thirst, worn-out draft-oxen, and attacks by lions, which either devoured their horses and cattle, or drove them to a distance from the encampment.

After several days' journey through Namaqua Land, and when they had reached the confines of that country, they arrived at the missionary station of Richterfeldt. Here they succeeded in procuring from Hans Larsen, a Dane, who was resident there, the requisite number of oxen, as also in inducing him to accompany them during their future wanderings. This was the greatest good fortune that travellers, unaccustomed to life in the wilderness, could have met with; for, independently of his having a perfect knowledge of the country, he could turn his hand to almost any thing—as for example, to train oxen either for riding or draft, repairing a waggon when needful, the discovery of water in the desert, &c. He was, moreover, a hunter of the first order, and possessed of herculean strength, an iron constitution, and a *sang-froid* that nothing could move. Andersson, indeed, admits that he and Galton could with difficulty have traversed the wilderness had they not been accompanied by Hans.

They now retraced their steps to Scheppmansdorf. The last preparations for the journey into the interior were made; and on the 13th of November they set forward. It had thus taken some three months from their first landing in Walvisch Bay to get all things in proper order.

Their course in the first instance was directed to Richterfeldt, from whence they made an excursion in a south-easterly direction to Barmen, another missionary station, at a distance of some days' journey from Richterfeldt.

On their return Andersson had his first serious adventure with a lion. At a short distance he had shot at the beast and lodged a ball in its body, when, finding itself wounded, it faced about and prepared to attack him. Andersson fell on one knee in readiness to fire his second barrel, but, being unable to obtain a proper view of its head, he delayed discharging his gun. The beast in the meanwhile suddenly made its spring, but, happily, overshot its mark, and passed clean over its opponent. With good reason Andersson called his escape "a lucky" one, as when making its bound the lion seldom miscalculates the distance. Some few days after the above occurrence the beast in question was found dead, and that at only a short distance from the scene of conflict.

From one cause or another their movements were slow. It was not before the beginning of March 1851 that Galton and Andersson left Schmelen's Hope, situated near the river Swakop, to set out in good earnest on their journey of discovery. Their first plan of visiting Ngami had been abandoned, because Galton determined on proceeding to another and nearer lake (the Omanbondé), of which he had received wonderful accounts from the natives. After somewhat more than a month's travel through part of Damara Land never previously explored, they reached the lake in question, which bitterly disappointed the expectations of both travellers. A dried-up mere, altogether without water, about an English mile in length, and partly overgrown with reeds, was the only reward for months of labour and anxiety.

After a short sojourn at this so-called lake, in the vicinity of which Andersson had, for the first time, the pleasure of seeing tracks of the elephant, the travellers resolved on attempting to make their way to Ovampo Land, the nearest country to the northward, of which, and of its inhabitants, the Damaras had given them extraordinary descriptions.

Their route lay through a wild and desolate country, of which they were the first to give an account. On the way it unfortunately happened that the axle-tree of their waggon snapped asunder, which compelled them to leave the vehicle, together with a number of their people, behind, and to prosecute the remainder of their journey with saddle- and pack-oxen.

In the beginning of May they arrived at Ovampo, the residence of Nangoro, the chief or king; and, during a six weeks' stay there, Andersson had opportunity to make his observations on the country and its inhabitants.

It was a question if they should extend their explorations to the not far-distant Cunéné River, then all but unknown; but, partly owing to the disabled state of their oxen, and partly to the many difficulties thrown in the way by the natives, this project was given up, and nothing remained for them but to retrace their steps to the disabled waggon that had been left under the charge of Hans, and which, during their absence, he had succeeded in repairing; and from thence they proceeded to the missionary station of Barmen, which they reached in the beginning of August.

Galton was now wearied out with the inconveniences attendant on the life of a traveller, and determined on proceeding to Europe by the first opportunity that offered from Walvisch Bay. As, however, no vessel was expected there until December, and they consequently had several months at their disposal, they determined on making an excursion to the eastward, where game was reported to be much more abundant than in the countries they had hitherto traversed.

After a fatiguing journey the travellers arrived on the 3rd of October at Tunobis, where Andersson for the first time became aware of the riches of the African animal world. Here was found, he says, an incredible number of wild animals; and he mentions, as an instance, that he, together with others, killed, in the course of a few days, thirty rhinoceroses, and that he himself, when quite alone, shot in one night no fewer than eight of those beasts, consisting of three different species, together with other large game.

After they had hunted to their hearts' content at Tunobis they returned to Walvisch Bay, where Galton, in the early part of 1852, engaged his passage in a ship bound for England, taking with him at the same time Andersson's collection of specimens of natural history, amongst which were the skins of about five hundred birds.

The plan abandoned by Galton of penetrating from the westward to Ngami was now taken up by Andersson. This, nevertheless, required a considerable outfit; and although Galton furnished him with many needful supplies, there were many more of which he stood in need. His waggon and such other things as he thought he could do without, he therefore exchanged for cattle, and with these proceeded by land to Cape Town, where he disposed of them, and with the proceeds purchased goods to barter with the natives, scientific instruments, and other requisites.

Prior to parting with Hans at Walvisch Bay it should be remarked he had entered into partnership with him, they agreeing to share alike in any profits that might accrue in the barter trade which they purposed carrying on with the tribes in the interior.

These preparations for his intended journey occupied Andersson a whole year: and therefore it was not before the end of January 1853 that he returned by sea to Walvisch Bay. In the month of May we find him again at Tunobis, the most easterly point previously reached by him when in company with Galton; but, to the great chagrin and disappointment both of himself and of his numerous and half-starved followers, no game was then to be found at that place.

During his stay there Andersson made disagreeable acquaintance with one of the deep pit-falls dug by the natives for the capture of large game, into which he and his horse tumbled headlong, but from which, happily, both rider and steed extricated themselves without very serious injury.

In consequence of the information Andersson received from the Damaras as to the difficulty of proceeding further with waggons, he left these vehicles at Tunobis and proceeded on his journey to the eastward with pack- and saddle-oxen. The supply of game now became abundant. At Kobis, one of the nearest stations to Ngami, Andersson had, he himself tells us, his surfeit of shooting. On this and many other occasions he adopted a system of hunting that in South-west Africa, during the dry season, is especially successful, namely, to lie in ambush at night near to some pool. During the daytime the larger animals are dispersed over a wide tract of country, sometimes of many miles in extent; but at night they resort to the water to quench their thirst; and if at such times the hunter knows his business he has the opportunity of obtaining much large game. These night hunts, however, are attended with greater peril than those by day. Andersson was accustomed to ensconce himself in a so-called skärm or screen—that is, a small circular enclosure six or eight feet in diameter, the walls usually consisting of loose stones, being about two feet in height; but this afforded him scarcely any protection, and he must besides, if he would count on a sure shot, allow the beast to approach to within a few paces before firing. We believe that the hunter is never so unprotected against savage animals as in such nocturnal combats. Andersson, indeed, on the first night of his stay at Kobis, was on three several occasions in imminent peril of his life. First came an elephant without his being aware of its approach, and with lowered trunk stood directly over him; that he could save himself as he did by throwing himself backwards on to the ground, and discharging his piece upwards at random is what could only happen once in a thousand times. A while afterwards he shot at and wounded a black rhinoceros; and when subsequently he left the skärm to look after another of those animals he had fired at and struck, he was fiercely attacked by the first rhinoceros, cast headlong to the earth, and had his right thigh ripped up. Lastly, when at sunrise he attempted to aid his boy, Kamapyu, who, whilst searching for his master, was attacked by the same beast, Andersson again escaped death as by a miracle; for just as he was on the point of being impaled on its sharp horn, the rhinoceros fell dead from its numerous wounds. For such a rapid succession of hair-breadth escapes it would be difficult to find a parallel.

A few days afterwards Andersson arrived at Lake Ngami,— "The object of my ambition," he writes*, "for a number of

---

* 'Lake Ngami,' p. 431. Mr. Andersson adds to the passage quoted in the text the following characteristic reflections:—

"The first sensation occasioned by this sight was very curious. Long as I had been prepared for the event, it now almost overwhelmed me. It was a mixture of pleasure and pain. My temples throbbed and my heart beat so violently that I was obliged to dismount, and lean against a tree for support, until the excitement had subsided. The reader will no doubt think that thus giving way to my feelings was very childish; but 'those who know that the first glimpse of some great object which we have read or dreamt of from earliest recollection is ever a moment of intensest enjoyment, will forgive the transport.' I felt unfeignedly thankful for the unbounded goodness and gracious assistance which I had experienced from Providence throughout the whole of this prolonged and perilous journey. My trials had been many; but, my dearest aspirations being attained, the difficulties were all forgotten. And here I could not avoid passing my previous life in review. I had penetrated into deserts almost unknown to civilized man; had suffered the extremity of hunger and thirst, cold and heat, and had undergone desperate toil, sometimes nearly in solitude, and often without shelter during dreary nights in vast wildernesses haunted by beasts of prey. My companions were mostly savages. I was exposed to numerous perils by land and by water, and endured torments from wounds inflicted by wild animals; but I was mercifully preserved by the Creator through the manifold dangers that hovered round my path. To Him are due all homage, thanksgiving, and adoration."

years, and for which I had forsaken home and friends, and staked my life."

Though he had so far succeeded, yet after all, he had only reached the starting-point of his explorations, Livingstone and Oswell having four years previously proceeded thus far. His ardent desire was to follow up discoveries which with them ended here, and as a naturalist and geographer to explore the lake. He therefore commenced the ascent of the Teoughe, a river that falls into Lake Ngami, where he found a vegetable and animal kingdom far exceeding in richness any thing he had previously met with in Africa. It was here that Andersson had first an opportunity of hunting the buffalo; and it was here he also met with the remarkable Tsetse-fly (*Glossina morsitans*), whose bite is fatal to domesticated horned cattle, horses, and sheep, but perfectly innocuous to animals in a wild state. In many districts, indeed, the fly prevents the keeping of other tame animals than the ass and the goat ; and it at the same time presents an almost insurmountable obstacle to journeys of discovery, the traveller losing both his saddle- and draft-oxen shortly after entering countries infested by this pernicious insect.

Andersson was not permitted to penetrate very far to the north of Ngami. For his supplies he was altogether dependent on the chief of the district he was traversing; and when he had proceeded eleven short days' journey up the Teoughe, all further support was withdrawn from him. He could neither obtain a boat, people, nor a guide, and was therefore compelled to return the way he came to the lake—a sad result of his labours, more especially as a rich and new field of discovery lay before him. More than once subsequently his projects were frustrated owing to similar obstacles being thrown in his way by the natives.

Andersson, during his sojourn at Ngami, devoted much of his time to studying the habits of the people and to researches in his favourite pursuit, natural history, and was thus enabled to make a large collection of zoological specimens.

When at length a waggon was required to convey these to the Cape, together with a large quantity of ivory which he had collected, he set off for Namaqua Land, accompanied by a single attendant, to procure one of these vehicles. It was a hurried journey of four months' duration; and probably neither on any former or subsequent occasion did he suffer more from hunger, thirst, the burning rays of the sun, and over fatigue than he did on this journey. Bearing in mind all these hardships and sufferings, which, thanks to the elasticity of his spirit and his still unbroken strength of body, he was happily enabled to endure, one cannot wonder at his exclaiming, in the words of a former traveller, "To journey in Africa requires the endurance of a camel and the strength of a lion."

This little expedition was not without its sporting results. One night he chanced to fall asleep in his skärm, when his mind became impressed with a confused sense of danger: whilst between sleeping and waking he could not make out the nature of the peril; but on coming fully to himself he distinctly heard the breathing of an animal immediately near his place of concealment, and a sound somewhat resembling the purring of a cat. A lion had crept close up to him as quietly as possible, but still not unnoticed by his dangerous neighbour. Andersson seized his gun, which was lying ready close by his side, aimed at the dark mass before him, and fired. The beast's roarings and convulsive movements showed plainly that the ball had told. It was not, however, until daylight that Andersson ventured forth from the skärm to ascertain the effect of his shot, when he found, to his great satisfaction, the lion lying dead at no great distance.

In the early part of 1854 Andersson repaired to Cape Town, and shortly afterwards visited Europe, where he published his well known work 'Lake Ngami.' At the end of 1856 he was back again at the Cape, for the purpose, as previously agreed upon, of joining his friend Frederick Green in an expedition intended as well for hunting as for the exploration of the river Cunéné.

As, however, that gentleman was then absent in the interior of Damara Land (in company with the late Mr. Wahlberg, who was shortly afterwards killed by an infuriated elephant), and was not expected to return for some time, Andersson undertook the superintendence of some recently discovered copper-mines on the confines of Namaqua Land; and when therefore his friend returned to the Cape, which was much sooner than had been anticipated, and set off on the journey to the Cunéné, he was unable to accompany him—a circumstance which at the time caused much annoyance to Andersson, but which, as the event proved, was little to be regretted, as, owing to the enmity of the Ovampo, the expedition proved altogether unsuccessful.

But at the expiration of a year, the mines in question having proved a failure, Andersson was free from his engagement; and he then determined on attempting to carry out the object that Green had failed in accomplishing.

In March 1858 he started from Objimbinque, situated on the river Swakop, which subsequently became his chief residence, after having by his own great exertions and the kind assistance he received from the missionaries fortunately succeeded in procuring the needful outfit. The course he took, which was through Western Damara Land, was not perhaps the most favourable, as he had to traverse a country where both game and water were known to be scarce; but it was shorter than the eastern route, and he hoped for the best.

For two months the expedition moved forward, though but slowly, owing partly to the rough nature of the ground and the dense brakes through which they were often obliged to cut their way with the axe, and partly to the frequent detours they were compelled to make in search of water. They persevered, nevertheless, and at length reached the Ovatjionba country, where they procured guides, who, however, soon lost themselves in the desert. They had been two days without water; and though the guides were threatened with death, they were unable to direct them to the precious liquid. On the third day Andersson and his people separated, and searched in different directions for water; but not a single drop was to be found in the then totally burnt-up country. To escape, therefore, from perishing by thirst no other alternative was left them but to retrace their steps, which they commenced doing without further delay; and at length, worn out with fatigue and languishing for water, of which they had then been destitute for six whole days, the caravan found its way back to Okova fountain.

Andersson now saw the impossibility of proceeding through western Damara Land to the river Cunéné. He nevertheless did not give up the project of penetrating to that river, and only altered his plans so far as to proceed in an easterly direction, where it was known that running water was to be met with.

Shortly afterwards, however, the waggon broke down and Andersson was under the necessity of sending it to Objimbinque for the needful repairs, with orders to the people in charge of it to rejoin the expedition with the vehicle at a distant point on the river Omuramba-Matako; and subsequently the journey was continued with saddle- and pack-oxen.

Game was now met with in abundance, and Andersson was very successful in hunting. One day he shot two giraffes, though not aware of having killed more than a single one. It happened thus:—The first he fired at fell within a very short distance; but as another giraffe started off from the very same spot he imagined it to be the same animal, and pursued it; and it was not until this had also succumbed, that he discovered the mistake.

He was now enabled to renew his night hunts. After long fatiguing marches during burning hot days, it was truly no child's play thus to lie out on the watch; but in this kind of chase he nevertheless took special delight. "A moonlight ambush by a pool well frequented by wild animals," he writes*, "is worth all the other modes of enjoying a gun put together. In the first place there is something mysterious and thrilling in finding one's self the secret and unsuspected spectator of the wild movements, habits, and propensities of the denizens of nature's varied and wonderful menagerie—no high feeding, no barred gates, no harsh and cruel keeper's voice having yet enervated, damped, or destroyed the elasticity, buoyancy, and frolicsomness of animal life. And then the intense excitement between each expected arrival! The distant footstep, now heard distinctly rattling over a rugged surface, now gently vibrating on the strained ear as it treads on softer ground; it may be that of a small antelope or an elephant, of a wild boar or rhinoceros, of a gnu or a giraffe, of a jackal or a lion. And then what opportunities present themselves of observing the habits and peculiarities of each species, and even of individuals, to say nothing of the terrible battles that take place and can so rarely be witnessed in the daytime. I have certainly learnt more of the untamed life of savage beasts in a single night's *tableau vivant* than during months of toilsome wanderings in the broad light of the sun."

By slow stages, during which Andersson had ample opportunity of using his rifle, they followed the course of the Matako River, and toward the end of August were rejoined by the people from Objimbinque with the repaired waggon. Andersson, however, derived little benefit from the vehicle; as it was soon found impossible to drag it through so rough a country as that they were then traversing. For a second time therefore he was obliged to leave it behind, and to proceed as before with saddle- and pack-oxen.

On September 16th Andersson reached Omanbondé, the so-called lake with which he made acquaintance on his first

* The Okavango River, p. 85.

journey in Africa, when in company with Galton; but it was with peculiar feelings that he revisited the spot. "It was now close upon eight years and a half," he says*, "since I was first here. Eight years and a half, the fifth part of a man's life in its full vigour! What was I at the beginning of this period, and what am I now? Where are the once ruddy cheeks? Where is that elasticity of foot and spirit that made me laugh at hardships and dangers? Where that giant health and strength that enabled me to vie with the natives in enduring the extremes of heat and cold? Gone! gone! ay, for ever! The spirit still exists unsubdued; but what with constant care, anxiety, and exposure, the power of performance has fled, leaving but the shadow of my former self. What have I accomplished during these long years? what is the result of all this toil, this incessant wear and tear of body and mind? The answer, if candid, must be, apparently, very little. This is a sad retrospect of the fifth part of a man's life, whilst still in the pride of manhood. And yet I feel I have not been idle, that I have done as much as any man under similar circumstances could have done; and so with this poor consolation I must rest content."

Omanbondé now contained much more water than when first visited by Andersson; and, as a consequence, much game was met with in the vicinity, such as the eland, the koodoo, the pallah, the quagga, the elephant, the lion, and the rhinoceros. During a hunt, however, one of his attendants was unfortunately killed by one of the last-named beasts, that had been previously wounded, which threw a damp over the whole party; with this exception all Andersson's hunts ended happily, and he was enabled to supply his numerous and hungry followers with abundance of food. He himself lived sumptuously. "One day" †, he says, " I dined on beef-steak, on lion, and hump 'de rhinocéros,' done in the ashes * * * *   I had never

* The Okavango River, p. 113.   † Ibid. p. 130.

before partaken of lion's flesh, but found it very palatable and juicy, not unlike veal, and very white. Rhinoceros-hump was a frequent and favourite dish of mine."

The stay of the expedition at Omanbondé was somewhat long, because Andersson, prior to plunging into an unknown country, wished to send to Objimbinque the large quantity of ivory he had collected, and to await the return of the messengers. It was therefore not until the commencement of January 1859 that he broke up the encampment and again set forward, but this time with the waggon in company.

During the journey, they came to a district where elephants very greatly abounded. "I had often heard the natives say, on being questioned as to the haunts of these animals," writes Andersson*, "that in certain localities they walked about as thick as cattle; but never till now had I been able to verify this apparently exaggerated statement. I did so at present; for the whole country in the neighbourhood of the vley (or piece of water) lately referred to, with the adjacent plain, was literally one network of elephant footprints. The trees and bushes, moreover, were so broken that one might fairly suppose a large army had just traversed the veldt. During the daytime they were not visible, but at night their shrill trumpetings would frequently startle us from our sleep. If, instead of exploring, I had turned my attention exclusively to elephant-hunting, I might have had magnificent sport, and profit too. The temptation was strong, but I considered it ignoble, however great the allurements I had to resist, to swerve from a predetermined purpose for the sake of gain and personal gratification." At length, and after a long and fatiguing journey, Andersson reached, not the Cunéné as he had intended, but another considerable river, the Okavango, which may be looked on as the greatest of his geographical discoveries, and the name of which serves as a title to the second of his published works. Having so long struggled through deep

* The Okavango River, p. 175.

sandy wastes, a broken country, and tangled underwood, it was a magnificent sight to him thus, in the heart of the continent, to view a noble stream, at least 1200 feet in breadth, but which, to his astonishment, he found to flow to the eastward instead of to the westward.

His ardent desire was now to follow up this mighty discovery; but the same adverse fate that had already more than once mocked his best endeavours even now crossed his path. He and nearly all his people were suddenly struck down with a dangerous fever, to which several of them shortly succumbed; and even when he himself began slowly to recover, many of his followers were still on the brink of the grave. He could not, therefore, longer hesitate in retreating from this most interesting, but to him most melancholy region. "To linger where I was," writes he*, "seemed certain death; and any visions of future success I might still entertain were too remote to justify me in so fearfully imperilling the lives of my fellow creatures. A precipitate retreat appeared therefore quite imperative. It cost, nevertheless, a severe struggle between duty and ambition before I could resolve upon it. I obeyed at last the monitions of conscience, and bade, with a sigh, farewell to the pursuit of fame and glory for ever. That this act of self-renunciation was not determined on without acute pangs it would be useless to deny. After such toils! such hardships! such sacrifices! and with the prospect of a final crowning success just dawning on me, it may well be imagined that I turned my back on the land of promise with drooping spirits and a heavy heart."

Andersson was thus unable to ascertain the further course of the Okavango—not even whether, as is most probable, it is connected with the river-system of the Lake Ngami; so far as we know this problem remains still unsolved.

On their return journey, which commenced in the beginning of June, and during which but slow progress was made, owing

* The Okavango River, p. 218.

to the weak state to which they were all reduced by fever, new difficulties arose. After proceeding five or six days in a southerly direction, Andersson was compelled to command a halt in consequence of intelligence having reached him, through parties he had sent forward to reconnoitre, that the country in front was untraversable owing to the scarcity of water; and subsequently information of a still more serious nature was brought him namely, that an attack on his encampment was imminent on the part of the Ovampo (the tribe that in 1857 made an onslaught on Green and his party). Stuck fast, so to say, as Andersson now was in the wilderness, his situation about the middle of November was very critical.

Fortunately the Ovampo had taken a long time in making their preparations; and intelligence of their murderous intentions in regard to Andersson had in the meantime reached Green, then on his return from an expedition to the eastward of the Ngami.

On learning there their evil designs, this gentleman instantly resolved on proceeding to the assistance of his friend, or, in the worst event, of avenging his death. That his movements might be more rapid he separated from his companions and started off all but alone; and though the dangers and difficulties of the route were great, they were overcome, and the travellers had once more the pleasure of embracing each other.

It now only remained to gain Green's encampment; for when their forces were united it was hardly probable the Ovampo would venture on an attack, as the prospect of success would then be small indeed. Without delay, therefore, Andersson broke up his camp, and commenced the march. They had, however, a five days' march before them through a fearful desert, but succeeded in accomplishing it without accident of any kind. "For that while"*, says Andersson, "we had little to satisfy hunger or thirst. Its happy issue was, under Providence,

* The Okavango River, p. 246.

mainly owing to the cheerful cooperation and indefatigable exertion of Green. Nothing less than his energy could have given us so prompt and, comparatively, so cheap a success."

Without any thing remarkable occurring our travellers afterwards proceeded in company to Objimbinque, which they reached in safety. * * * *

Arrived at Objimbinque, Andersson occupied himself in preparing for the press his second work, 'The Okavango River,' and afterwards proceeded to the Cape, where, in 1861, he was married to Miss Aitchison.

His health being now much impaired, owing to the hardships he had undergone during his several journeys, and more especially by the fever he caught near the river Okavango, there was no longer a possibility of his resuming a wandering hunter's life. He therefore determined on devoting himself to trading-pursuits, for which purpose he settled at Objimbinque, where he provided the native and other hunters with what they needed for the chase, and received from them in return ivory, ostrich-feathers, and cattle. At first he prospered, but after a time, owing to a deadly feud between the Damaras and Namaquas, the tide turned against him. Andersson, who dwelt in the country of the former, took their part and assisted them both by word and deed. The Namaquas, to revenge themselves on him, seized his convoys one after the other, and at length, with arms in their hands, made an attack on Objimbinque; Andersson, who had put himself at the head of the Damaras, was wounded during the battle that ensued, by a ball that shattered one of his knees—and being deserted by his cowardly Damara allies, with difficulty saved his life. His wound confined him long to his bed, and ultimately rendered him a cripple for the remainder of his days.

Some months afterwards he repaired to Cape Town, where he chiefly occupied himself in the composition of a new work, entitled "The Avifauna of South-west Africa." During his many years' abode there he had devoted much of his time and attention to this object; and numerous beautiful drawings of birds, principally from specimens in his own collection, were made by himself and by his friend, the well-known traveller and artist Mr. Baines, for the illustration of his intended work. * * * *

In May 1866, Andersson again returned to Damara Land, and resumed his barter-trade with the natives and others. But, as the Namaquas continued their persecutions, he determined, with his usual indomitable resolution (although then a cripple, and although his constitution was fairly broken down), on making a journey to the Cunéné for the purpose of ascertaining if further to the northward a suitable locality for a trading-station might be found, where he would at least be free from the robberies of his enemies.

Accompanied by a young Swede named Axel Ericson, Andersson, in May 1867, left Objimbinque and set off on this long and perilous journey; and though he was then so worn out as hardly to be able to sit on his horse, he, together with several followers, succeeded, though deserted by his native guide, in reaching the river in question.

Hardly had he arrived there, however, before his illness became so serious as to cause him to determine on retracing his steps. But it was not ordained that he should ever again rejoin his friends; for on his way through Ovampo Land, on July 5th, 1867, his eyes were closed in death; and in the country of the tribe Wagaambi his remains were interred in a sand-hill by his Swedish friend and follower Axel Ericson.

As an appropriate supplement to the foregoing biographical sketch, I may here add an interesting extract from the annual address of the Chairman of the Cape-Town Public Library for the year 1868:—

"Whilst Livingstone and others have been upraising the

veil which has so long obscured South-eastern Africa, Charles John Andersson and others have been performing similar work for South-western Africa: Andersson, in his quiet, modest, but most persistent and (considering all the difficulties and obstacles he had to encounter), I will add, heroic manner, has made us as familiar with the maps of Damara Land and Ovampo Land and onwards towards the Cunéné, as with those of the Free State, or Natal, or the Cape Colony itself.

What could be more touching than the record of his last fatal journey, told so well a few months ago by his friend and fellow-explorer Mr. Frederick Green? There amid the wilds of Ondonga, tended only by his faithful servant, baffled and prostrate, he calmly prepared himself to die; and, after farewell thoughts of his wife and children and home, he begged to have read to him the Psalms in his native Swedish, which in childhood he had learnt, and which in the supreme moment were now his consolation and his stay.

The University of Lund have only by this last mail conferred on him the Degree of Doctor of Philosophy, a fitting acknowledgment of the service he has rendered to the march of civilization and of science; but months before, he had already passed beyond the reach of earthly distinction, and finished a career which will, however, still give honour to his memory as one of the best and bravest of the pioneers of progress in South Africa. I speak of him as I knew him; and I could not make mention of his name at all without offering this tribute at once of warm affection and high esteem."

The extract above cited fitly closes the narrative of Charles John Andersson's career; and I will only add to it the expression of my persuasion that he was one of whom it may with truth be said that he habitually

"Looked through Nature, up to Nature's God,"

thus cultivating a tone of thought which most effectually enhances the delight of studying the works of the Almighty Creator, who has stamped upon the face of nature the impress of his own infinity, and of whom we are assured that "his tender mercies are over all his works"*, and that before him not even one Sparrow "is forgotten"†.

J. H. G. (1872).

* Psalm cxlv. ver. 9.　　　† Luke, ch. xii. ver. 6.

PREFACE BY THE EDITOR.

The four hundred and twenty eight species of birds enumerated in the following pages have been observed either as residents, as migrants, or as accidental visitors in that part of South-western Africa of which Damara Land is the central portion, but which also comprises Ovampo Land to the north, the district adjacent to Lake Ngami to the north-east, a portion of the Kalahari Desert to the south-east, Great and Little Namaqua Land to south, and the coast and adjacent islets of the Atlantic to the west.

All the species included in this volume have been identified by myself from an examination of specimens obtained in the above-mentioned countries, except in those cases where the contrary is specifically mentioned*.

The specimens so identified were chiefly contained in the collection left by Mr. Andersson at his decease, and subsequently sent to London for the purpose of being disposed of; but some were seen by me in other collections of Mr. Andersson's which reached England in previous years.

In this work of identification I have been exceedingly indebted to the assistance of my friends Mr. R. B. Sharpe, Mr. J. E. Harting, M. Jules Verreaux, Mr. Layard, my late much-regretted friend Mr. G. R. Gray, and other eminent ornithologists, whose kind aid is more particularly acknowledged in the following pages. To the two first-named

---

* A memorandum to the above effect, however, was accidentally omitted in the case of *Circus maurus* and of *Circus ranivorus*, pp. 33 and 34.

gentlemen I am under especial obligation for a free access to their rich collections, as well as for much valuable information; and, indeed, but for the help of my friend Mr. Sharpe, I should have despaired of effecting a satisfactory identification of many of the smaller insessorial species; but with the aid which I have received, this has, I believe, been accomplished with considerable accuracy.

Most of Mr. Andersson's specimens were numbered by him as they were obtained; and many of these numbers corresponded with memoranda in his note-book, which greatly facilitated the work of identification.

By far the greater number of the species referred to in the following pages were obtained by Mr. Andersson himself; but a few which he did not meet with, but which have been satisfactorily ascertained on other authority to have occurred within the districts to which the present volume refers, are here included, for the purpose of making the entire catalogue of the birds known to inhabit these countries as complete as possible.

Mr. Andersson's notes as here given, have been extracted partly from his rough note-book and partly from the MS. (which he had partially prepared at the time of his decease) for his projected work on the avifauna of South-western Africa, to both which documents I have had full access by the kind permission of Mr. Andersson's widow.

To have given the whole of these MSS. in full would have extended the present work to a very inconvenient size; and I have therefore only extracted such portions as appeared to me to embody Mr. Andersson's personal and original observations.

The operation of combining these extracts into a consecutive account has made it necessary to have recourse to some verbal emendations; but I have endeavoured scrupulously to avoid any alterations which could in any degree vary the meaning of the author.

Mr. Andersson's MSS. contained minute descriptions of the plumage of most of the birds of South-west Africa; but to have introduced these into the present volume would have unduly increased its bulk; and their introduction is the less requisite as, probably, almost all ornithologists who may peruse these pages are possessed of Mr. Layard's most useful 'Catalogue of the Birds of South Africa,' where a sufficient description of each species will readily be found.

I have, however, transcribed Mr. Andersson's account of the colouring of those parts which fade after death, in all cases where he has recorded these particulars from the examination of freshly killed specimens; and I have also inserted descriptions of plumage in a few instances where it has appeared to me desirable to do so.

Of the measurements of such newly killed specimens as Mr. Andersson has recorded, I have given examples in the case of all those species in respect of which he has also recorded the sex of the individuals measured.

For the nomenclature and arrangement of the species contained in this volume Mr. Andersson is not answerable; that duty having fallen to my lot as editor, I am responsible for the manner in which it has been carried out.

The generic and the specific names are, for the most part, those used in Gray's 'Hand-list of Birds;' but I have not scrupled to deviate from that high authority in some cases in which it appeared to me that adequate reasons existed for making such a deviation.

I have not intercalated in the text the names of the various subfamilies, but such of them as appear to me to have a sound and natural foundation are given in the preliminary table of species at the commencement of the volume.

I have not attempted to give a full synonymy of each species; but I have given a reference to Mr. Layard's volume in the case of every bird recorded in that work, and in the case of other species I have referred to such authors as have noticed the occurrence of these additional species in Southern Africa. I have also, in all cases, referred to those authors who have recorded the occurrence of any of the birds here mentioned in the countries to which the present volume refers, and occasionally to other authors whose works elucidate questions of interest as to the synonymy or economy of various species referred to in the following pages.

I have also given a reference to a figure of each species in all cases where it was known to me that the species had been figured.

It may be right to add that certain extracts from Mr. Andersson's ornithological notes have been already published, some having been contributed by Mr. Andersson to the ornithological appendix to Chapman's 'Travels in South Africa,' and others to the 'Proceedings of the Zoological Society' for 1864; but this circumstance, it is apprehended, does not materially diminish the desirableness of preserving in a more complete form the record of the personal observations of so eminent a traveller and so persevering an ornithologist as the late Mr. Andersson unquestionably was.

At the commencement of this volume will be found a map of Damara Land and the countries adjacent; and at its close a scale of British inches and lines, which latter may be useful to any continental naturalists who have occasion to refer to Mr. Andersson's measurements; also lithographs of the sternum and larynx of *Machæramphus Anderssoni* from Mr. Andersson's original drawings. These lithographs are not referred to in the article on this very interesting bird of prey, as the original drawings had been accidentally mislaid, and were not found again until after that article had been printed.

A list of the various works referred to in this volume will be found at the end of this preface, as it is thought that this may prove convenient to readers desirous of consulting any of these authorities.

EDITOR'S PREFACE.

The following memorandum from the pen of Mr. Andersson is here inserted as giving some interesting information relating to the nidification and migration of the birds of Damara Land and the neighbouring countries:—

"The pairing-and-breeding season of birds in Damara, Namaqua, and parts adjacent depends much, if not entirely, on the falling of the rains; that is, the breeding-season is late or early according to late or early rains. From November to May is probably the chief period of incubation; but very many birds pair as early as September: Owls, Bee-eaters, and Grouse are amongst the earliest breeders.

"Near the sea-coast, or rather those portions of it where the periodical rivers have their embouchures, the breeding-season is somewhat different, or, perhaps it would be more correct to say, occurs later in the year. The cause is simple: rain rarely or never falls in those parts; and it is not until long after the rivers (having their sources and origin in the distant interior) have subsided that the scanty vegetation recovers from its 'torpor;' and with it returns the insect-life which enables the parent birds to seek and obtain suitable sustenance for their tender broods.

"The moulting-season begins with the return of the wet season.

"It is during the rainy time of the year that the greatest variety of birds is to be observed; for, though all but deserts during the dry season, Damara and Namaqua Land, from their peculiar positions &c., are then a regular paradise to the feathered tribes, the insect- and reptile-life being at that period exceedingly prolific. Swarms of migratory Hawks and Kites may then be observed in pursuit of the myriads of *Termites* which at this season infest the air, but at the same time brighten it, as it were, with innumerable silvery dots and streaks as their gorgeous wings and white bodies encounter the fiery sunlight. Here and there a flock of Storks may be observed busily chasing the devastating locusts, or performing graceful gyrations in the air; and whilst the temporary rain-pools often abound with rare and handsome Waterfowl, the shores are frequented by the elegant Heron, the lively Sandpiper, the graceful Avocet, and the gorgeous Flamingo. The Atlantic on the west, the Orange River to the south, the Okavango River and the Lake Ngami with the watersheds to the north and east, contribute chiefly to these large and varied annual incursions and migrations."

In conclusion it may be desirable to refer to the synonymy of the Plover which is included in the present volume under the name of *Eudromias asiaticus* (No. 316), in order to add an observation respecting it, which, as the article upon that species is already printed, cannot be inserted in its proper place. In 'The Ibis' for 1872, p. 144, Dr. Finsch gives it as his opinion that this Plover should bear the specific name of "*damarensis*," proposed for it by Strickland in the 'Contributions to Ornithology' for 1852, p. 158, it being Dr. Finsch's view that the older specific names of "*asiaticus*" and "*caspius*" were intended by their author (Pallas) to apply to the nearly allied species figured by Mr. Harting in 'The Ibis' for 1870, pl. 6, under the title of *Eudromias veredus*. The synonymy of both these Plovers is fully discussed in Dr. Finsch's paper above cited, to which, as well as to Mr. Harting's paper in 'The Ibis' for 1870, p. 201, I beg to refer my readers.

J. H. G. (1872).

ERRATA.

Page 13 line 3 *for* "preceding" *read* "succeeding."
,, 266 ,, 14 *for* "uncommon" *read* "not uncommon."
,, 304 ,, 7 *omit* "shading lighter on the inner vanes."

LIST OF WORKS

REFERRED TO IN THIS VOLUME.

ALEXANDER, SIR J. E. Expedition of Discovery into the Interior of Africa. Description of New Species of Birds, by G. R. Waterhouse. 8vo. London, 1838. (See also letter W.)

ANDERSSON, C. J. Lake Ngami. 8vo. London, 1856.

——. The Okavango River. 8vo. London, 1861.

ANNALS AND MAGAZINE OF NATURAL HISTORY. 8vo. London (second series), 1848 to 1857.

BAINES, T. Explorations in South-west Africa. 8vo. London, 1864.

BLANFORD, W. T. Observations on the Geology and Zoology of Abyssinia. 8vo. London, 1870.

BONAPARTE, C. L. Conspectus Generum Avium. 8vo. Leyden, 1850.

——. Iconographie des Pigeons. Fol. Paris, 1857.

BOURJOT ST.-H., A. Collection de Perroquets. Fol. Paris and Strasbourg, 1835 to 1839.

BREE, C. R. A History of the Birds of Europe not observed in the British Islands. 8vo. London, 1863 to 1867.

BUFFON, G. L. L. DE. Planches Enluminées. Fol. Paris, 1770 to 1786.

BURCHELL, W. J. Travels in the Interior of Southern Africa in 1811 to 1815. 4to. London, 1822 to 1824.

CHAPMAN, J. Travels in the Interior of South Africa. 8vo. London, 1868.

CUVIER, LE BARON. Le Règne Animal, distribué d'après son organisation. 8vo. Paris: ed. 1, 1817; ed. 2, 1829.

DAUDIN, F. M. Traité élémentaire et complet d'Ornithologie, ou Histoire Naturelle des Oiseaux. 4to. Paris, 1800.

DECKEN, BARON C. C. VON DER. Reisen in Ost-Afrika. Vol. III. Säugethiere, Vögel, Amphibien, Crustaceen, Mollusken, und Echinodermen. Bearbeitet W. C. H. Peters, J. Cabanis, F. Hilgendorf, Ed. Villartens und C. Semper. 8vo. Leipzig, 1869.
——. Ditto. Vol. IV. Vögel Ost-Afrika's. Von O. Finsch und G. Hartlaub. (See also under letter F.) 8vo. Leipzig, 1870.
DES MURS, O. Iconographie Ornithologique. Fol. Paris, 1849.
DES MURS, O., LEFEBVRE, T., and DILLON, Q. Voyage en Abyssinie par une Commission Scientifique composée de M. Lefebvre, Lieutenant de Vaisseau, A. Petit et Quartin Dillon, Naturalistes du Museum. Zoologie par O. Des Murs. Text 8vo, Atlas fol. Paris, 1848.
EDINBURGH JOURNAL OF NATURAL AND GEOGRAPHICAL SCIENCE. 8vo. Edinburgh, 1819 to 1831.
EDWARDS, G. Natural History of Birds. 4to. London, 1743 to 1751.
——. Gleanings of Natural History. 4to. London, 1758 to 1764.
EYTON, T. C. Monograph of the Anatidæ. 4to. London, 1838.
EXPLORATION SCIENTIFIQUE DE L'ALGÉRIE pendant les Années 1840, 1841, 1842, publiée par ordre du Gouvernement et avec le concours d'une Commission Académique. Sciences Physiques, Zoologie. Vol. IV. Fol. Paris, 1849.
FINSCH, O., and HARTLAUB, G. Die Vögel Ost-Afrika's, in Baron Carl Claus von der Decken's Reisen in Ost-Afrika, vol. iv. 8vo. Leipzig, 1870. (See also under letter D.)
GMELIN, J. F. Systema Naturæ, per Regna tria Naturæ secundum Classes, Ordines, Genera, Species, cum characteribus, differentiis, synonymis, locis. 8vo. Leyden, 1789.
GOULD, J. Birds of Australia. Fol. London, 1848.
——. Birds of Europe. Fol. London, 1837.
——. Birds of Great Britain. Fol. London, 1862.
——. Icones Avium. Fol. London, 1837.
GRAY, G. R. Appendix to Trotter, Allen, and Thomson's Expedition to the Niger in the year 1841. 8vo. London, 1841. (See also letter T.)
——. Genera of Birds, comprising their generic characters, illustrated by D. Mitchell. Fol. London, 1844 to 1849.
——. Hand-list of Genera and Species of Birds, distinguishing those contained in the British Museum. 8vo. London, 1869 to 1871.
GRAY, J. E. Gleanings from the Menagerie and Aviary at Knowsley Hall. Fol. Knowsley, 1846.
HARTLAUB, G. Beitrag zur Ornithologie Westafrica's, in 'Abhandlungen aus dem Gebiete der Naturwissenschaften.' 4to. Hamburg, 1852.

LIST OF WORKS REFERRED TO.   xxxiii

HARTLAUB, G.   System der Ornithologie Westafrica's.   8vo.   Bremen, 1857.
HEMPRICH, F. G., and EHRENBERG, C. G.   Symbolæ Physicæ.   Fol. Berlin, 1828.
HEUGLIN, T. VON.   Ornithologie Nord-Ost Afrika's.   8vo. Cassel, 1869 to 1871.
——.   Systematische Uebersicht der Vögel Nord-Ost Afrika's.   8vo. Vienna, 1855.
IBIS.   The Ibis, a Magazine of General Ornithology.   8vo. London, 1859 to 1872.
ISIS.   Isis oder encyclopädische Zeitung.   4to. Jena, 1817.
JARDINE, SIR W.   Sun-birds, in Jardine's ' Naturalist's Library.'   12mo. Edinburgh, n. d.
JARDINE, SIR W., and SELBY, P. J.   Illustrations of Ornithology.   Fol. Edinburgh, 1826 to 1843.
JORNAL DE SCIENCIAS Mathem. Phys. e Naturaes.   Publ. da Academia R. de Lisboa.   8vo. Lisbon, 1867 to 1871.
JOURNAL FÜR ORNITHOLOGIE.   8vo. Cassel, 1853 to 1871.
LATHAM, J.   General Synopsis of Birds.   4to. London, 1781 to 1785.
LAYARD, E. L.   The Birds of South Africa, a Descriptive Catalogue of all the known Species occurring south of the 28th parallel of South Latitude.   First Edition.   8vo. Cape Town, 1867.
LEVAILLANT, F.   Histoire Naturelle des Oiseaux d'Afrique.   Fol. Paris, 1799 to 1808.
——.   Histoire Naturelle des Oiseaux de Paradis et des Rolliers, suivie de celle des Toucans et des Barbus.   Fol. Paris, 1806 and 1807.
——.   Histoire Naturelle des Promerops et des Guêpiers.   Fol. Paris, 1807.
——.   Histoire Naturelle des Perroquets.   Fol. Paris, 1801 to 1805.
LICHTENSTEIN, H.   Verzeichniss der Doubletten des zoologischen Museums der Königl. Universität zu Berlin.   4to. Berlin, 1823.
LIVINGSTONE, D.   Missionary Travels and Researches in South Africa. 8vo. London, 1857.
MALHERBE, A.   Monographie des Picidées.   Fol. Metz, 1861 to 1863.
MARSHALL, C. H. T., and MARSHALL, G. F. L.   A Monograph of the Capitonidæ or Scansorial Barbets.   4to. London, 1870 and 1871.
MÜLLER, BARON J. W. DE.   Description de nouveaux Oiseaux d'Afrique. Fol. Stuttgart, 1853.
NAUMANN, J. A.   Naturgeschichte der Vögel Deutschlands.   8vo. Leipzig, 1822.
NEWTON, A.   A History of British Birds by the late William Yarrell.

c

Edited by Alfred Newton. 8vo. London, 1871. (See also under Y.)

ÖFVERSIGT af Kongl. Vetenskaps Akademiens Förhandlingar. 8vo. Stockholm, 1844 to 1871.

PALLAS, P. S. Reise durch verschiedene Provinzen des Russischen Reichs. 4to, and Atlas fol. Petersburg, 1771 to 1776.

——. Zoographia Rosso-Asiatica. 4to. Petersburg, 1811.

RADDE, G. Reisen im Süden v. Ost-Sibirien in d. J. 1855–59. Säugethiere u. Vögel. 4to. Petersburg, 1862 and 1863.

REICHENBACH, A. B. Praktische Naturgeschichte der Vögel. Leipzig, 1847.

REICHENBACH, L. Synopsis Avium. 8vo. Dresden and Leipzig, 1851.

REVUE ET MAGASIN DE ZOOLOGIE pure et appliquée. 8vo. Paris, 1849 to 1870.

RÜPPELL, E. Atlas zu der Reise im nördlichen Afrika. Fol. Frankfurt am Main, 1826.

——. Monographie der Gattung *Otis* im Museum Senckenbergianum. 4to. Frankfurt am Main, 1837.

——. Neue Wirbelthiere, zu der Fauna von Abyssinien gehörig. Fol. Frankfurt am Main, 1835 to 1840.

——. Systematische Uebersicht der Vögel Nord-Ost Afrika's. 8vo. Frankfurt am Main, 1845.

SCHLEGEL, H. Museum d'Histoire Naturelle des Pays-Bas. Revue Méthodique et Critique de la Collection des Oiseaux. Leyden, 1862 to 1866.

SHARPE, R. B. Catalogue of African Birds in the Collection of R. B. Sharpe. 8vo. London, 1871.

——. A Monograph of the Alcedinidæ or Kingfishers. 4to. London, 1868 to 1871.

SHARPE, R. B., and DRESSER, H. E. A History of the Birds of Europe, including all the Species inhabiting the Western Palæarctic Region. Fol. London, 1871.

SHAW, G. General Zoology, or Systematic Natural History. 8vo. London, 1809 to 1812.

SHAW, G., and NODDER, F. P. Naturalist's Miscellany. 8vo. London, 1790 to 1813.

SHELLEY, G. E. The Birds of Egypt. 8vo. London, 1872.

SMITH, A. Illustrations of the Zoology of South Africa. Aves. 4to. London, 1849.

——. Report of the Expedition for Exploring Central Africa from the Cape of Good Hope. 8vo. Cape Town, 1836.

SOUTH-AFRICAN QUARTERLY JOURNAL. 8vo. Cape Town, 1829 to 1835.

SOUTH-AFRICAN MUSEUM. Catalogue of the South-African Museum now Exhibiting in the Egyptian Hall, Piccadilly. 8vo. London, 1837.
SPARRMAN, A. Museum Carlsonianum. 4to. Holmiæ, 1786 to 1788.
STRICKLAND, H. E. Ornithological Synonyms. 8vo. London, 1855.
STRICKLAND, H. E., and SCLATER, P. L. List of a Collection of Birds procured by C. J. Andersson in the Damara Country, with notes. In Jardine's 'Contributions to Ornithology' for 1852. 8vo. Edinburgh, 1853.
SUNDEVALL, C. J. Conspectus Avium Picinarum. 8vo. Stockholm, 1866.
SWAINSON, W. Birds of Western Africa. In Jardine's 'Naturalist's Library.' 12mo. Edinburgh, 1837.
——. Zoological Illustrations. 8vo. London, 1820 to 1833.
TEMMINCK, C. J. Histoire Naturelle Générale des Pigeons et des Gallinacés. 8vo. Amsterdam, 1813 to 1815.
——. Manuel d'Ornithologie. 2nd edition, 8vo. Paris, 1820 to 1835.
——. Histoire Naturelle Générale des Pigeons, avec figures en couleurs peintes par Madame Knip, née Pauline de Courcelles. Le texte par C. J. Temminck. Fol. Paris, 1807.
TEMMINCK, C. J., and LAUGIER, LE BARON. Nouveau Recueil de Planches Coloriées d'Oiseaux. Fol. Paris, 1820 to 1836.
TROTTER, H. D., ALLEN, W., and THOMSON, T. R. H. An Expedition to the Niger in the year 1841. Appendix by G. R. Gray. 8vo. London, 1848. (See also letter G.)
VIEILLOT, L. P. Galerie des Oiseaux. 4to. Paris, 1825.
——. Histoire Naturelle des plus beaux Oiseaux Chanteurs de la zône torride. Fol. Paris, 1805.
——. Nouveau Dictionnaire d'Histoire Naturelle. 8vo. Paris, 1816 to 1819.
WAGLER, J. C. Systema Avium. 12mo. Stuttgard, 1827.
WATERHOUSE, G. R. Descriptions of New Species of Birds. In Sir J. E. Alexander's 'Expedition of Discovery into the Interior of Africa.' 8vo. London, 1838. (See also letter A.)
YARRELL, W. A History of British Birds. Edited by Alfred Newton. 8vo. London, 1871. (See also under N.)
ZOOLOGICAL JOURNAL. 8vo. London, 1824 to 1834.
ZOOLOGICAL SOCIETY. Proceedings of the Scientific Meetings of the Zoological Society of London. 8vo. London, 1830 to 1871.
——. Transactions of the Zoological Society of London. 4to. London, 1835 to 1871.

TABLE OF SPECIES.

ACCIPITRES.

Family VULTURIDÆ.

Subfamily NEOPHRONINÆ.
No.
1. Neophron percnopterus.
2. —— pileatus.

Subfamily VULTURINÆ.
3. Otogyps auricularis.
4. Vultur occipitalis.
5. Gyps Kolbii.
6. —— Rüppelli.

Family FALCONIDÆ.

Subfamily AQUILINÆ.
7. Aquila vulturina.
8. —— nævioides.
9. Hieraëtus pennatus.
10. Pseudaëtus spilogaster.
11. —— bellicosus.
12. Haliaëtus vocifer.
13. Circaëtus pectoralis.
14. Helotarsus ecaudatus.

Subfamily BUTEONINÆ.
No.
15. Buteo jackal.
16. —— desertorum.

Subfamily FALCONINÆ.
17. Falco minor.
18. —— cervicalis.
19. Chicquera ruficollis.
20. Hypotriorchis subbuteo.
21. Erythropus vespertinus.
22. —— amurensis.
23. Tinnunculus cenchris.
24. —— alaudarius.
25. —— rupicolus.
26. —— rupicoloides.
27. Polihierax semitorquatus.

Subfamily MILVINÆ.
28. Elanus cæruleus.
29. Milvus migrans.
30. —— Forskahli.
31. Machærhamphus Anderssoni.

Subfamily ACCIPITRINÆ.

No.
32. Kaupifalco monogrammicus.
33. Melierax musicus.
34. —— polyzonus.
35. —— gabar.
36. —— niger.
37. Accipiter tachiro.
38. —— polyzonoides.
39. —— minullus.
40. —— rufiventris.

Subfamily CIRCINÆ.

41. Circus Swainsoni.
42. —— cinerarius.
43. —— maurus.
44. —— ranivorus.

Family SERPENTARIIDÆ.

Subfamily SERPENTARIINÆ.
45. Sagittarius secretarius.

Family STRIGIDÆ.

Subfamily STRIGINÆ.
No.
46. Strix poensis.

Subfamily SURNINÆ.

47. Athene perlata.
48. Tænioglaux capensis.

Subfamily BUBONINÆ.

49. Scops capensis.
50. —— leucotis.
51. Huhua Verreauxi.
52. Bubo maculosus.

Subfamily SYRNIINÆ.

53. Phasmoptynx capensis.

PASSERES.

Division *FISSIROSTRES*.

Family CAPRIMULGIDÆ.

Subfamily CAPRIMULGINÆ.
54. Caprimulgus rufigena.
55. —— pectoralis.
56. —— lentiginosus.
57. Cosmetornis vexillarius.

Family CYPSELIDÆ.

Subfamily CYPSELINÆ.
58. Cypselus gutturalis.

59. Cypselus barbatus.
60. —— parvus.

Family HIRUNDINIDÆ.

Subfamily HIRUNDININÆ.

61. Hirundo Monteiri.
62. —— rustica.
63. —— cucullata.
64. —— dimidiata.
65. Cotile fuligula.

TABLE OF SPECIES.

Family CORACIADÆ.

Subfamily CORACIANÆ.

No.
66. Coracias caudata.
67. —— nævia.
68. —— garrula.

Family ALCEDINIDÆ.

Subfamily DACELONINÆ.

69. Halcyon cyanolouca.
70. —— semicærulea.
71. —— chelicutensis.

Subfamily ALCEDININÆ.

72. Alcedo semitorquata.
73. Ceryle maxima.
74. —— rudis.
75. Corythornis cyanostigma.

Family MEROPIDÆ.

Subfamily MEROPINÆ.

76. Merops apiaster.
77. —— superciliosus.
78. —— nubicoides.
79. Melittophagus pusillus.
80. Dicrocercus hirundinaceus.

Division *TENUIROSTRES*.

Family UPUPIDÆ.

Subfamily UPUPINÆ.

81. Upupa minor.

Subfamily IRRISORINÆ.

82. Irrisor erythrorhynchos.

No.
83. Irrisor cyanomelas.
84. —— aterrimus.

Family PROMEROPIDÆ.

Subfamily NECTARININÆ.

85. Nectarinia famosa.
86. Cinnyris chalybea.
87. —— afra.
88. —— bifasciata.
89. —— fusca.
90. —— talatala.
91. Chalcomitra gutturalis.
92. Anthobaphes violacea.

Family MELIPHAGIDÆ.

Subfamily MELIPHAGINÆ.

93. Zosterops capensis.
94. —— senegalensis.

Family TROGLODYTIDÆ.

95. Sylvietta rufescens.

Division *DENTIROSTRES*.

Family PARIDÆ.

Subfamily PARINÆ.

96. Parisoma subcæruleum.
97. —— Layardi.
98. Anthoscopus minutus.
99. —— Caroli.
100. Parus afer.
101. —— niger.

TABLE OF SPECIES.

Family LUSCINIDÆ.

Subfamily CALAMODYTINÆ.

No.
102. Drymoica maculosa.
103. —— affinis.
104. —— flavicans.
105. —— ocularius.
106. —— Smithii.
107. —— chiniana.
108. —— subruficapilla.
109. —— rufilata.
110. —— Levaillantii.
111. Cisticola terrestris.
112. Aëdon subcinnamomea.
113. —— fasciolata.
114. —— leucophrys.
115. —— paena.
116. Thamnobia coryphæus.
117. Camaroptera olivacea.
118. Dryodromas damarensis.
119. —— flavida.
120. Eremomela flaviventris.
121. —— usticollis.
122. Calamodyta arundinacea.
123. —— bæticata.
124. Calamodus schœnobænus.

Subfamily SYLVIANÆ.

125. Sylvia hortensis.
126. Phyllopneuste hypolais.
127. —— trochilus.

Subfamily SAXICOLINÆ.

128. Pratincola torquata.
129. Saxicola familiaris.
130. —— Schlegelii.
131. —— Stricklandii.

No.
132. Saxicola infuscata.
133. —— pileata.
134. —— leucomelæna.
135. —— Atmorii.
136. Myrmecocichla formicivora.

Family MOTACILLIDÆ.

Subfamily MOTACILLINÆ.

137. Motacilla capensis.
138. —— Vaillantii.
139. Budytes flava.

Subfamily ANTHINÆ.

140. Anthus Raaltcni.
141. —— caffer.
142. —— pyrrhonotus.
143. —— campestris.

Family TURDIDÆ.

Subfamily TURDINÆ.

144. Turdus letsitsirupa.
145. —— libonyanus.
146. —— olivaceus.
147. Monticola brevipes.
148. Chætops pycnopygius.
149. Cossypha caffra.
150. —— bicolor.

Family PYCNONOTIDÆ.

Subfamily PYCNONOTINÆ.

151. Pycnonotus nigricans.
152. —— tricolor.

Subfamily PHYLLORNITHINÆ.

153. Phyllastrephus capensis.
154. Criniger flaviventris.

… TABLE OF SPECIES. xli

Subfamily CRATEROPODINÆ.
No.
155. Crateropus bicolor.
156. —— melanops.
157. —— Jardinii.
158. —— Hartlaubi.

Family ORIOLIDÆ.

Subfamily ORIOLINÆ.
159. Oriolus galbula.
160. —— notatus.

Family DICRURIDÆ.

Subfamily DICRURINÆ.
161. Dicrurus musicus.

Family MUSCICAPIDÆ.

Subfamily MUSCICAPINÆ.
162. Melanopepla pammelæna.
163. Bradornis mariquensis.
164. Muscicapa griseola.
165. Tchitrea viridis.
166. Platysteira pririt.
167. —— affinis.
168. —— torquata.

Subfamily CAMPEPHAGINÆ.
169. Campephaga nigra.
170. Ceblepyris pectoralis.

Family LANIIDÆ.

Subfamily LANIINÆ.
171. Lanius minor.
172. Enneoctonus collurio.
173. Fiscus collaris.
174. —— subcoronatus.

Subfamily MALACONOTINÆ.
No.
175. Urolestes melanoleucus.
176. Nilaus brubru.
177. Eurocephalus anguitimens.
178. Prionops talacoma.
179. —— Retzii.
180. Laniarius atrococcineus.
181. —— major.
182. —— sticturus.
183. Dryoscopus cubla.
184. Telophorus gutturalis.
185. Chlorophoneus similis.
186. Pomatorhynchus erythropterus.
187. —— trivirgatus.

Division CONIROSTRES.

Family CORVIDÆ.

Subfamily CORVINÆ.
188. Corvultur albicollis.
189. Corvus scapulatus.
190. —— capensis.

Family STURNIDÆ.

Subfamily JUIDINÆ.
191. Cinnyricinclus Verreauxi.
192. Juida australis.
193. —— Mevesii.
194. Lamprocolius phœnicopterus.
195. Spreo bicolor.
196. Amydrus caffer.

Subfamily STURNINÆ.
197. Dilophus carunculatus.

Subfamily BUPHAGINÆ.
No.
198. Buphaga africana.

Family FRINGILLIDÆ.

Subfamily PLOCEINÆ.
199. Bubalornis erythrorhynchus.
200. Plocepasser mahali.
201. Philetærus socius.
202. Hyphantornis spilonotus.
203. —— velatus.
204. Euplectes capensis.
205. —— taha.
206. Pyromelana oryx.
207. Quelea sanguinirostris.

Subfamily SPERMESTINÆ.
208. Amadina erythrocephala.
209. Alario aurantia.
210. Hypochera ultramarina.
211. Pytelia melba.
212. Lagonosticta minima.
213. Sporopipes squamifrons.
214. Estrelda astrild.
215. —— erythronota.
216. Mariposa cyanogastra.
217. Uræginthus granatinus.

Subfamily VIDUANÆ.
218. Vidua regia.
219. —— principalis.
220. —— paradisea.

Subfamily PYRRHULINÆ.
221. Crithagra angolensis.
222. —— chrysopyga.
223. —— chloropsis.

Subfamily FRINGILLINÆ.
No.
224. Poliospiza gularis.
225. —— crocopygia.
226. Petronia petronella.
227. Passer arcuatus.
228. —— motitensis.
229. —— diffusus.

Subfamily EMBERIZINÆ.
230. Fringillaria flaviventris.
231. —— impetuani.

Family ALAUDIDÆ.

Subfamily ALAUDINÆ.
232. Pyrrhulauda Smithi.
233. —— verticalis.
234. —— australis.
235. Alauda conirostris.
236. —— Grayi.
237. —— erythrochlamys.
238. Calendula crassirostris.
239. Megalophonus sabota.
240. —— cinereus.
241. —— Anderssoni.
242. —— africanoides.
243. —— nævius.
244. —— lagepa.
245. Certhilauda rufula.
246. —— semitorquata.

Family MUSOPHAGIDÆ.

Subfamily COLIINÆ.
247. Colius capensis.
248. —— erythromelon.

Subfamily MUSOPHAGINÆ.
249. Schizorhis concolor.

TABLE OF SPECIES.

Family BUCEROTIDÆ.

Subfamily BUCEROTINÆ.

No.
250. Bucorvus abyssinicus.
251. Tockus nasutus.

No.
252. Tockus melanoleucus.
253. —— Monteiri.
254. —— flavirostris.
255. —— erythrorhynchus.

SCANSORES.

Family PSITTACIDÆ.

Subfamily PSITTACINÆ.

256. Poicephalus robustus.
257. —— Meyeri.
258. —— Rüppelli.
259. Psittacula roseicollis.

Family CAPITONIDÆ.

Subfamily POGONORHYNCHINÆ.
260. Pogonorhynchus leucomelas.

Family PICIDÆ.

Subfamily PICINÆ.
261. Thripias namaquus.
262. Dendropicus Hartlaubii.
263. —— cardinalis.
264. Ipagrus capricorni.

265. Ipagrus Brucei.
266. —— variolosus.

Family CUCULIDÆ.

Subfamily INDICATORINÆ.
267. Indicator minor.

Subfamily CENTROPODINÆ.
268. Centropus senegalensis.

Subfamily CUCULINÆ.
269. Coccystes glandarius.
270. Oxylophus jacobinus.
271. —— caffer.
272. —— serratus.
273. Cuculus clamosus.
274. —— canorus.
275. —— gularis.
276. Chrysococcyx cupreus.
277. —— Klaasi.

COLUMBÆ.

Family COLUMBIDÆ.

Subfamily TRERONINÆ.
278. Phalacrotreron calva.

Subfamily COLUMBINÆ.
279. Stictœnas phæonotus.

280. Turtur senegalensis.
281. Streptopelia damarensis.
282. —— semitorquata.
283. Œna capensis.
284. Chalcopelia afra.

GALLINÆ.

Family MELEAGRIDÆ.

Subfamily NUMIDINÆ.

No.
285. Numida cornuta.

Family PTEROCLIDÆ.

Subfamily PTEROCLINÆ.

286. Pterocles bicinctus.
287. —— variegatus.
288. Pteroclurus namaqua.

Family TETRAONIDÆ.

Subfamily PERDICINÆ.

289. Pterniscs nudicollis.

No.
290. Pternises Swainsonii.
291. Scleroptera gariepensis.
292. —— subtorquata.
293. —— pileata.
294. —— adspersa.
295. Coturnix communis.
296. —— Delegorguei.

Subfamily TURNICINÆ.

297. Turnix lepurana.

STRUTHIONES.

Family STRUTHIONIDÆ.

Subfamily STRUTHIONINÆ.

No.
298. Struthio australis.

GRALLÆ.

Family OTIDIDÆ.

Subfamily OTIDINÆ.

No.
299. Eupodotis kori.
300. —— ruficrista.
301. —— Rüppellii.
302. —— afra.
303. —— afroides.

Family CHARADRIADÆ.

Subfamily CURSORINÆ.

No.
304. Cursorius senegalensis.
305. —— bicinctus.
306. —— cinctus.
307. —— chalcopterus.

TABLE OF SPECIES.

Subfamily GLAREOLINÆ.
No.
308. Glareola melanoptera.
309. —— pratincola.

Subfamily ŒDICNEMINÆ.
310. Œdicnemus capensis.
311. —— vermiculatus.

Subfamily CHARADRINÆ.
312. Lobivanellus lateralis.
313. Hoplopterus speciosus.
314. Chettusia coronata.
315. Squatarola varia.
316. Eudromias damarensis *.
317. Ægialites alexandrinus.
318. —— marginatus.
319. —— pecuarius.
320. —— tricollaris.
321. —— hiaticula.

Subfamily CINCLINÆ.
322. Cinclus interpres.

Subfamily HÆMATOPINÆ.
323. Hæmatopus Moquini.

Family GRUIDÆ.

Subfamily GRUINÆ.
324. Bugeranus carunculatus.
325. Tetrapteryx paradisea.
326. Balearica regulorum.

Family CICONIIDÆ.

Subfamily CICONIINÆ.
327. Ciconia alba.
328. Sphenorrhynchus Abdimii.

No.
329. Ephippiorhynchus senegalensis.
330. Leptoptilus crumeniferus.

Family ANASTOMATIDÆ.

Subfamily ANASTOMATINÆ.
331. Anastomus lamelligerus.

Family ARDEIDÆ.

Subfamily ARDEINÆ.
332. Ardea cinerea.
333. —— melanocephala.
334. —— goliat.
335. —— purpurea.
336. —— rufiventris.
337. Bubulcus ibis.
338. Ardeola comata.
339. Herodias alba.
340. —— intermedia.
341. —— garzetta.
342. Ardeiralla Sturmii.
343. Butorides atricapilla.
344. Ardetta minuta.
345. Nycticorax ægyptius.

Family SCOPIDÆ.

Subfamily SCOPINÆ.
346. Scopus umbretta.

Family PLATALEIDÆ.

Subfamily PLATALEINÆ.
347. Platalea tenuirostris.

* *Vide* Preface, p. xxx.

TABLE OF SPECIES.

Family TANTALIDÆ.

Subfamily TANTALINÆ.

No.
348. Tantalus ibis.

Subfamily IBIDINÆ.

349. Geronticus calvus.
350. Ibis æthiopica.
351. Hagedashia caffrensis.

Family SCOLOPACIDÆ.

Subfamily NUMENIINÆ.

352. Numenius arquatus.
353. —— phæopus.

Subfamily TOTANINÆ.

354. Totanus calidris.
355. —— glottis.
356. —— stagnatilis.
357. —— glareola.
358. Actitis hypoleucus.
359. Terekia cinerea.

Subfamily TRINGINÆ.

360. Philomachus pugnax.
361. Tringa canutus.
362. —— subarquata.
363. —— Bairdii.
364. —— minuta.
365. Calidris arenaria.

Subfamily SCOLOPACINÆ.

366. Gallinago major.
367. Rhynchæa capensis.

Family RECURVIROSTRIDÆ.

Subfamily AVOCETTINÆ.

No.
368. Recurvirostra avocetta.

Subfamily HIMANTOPODINÆ.

369. Himantopus autumnalis.

Family RALLIDÆ.

Subfamily RALLINÆ.

370. Rallus cærulescens.
371. Ortygometra pygmœa.
372. —— marginalis.
373. Alecthelia dimidiata.
374. Limnocorax niger.

Family GALLINULIDÆ.

Subfamily GALLINULINÆ.

375. Gallinula angulata.
376. —— chloropus.
377. Porphyrio smaragnotus.
378. —— Alleni.

Subfamily FULICINÆ.

379. Fulica cristata.

Family PARRIDÆ.

Subfamily PARRINÆ.

380. Parra africana.
381. —— capensis.

ANSERES.

Family PHŒNICOPTERIDÆ.

Subfamily PHŒNICOPTERINÆ.
No.
382. Phœnicopterus erythræus.
383. —— minor.

Family ANATIDÆ.

Subfamily PLECTROPTERINÆ.
384. Plectropterus gambensis.
385. Sarkidiornis melanotus.
386. Chenalopex ægyptiacus.

Subfamily ANSERINÆ.
387. Nettapus auritus.

Subfamily ANATINÆ.
388. Dendrocygna viduata.
389. Mareca capensis.
390. Pœcilonetta erythrorhyncha.
391. Nettion hottentota.
392. Anas sparsa.
393. —— xanthorhyncha.
394. Spatula capensis.
395. Aythia capensis.
396. Thalassornis leuconota.
397. Erismatura maccoa.

Family PODICIPIDÆ.

Subfamily PODICIPINÆ.
398. Podiceps cristatus.
399. —— nigricollis.
400. —— minor.

Family SPHENISCIDÆ.

Subfamily SPHENISCINÆ.
No.
401. Spheniscus demersus.

Family PROCELLARIIDÆ.

Subfamily PROCELLARIINÆ.
402. Puffinus major.
403. Procellaria pelagica.
404. —— oceanica.
405. Pseudoprion turtur.
406. Daption capensis.
407. Ossifraga gigantea.

Subfamily DIOMEDEINÆ.
408. Diomedea exulans.
409. —— melanophrys.

Family LARIDÆ.

Subfamily STERCORARIINÆ.
410. Stercorarius pomarinus.
411. —— parasiticus.

Subfamily LARINÆ.
412. Larus vetula.
413. Cirrhocephalus poiocephalus.

Subfamily STERNINÆ.
414. Sterna caspia.
415. —— Bergii.
416. —— cantiaca.
417. —— fluviatilis.
418. Sternula balænarum.

No.
419. Pelodes hybrida.
420. Hydrochelidon nigra.

Subfamily RHYNCHOPSINÆ.
421. Rhynchops flavirostris.

Family PELECANIDÆ.

Subfamily SULARINÆ.
422. Sula capensis.

Subfamily PLOTINÆ.
No.
423. Plotus Levaillantii.

Subfamily GRACULINÆ.
424. Graculus carbo.
425. —— capensis.
426. —— neglectus.
427. —— africanus.

Subfamily PELECANINÆ.
428. Pelecanus minor.

THE BIRDS OF DAMARA LAND

AND THE ADJACENT COUNTRIES OF

SOUTH-WEST AFRICA.

ACCIPITRES.

VULTURIDÆ.

1. Neophron percnopterus (Linn.). Egyptian Vulture.

Neophron percnopterus, Gould's Birds of Europe, pl. 3.
    ,,   ,,   Layard's Cat. No. 2.
    ,,   ,,   Finsch & Hartlaub's Vögel Ost-Afrika's, p. 33.

This Vulture is not uncommon in Damara Land and Great Namaqua Land and the parts adjacent, more especially in the neighbourhood of the coast. It is usually found in pairs, and is a regular scavenger, being generally seen in search of the filthiest food.

The irides are reddish yellow; the bill and naked patch on the lower part of the throat are bright orange, darkest above, the tips of both mandibles pale liver-brown; the feet and legs yellowish white.

[Mr. Andersson, in a note to p. 22 of his work entitled 'Lake Ngami,' says that he has seen this Vulture feed on the fruit of a wild gourd found near the coast of Damara Land, and called "Naras."

I have not myself seen a Damara example of this Vulture, but the species is too well known to admit any doubt as to its identification.—ED.]

2. Neophron pileatus (Burch.). Pileated Vulture.

Cathartes monachus, Temminck's Pl. Col. pl. 222.
Neophron pileatus, Layard's Cat. No. 3.
  „   „   Finsch & Hartlaub's Vögel Ost-Afrika's, p. 37.

This species is not so common as the preceding one in Damara Land, but becomes more numerous as one approaches the Orange River. Its habits are similar to those of *N. percnopterus*, and it is comparatively fearless where it is not disturbed. I have observed it single and also in small families.

[Mr. Andersson's identification of this species is confirmed by an excellent coloured drawing from the pencil of Mr. Baines; but I have not personally had the opportunity of examining a Damara specimen.

If Burchell's description of his *Vultur pileatus*, given in his 'Travels in South Africa,' vol. ii. p. 195, was really intended to apply to this species, it is not very accurate, as has been already pointed out by a subsequent author, *vide* Blanford's 'Geology and Zoology of Abyssinia,' p. 287, note.—ED.]

3. Otogyps auricularis, Daud. Sociable Vulture.

L'Oricou, Levaillant's Ois. d'Afr. pl. 9.
Otogyps auricularis, Layard's Cat. No. 5.

This is the commonest Vulture in Damara and Great Namaqua Land, and is also found in all the parts bordering on those countries. It is a very powerful bird, and is the first to appear about a carcase.

I believe naturalists are not quite agreed as to whether Vultures hunt by sight, by scent, or by both faculties combined. I have myself no doubt that they employ the one sense as well as the other in finding their prey, though I feel inclined to give sight the preference; and I had once a very striking proof of how they employ their vision in guiding them to carrion—in this instance, however, not so much by the actual sight of the carrion (though the first discovery probably originated in that way) as by another singular contrivance. Early one morning as I was toiling up the ascent of a somewhat elevated ridge of hills, with the view of obtaining bearings for my travelling map, and before arriving at the summit, I observed several Vultures descending near me; but thinking I had merely disturbed them from their lofty perch, I did not take any particular notice of their appearance, as the event was one of usual occurrence; but on gaining my destination, I found that the birds were not coming merely from the hill summit, but from an indefinite distance on the other side. This circumstance, coupled with the recollection that I had wounded a zebra on the preceding day in the direction towards which the Vultures were winging their way, caused me to pay more attention. The flight of the Vultures was low, at least five hundred to a thousand feet below the summit of the mountain; and on arriving near the base they would abruptly rise without deviating from their direct course; and no sooner was the obstacle in their way thus surmounted than they again depressed their flight. Those Vultures which I saw could not have themselves seen the carrion, but simply hunted in direct sight of one another. There was a numerous arrival; and although I could not always detect the next bird as soon as I lost sight of the previous one, yet, when at length it did come into view, it never seemed uncertain about its course. Having finished my observations I descended, and proceeded in the direction which the Vultures had pursued, and after about half an hour's rapid walking I found, as I anticipated, the carcase of a zebra, with a numerous company of Vultures busily discussing it.

[The drawings of birds from Damara Land and the adjoining countries, which were executed by Mr. Baines for the late Mr. Andersson, include this species. I have not myself had the opportunity of examining a Damara specimen.—ED.]

4. **Vultur occipitalis,** Burch.   White-headed Vulture.

Vultur eulophus, Hemprich & Ehrenberg's Aves, pl. 14.
Vultur occipitalis, Layard's Cat. No. 4.

I do not remember to have met with this fine Vulture in Damara Land, but have observed it (though only at a distance) on a few occasions in Great Namaqua Land.

[Mr. Andersson's collection of ornithological drawings contains an excellent portrait, from the pencil of Mr. Baines, of an adult bird of this species, which appears, by a memorandum attached to the drawing, to have been obtained "near Seeo-Kaama Hill, the Koppes, S.W. of Lake Ngami, March 2, 1862." I have not myself seen it from Damara Land.—ED.]

5. Gyps Kolbii (Daud.). South-African Griffon-Vulture.

Le Chassefiente, Levaillant's Ois. d'Afr. pl. 10.
Gyps fulvus, Layard's Cat. No. 6.
Gyps Kolbii, Gray's Hand-list of Birds, No. 9.
Vultur fulvus (part.), Finsch & Hartlaub's Vögel Ost-Afrika's, p. 1.

This Vulture is sparingly found in Damara Land. I have chiefly observed it in the vicinity of the sea, above Oosop rocks on the lower course of the Swakop River.

[I have had some hesitation in treating this Vulture, which is evidently the southern representative of *Gyps fulvus*, as specifically distinct from its more northern congener, from which however, it constantly differs in being less fulvous when in immature plumage, and also in having its head and neck less completely clothed with downy hairs and plumelets—differences which may perhaps suffice to constitute a specific distinction.

I have seen specimens from various parts of South Africa, but have not personally examined one from Damara Land.—ED.]

6. Gyps Rüppelli, Bon. Rüppell's Griffon-Vulture.

Vultur Kolbii, Rüppell's Atlas Reise nörd. Afr. pl. 32.
Gyps vulgaris, Layard's Cat. No. 7.
Gyps Rüppellii, Gray's Hand-list of Birds, No. 12.

[Mr. Andersson's last collection contained a Vulture of this species shot in Ondonga, Ovampo Land, on the 14th November, 1866.—ED.]

FALCONIDÆ.

7. Aquila vulturina, Daud. Verreaux's Eagle.

Aquila Verreauxii, Des Murs, Zool. de la Voyage en Abyssinie, pl. 5.
   „   „   Layard's Cat. No. 13.
Pteroaëtus vulturina, Gray's Hand-list of Birds, No. 98.

This Eagle nests in Little Namaqua Land on lofty

rocks; but I cannot specify an instance of its occurrence to the northward of the Orange River. It is at once a friend and a foe to the farmer, as, though it occasionally devours the young lambs and kids, it is said to warn the Boer of another enemy, the leopard, to which it attracts notice by its piercing cries and by circling over the spot where the intruder has appeared. It generally selects some projecting inaccessible ledge, on which it constructs a large eyrie composed of sticks. The egg is nearly oval and of a whitish colour, sparsely spotted with brown, more especially towards the thicker end; its length is 3" 8''', and breadth 2" 7'''.

[The identification of this Eagle rests on Mr. Andersson's authority, as his collection did not contain an example of it.—ED.]

8. Aquila nævioides, Cuv. Tawny Eagle.

Aquila nævioides, Cuvier's Règne An. ed. 2, vol. i. p. 326.
Falco belisarius, Levaillant jeune, in Expl. de l'Algérie, Ois. pl. 2 (adult in worn plumage).
Aquila nævioides, Lord Lilford in Ibis, 1865, pl. 5 (adult newly moulted, and young in pale plumage).
Aquila mogilnik, Alléon in Rev. de Zool. 1866, pl. 20 (young in dark plumage).
Aquila senegalla, Layard s Cat. No. 11.
Aquila nævioides, Chapman's Travels in S. Afr., App. p. 389.
 ,, ,, Elwes & Buckley in Ibis, 1870, p. 67.

The Tawny Eagle is not uncommon in Damara Land and Great Namaqua Land. It perches usually on the topmost branches of lofty trees, often remaining stationary for hours together; it builds on the top of high and generally of inaccessible trees, and constructs a large nest of dry sticks. It is very destructive to the young of diminutive antelopes, and to hares, bustards, and plovers; but it also feeds largely (and, I fancy, by choice) on carrion, besides which it devours fish, frogs, and earthworms; it pursues and plunders other less powerful birds of prey, and also robs the sportsman of wounded game.

In immature birds the iris is brown, in adults pale yellow freckled with brown, and with a narrow exterior edging of dark brown; the bill is dark horn-colour, but livid at the base and with the under mandible yellow; the gape, cere, and feet are also yellow.

9. **Hieraëtus pennatus** (Gmel.). Booted Eagle.

Aquila pennata, Gould's Birds of Europe, pl. 9.
„ „ Layard's Cat. No. 10.
Hieraëtus pennatus, Gray's Hand-list of Birds, No. 100.

[Mr. Andersson's last collection contained one of these Eagles obtained in Ondonga, Ovampo Land, on 14th November, 1866. —Ed.]

10. **Pseudaëtus spilogaster** (Dub.). Spotted-breasted Hawk-Eagle.

Spizaëtos zonurus, Müller's Ois. d'Afrique, pl. 1 (adult).
Spizaëtus Ayresii, Gurney in Ibis, 1862, pl. 4 (immature).
Spizaëtus spilogaster, Layard's Cat. No. 17.
Aquila Bonellii, Layard's Cat. No. 12.
„ „ Chapman's Travels in S. Afr., App. p. 389.
Aquila fasciata, Gurney in Ibis, 1868, p. 138.
Aquila spilogaster, Finsch & Hartlaub's Vögel Ost-Afrika's, p. 48.

I have obtained examples of this species at Objimbinque, the mouth of the Onanès River, Bull's Port, the Omaruru River, and Ondonga. Its flight is heavy; but when it has once risen to a certain height it soars powerfully; it perches on trees and rocks, and is fond of returning to the same tree or other post of observation. Its food consists of small quadrupeds and birds of all kinds.

The irides are yellow, but paler in the immature than in the adult bird; the toes, cere, and basal part of the mandibles greenish yellow; the anterior part of the bill dark horn-colour, almost black at the point.

Measurements of a male:—

	in.	lin.
Entire length	23	16
Length of folded wing	15	7
,, tarsus	3	10
,, middle toe	2	6
,, tail	10	9
,, bill	1	10

[It appears to me that this and the succeeding species belong most properly to the genus *Pseudaëtus* of Hodgson, of which the type is the Bonelli's Eagle (*Aquila fasciata* of Vieillot), to which the present species, in particular, is very closely allied.

An individual of this species, which was incorrectly identified by me some years since as *Aquila Bonellii* (= *fasciata*), led Mr. Layard into the error of introducing the latter species into his Catalogue of South-African Birds, *loc. cit.*—Ed.]

11. Pseudaëtus bellicosus (Daud.). Martial Hawk-Eagle.

Aquila bellicosa, Smith's Zool. of S. Afr. pl. 42.
Spizaëtus bellicosus, Layard's Cat. No. 15.
Aquila bellicosa, Chapman's Travels in S. Afr., App. p. 389.

I never identified but one pair of these Eagles, which I found close to Objimbinque, and the female of which I killed; they were both very wild, always perching on the topmost branches of the loftiest trees.

Measurements of the female above referred to:—

	in.	lin.
Entire length	32	10
Length of folded wing	25	0
,, tarsus	5	5
,, middle toe	3	3
,, tail	13	0
,, bill	2	9

[The female bird above alluded to, a fine adult specimen, formed a part of Mr. Andersson's collection of specimens; and his portfolio of coloured drawings comprised one by Mr. Baines of an adult of this species, to which is attached the following note:—" Eagle supposed to be living on a brood of young ostriches, having killed one on the morning of this day, March 18th, 1862, between Mount Lubels and Mount Nguiba, twenty or twenty-five miles south of Lake Ngami "*. Mr. Andersson states in his notes that he believes that he also observed the Crowned Eagle (*Spizaëtus coronatus*) at least twice during his travels, but, not having succeeded in obtaining a specimen, he was unable to include it as an ascertained Damara-Land species. —Ed.]

12. **Haliaëtus vocifer** (Daud.). African Sea-Eagle.

Haliaëtus vocifer, Des Murs, Icon. Orn. pl. 8.
    ,,   ,, Layard's Cat. No. 21.
    ,,   ,, Chapman's Travels in S. Afr., App. p. 389.

This very handsome species is not an inhabitant of either Damara or Great Namaqua Land, but is tolerably common in the Lake-region and its watersheds and also along the course of the Okavango.

[Mr. Andersson's portfolio contained a portrait by Mr. Baines of an adult Eagle of this species obtained on the Botletlé River on 30th April, 1862, but I have not personally examined a Damara specimen.—Ed.]

* *Conf.* Baines, ' Explorations in South-West Africa,' pp. 401, 402.

13. Circaëtus pectoralis, Smith. Black-breasted Harrier-Eagle.

Circaëtus pectoralis, Strickland's Orn. Syn. No. 75.
„ *thoracicus,* Layard's Cat. No. 18.

On March 1st, 1865, I observed an adult of this species soaring very low, just in front of my window*; and I have subsequently killed this bird in Damara Land.

[I am not aware that any plate of this fine species has as yet been published. *Circaëtus cinereus* of Vieillot, of which *C. funereus* of Rüppell is a synonym, is a somewhat larger species, and is darker in its colouring than the immature bird of *C. pectoralis,* with which some naturalists have supposed it to be identical. *C. pectoralis* has been ascertained by Mr. Ayres to feed chiefly on reptiles.—ED.]

14. Helotarsus ecaudatus (Daud.). Rufous-backed Bateleur Eagle.

Le Bateleur, Levaillant's Ois. d'Afr. pls. 7 & 8.
Helotarsus ecaudatus, Layard's Cat. No. 23.
„ „ Chapman's Travels in S. Afr., App. p. 361.

This is probably the most common Eagle in Damara and Great Namaqua Land; it is usually found in plains, and builds its nest on trees. The nest consists of a large mass of sticks pretty firmly bound together without any kind of lining. Several pairs bred in the neighbourhood of my place at Objimbinque; but their nests were always exceedingly difficult of access, on account of the terribly thorny nature of the trees on which they constructed their eyries.

The iris of this species is of a magnificently brilliant and transparent brown; the legs are a light orange; the basal part of the bill a rich dark orange, the tip black, and the intermediate part yellow.

* Probably at Objimbinque.

[The notes of coloration given above by Mr. Andersson have probably been taken from a bird not fully adult, as in old birds the feet and bare skin between the eye and the bill are of a coral-red. It is, however, remarkable (as I am informed by Mr. Bartlett of the Zoological Society's Gardens in the Regent's Park) that, if the bird be irritated, the coral-red of all these parts fades temporarily to an orange-yellow, as he has frequently observed in the case of living specimens which he has had under his care. Mr. Chapman (*loc. cit.*) states that this species is found at Lake Ngami, and gives a remarkable account of his being fiercely and pertinaciously attacked by one of these birds, at which he had fired as it was sitting on its nest: this nest, like those observed by Mr. Andersson, was on a tree; but an instance of this Eagle nesting on a rocky cliff is recorded in 'The Ibis,' 1868, p. 140.—ED.]

15. Buteo jackal (Daud.). Jackal Buzzard.

Le Rounoir, Levaillant's Ois. d'Afr. pl. 16.
Buteo jackal, Layard's Cat. No. 8.

To the best of my belief this Buzzard has never been observed in Damara Land; and it is scarce in Great Namaqua Land, where, indeed, I have only occasionally observed it in the southern parts, usually in the neighbourhood of rocks, on which it perches.

When hunting, it soars steadily aloft, swooping down suddenly with great force and velocity.

Its chief food consists of moles, rats, mice, lizards, and insects.

Measurements of a male and a female:—

	Male.		Female.	
	in.	lin.	in.	lin.
Entire length	20	3	21	0
Length of folded wing	16	0	17	10
,, tarsus	3	6	3	6
,, middle toe	1	6	1	9
,, tail	8	6	8	5
,, bill	1	9	1	10

16. Buteo desertorum (Daud.). Desert Buzzard.

Falco cirtensis, Levaillant jeune, in Expl. de l'Algérie, *Ois.* pl. 3.
Falco tachardus, Bree's Birds of Europe, vol. i. p. 97 (pl.).
Buteo desertorum, Layard's Cat. No. 9.

This species is not uncommon in Ondonga, but it is very wary; the stomach of a female shot by my servant contained two fully fledged Doves (*Turtur senegalensis*); this specimen was excessively fat.

Measurements of a female:—

	in.	lin.
Entire length	19	5
Length of wings when folded	14	2
„ tarsus	3	0
„ middle toe	1	7
„ tail	8	3
„ bill	1	7

[The only difference which I can detect between the South African and the northern races of this Buzzard is, that in the southern race immature specimens usually have more white mingled with the brown plumage of the abdomen than occurs in young birds from more northern localities; but this does not seem to me to be an adequate ground for specific distinction, and I therefore consider Daudin's specific name cited above to be applicable to the northern as well as to the southern examples of this species.—ED.]

17. Falco minor, Bon. South-African Peregrinoid Falcon.

Falco minor, Layard's Cat. No. 25.

I never observed but one individual of this Falcon, a young female, which I obtained at Objinere, about two days journey from Objimbinque.

[Mr. Andersson's last collection contained an adult male of this species, obtained in Ondonga on 30th January, 1867, which was doubtless subsequent to the date of the note above recorded. This species has not as yet been figured. In the English names which I have appended to this and the preceding species, I have aimed at indicating their near affinity to the two more northern Falcons with which they are respectively most closely allied.—ED.]

18. Falco cervicalis, Licht. South-African Lanneroid Falcon.

Falco biarmicus, Temminck's Pl. Col. pl. 324.
   ,,   ,,   Gurney, Birds Damar., Proc. Zool. Soc. 1864, p. 2.
   ,,   ,,   Andersson, Birds Damar., Proc. Zool. Soc. 1864, p. 3.
   ,,   ,,   Layard's Cat. No. 27.
   ,,   ,,   Chapman's Travels in S. Afr., App. p. 390.
Falco cervicalis, Gray's Hand-list of Birds, No. 172.

This Falcon (called in Damara Land "Onikothé") is to be met with from the Cape Colony in the south to the Okavango River in the north, and as far eastward as Lake Ngami; it is particularly numerous in Little Namaqua Land and also in the neighbourhood of the Okavango, and it is occasionally seen along the seacoast. It preys chiefly on birds and has a very powerful and sweeping flight.

Measurements of a male and a female:—

	Male.	Female.
	in. lin.	in. lin.
Entire length	16 3	18 0
Length of folded wing	12 3	13 6
,, tarsus	2 0	2 2
,, middle toe	1 7	1 11
,, tail	6 10	6 0
,, bill	1 0	1 3

The female bird, of which the measurements are here given, flew fiercely above me, and so near that I felt the air driven by its wings against my face.

[The tail in this female specimen, which formed part of the

collection left by Mr. Andersson at his decease, appears to have been recently moulted and not fully re-grown.

Mr. Andersson mentions another female specimen in which the tail measured 7" 6'''.—ED.]

19. Chicquera ruficollis (Swains.). Rufous-necked Falcon.

Falco ruficollis, Swainson's Birds of West Africa, vol. i. pl. 2.
Hypotriorchis ruficollis, Layard's Cat. No. 28.
Falco ruficollis, Chapman's Travels in S. Afr., App. p. 390.
Chicquera ruficollis, Gray's Hand-list of Birds, No. 196.
Falco ruficollis, Finsch & Hartlaub's Vögel Ost-Afrika's, p. 72.

This beautiful Falcon is rare in Damara and Great Namaqua Land and in all other parts of South Africa which I have traversed; it is invariably found in pairs, and perches on the tops of trees, from whence it watches by the hour for its prey, which consists of birds and insects. It feeds voraciously on white ants whenever it has the opportunity.

The iris is a deep brown, the legs, cere, and base of bill bright yellow, point of bill bluish.

Measurements of a male and a female :—

	Male. in. lin.	Female. in. lin.
Entire length	11 11	13 7
Length of folded wing	8 1	9 1
,, tarsus	1 6	1 7
,, middle toe	1 3	1 4
,, tail	5 4	6 2
,, bill	0 10	1 0

20. Hypotriorchis subbuteo (Linn.). British Hobby.

Falco subbuteo, Gould's Birds of Europe, pl. 22.
Hypotriorchis subbuteo, Layard's Cat. No. 29.
Falco subbuteo, Chapman's Travels in S. Afr., App. p. 390.

This handsome species occasionally makes its appearance in Damara Land during the rainy season, and is then often found in company with the myriads of *Erythropus vespertinus, Tinnunculus rupicolus, Milvus migrans,* &c. which appear nearly simultaneously. During its temporary visit to Damara Land its chief food consists of locusts and white ants, which usually abound during the rainy season.

The irides are dark brown, the cere greenish yellow, the bare skin round the eye yellowish, the legs and toes clear yellow.

Average dimensions of two females :—

	in.	lin.
Entire length	13	4
Length of folded wing	10	8
,, tarsus	1	5
,, middle toe	1	5
,, tail	5	11
,, bill	0	11

[The Hobby appears to occur in Ovampo Land as well as in Damara Land, a specimen from Ondonga being comprised in Mr. Andersson's last collection.—ED.]

21. Erythropus vespertinus (Linn.). Western Red-footed Hobby.

Falco rufipes, Gould's Birds of Europe, pl. 23.
Falco vespertinus, Hartlaub's Orn. West-Africa's, No. 738.
Erythropus vespertinus, Gurney, Birds Damar., Proc. Zool. Soc. 1864, p. 2.
Falco rufipes, Chapman's Travels in S. Afr., App. p. 391.
Erythropus vespertinus, Gray's Hand-list of Birds, No. 213.
Falco vespertinus, Sharpe & Dresser's Birds of Europe, pl. 1.

This pretty Falcon strongly resembles the Hobby both in form and habits, but is much more numerous; it usually arrives in Damara and Great Namaqua Land about the rainy season, and again retires northwards upon the approach of the dry season. During these annual visits it is exceedingly abundant and may be counted by hundreds and by thousands; nay, their numbers at times exceed all belief. On one particular occasion a friend of mine and myself attempted to form a rough approximation to the number of these birds actually within sight, and of the black and yellow-billed Kites with which they appeared to be mixed up in about equal proportions. Taking a small section of the sky, we came to the conclusion, by counting and estimating, that there were at least ten thousand individuals; and as the heavens above and all around us appeared to be darkened by a living mass of Kites and Hawks, we set down the aggregate number immediately within our view at fifty thousand, feeling at the same time that we were probably below the mark.

These birds, during their stay in Damara Land, feed almost exclusively on white ants, on which they fatten amazingly, as does every other bird and animal that diets upon this insect; locusts are another favourite kind of food, but less certain.

The irides in this Falcon are dark brown, the upper part of the base of the bill, the ring round the eyes, the legs and the toes are all reddish orange; the points of the mandibles are bluish.

[Mr. Andersson's last collection contained examples of this species obtained in Ovampo Land as well as in Damara Land. —Ed.]

22. Erythropus amurensis (Radde). Eastern Red-footed Hobby.

Falco vespertinus, var. *amurensis*, Radde, Sibir. Reise, p. 102, pl. 1. fig. 2.
Falco amurensis, Von Homeyer, in Journal für Orn. 1868, p. 251.
Erythropus amurensis, Gurney, in Ibis, 1868, pl. 2.
„   „   Gray's Hand-list of Birds, No. 214.
Falco Raddei, Finsch & Hartlaub's Vögel Ost-Afrika's, p. 74.

[An example of this species, which was obtained by Mr. Andersson in Damara Land, is recorded in 'The Ibis' for 1868, p. 42. An adult male and an immature male, but both without tickets attached, were comprised in Mr. Andersson's last collection.—ED.]

23. Tinnunculus cenchris (Frisch). Western Grey-winged Kestrel.

Falco tinnunculoides, Gould's Birds of Europe, pl. 27.
Tinnunculus cenchris, Gurney, Birds Damar., Pr. Zool.Soc.1864,p. 2.
„   „   Layard's Cat. No. 31.
Falco cenchris, Chapman's Travels in S. Afr., App. p. 390.
Tinnunculus cenchris, Sharpe & Dresser's Birds of Europe, pl. 23.

This species is rather scarce in Damara Land, and only makes its appearance during the rainy season, when it is found in company with *Tinnunculus rupicolus, Erythropus vespertinus, Milvus migrans,* &c.

The irides are brown, the basal part of the bill greenish yellow merging into bluish black, bare skin round the eye yellow, legs and toes pale yellow.

Measurements of a male :—

		in.	lin.
Entire length		12	0
Length of folded wing		9	7
„	tarsus	1	5
„	middle toe	1	0
„	tail	5	10
„	bill	0	10

24. Tinnunculus alaudarius (Gmel.).   British Kestrel.
Falco tinnunculus, Gould's Birds of Europe, pl. 26.
„   „   Hartlaub's Orn. West-Africa's, No. 20.
Tinnunculus alaudarius, Gray's Hand-list of Birds, No. 203.
Falco tinnunculus, Sharpe & Dresser's Birds of Europe, pl. 12.

[The only South-African example of this Kestrel which has come under my notice is a female killed at Objimbinque, Damara Land, on 1st February, 1865, and comprised in Mr. Andersson's last collection. This specimen, which is preserved in the Norwich Museum, is of the ordinary European type.—ED.]

25. Tinnunculus rupicolus (Daud.).   Lesser South-African Kestrel.
Le Montagnard, Levaillant's Ois. d'Afr. pl. 35.
Tinnunculus rupicolus, Strickland & Sclater, Birds Dam., Cont. Orn. 1852, p. 142.
„   „   Layard's Cat. No. 32.

Next to *Erythropus vespertinus*, this is the commonest species of Falcon in Damara and Great Namaqua Land; it is one of the few Hawks which remain in Damara Land during the dry season; but its numbers are then lessened by a partial migration to more favoured regions. These Kestrels perch on rocks and trees indiscriminately; but I have found that, when they have the choice of both, they generally rest on the trees during the heat of the day, but on the least approach of danger retreat to the hills. They usually nest on rocks; but I have found exceptions to this rule. The nest is composed of sticks, and lined internally with hair and feathers. They lay from six to eight eggs, very similar to those of the European Kestrel.

This species feeds on mice, lizards, beetles, &c. There does not seem to be any very marked difference in size between the male and female.

The iris is dull brown, the bare skin round the eye and the legs are yellow.

26. Tinnnunculus rupicoloides (Smith). Greater South-African Kestrel.

> *Falco rupicoloides*, Smith's Zool. of S. Africa, pl. 92.
> *Tinnunculus rupicoloides*, Gurney, Birds Damar., Proc. Zool. Soc. 1864, p. 2.
> „   „   Layard's Cat. No. 33.
> *Falco rupicoloides*, Chapman's Travels in S. Afr., App. p. 390.
> „   „   Finsch & Hartlaub's Vögel Ost-Afrika's, p. 76.

This species is very sparingly met with in Damara Land, but is a little more frequent as one approaches the Orange River; it is also found at Lake Ngami.

The iris is whitish, deeply impregnated with brown; the eyelid and cere are yellow; the bare space round the eye and the basal part of the bill livid yellowish, the remainder of the bill horn-colour; the legs and toes bright yellow.

[In addition to the localities above given, this species occurs in Ondonga, Ovampo Land, Mr. Andersson's last collection having contained specimens from that locality.—ED.]

27. Polihierax semitorquatus (Smith). African Dwarf Falcon.

> *Falco semitorquatus*, Smith's Zool. of S. Africa, pl. 1 (male, adult).
> *Poliohierax semitorquatus*, Andersson, in Proc. Zool. Soc. 1864, p. 4.
> *Hypotriorchis castanonotus*, Heuglin, in Ibis, 1860, p. 407, and Sclater, in Ibis, 1861, pl. 12 (female, adult).
> *Falco semitorquatus*, Chapman's Travels in S. Afr., App. p. 391.
> „   „   Heuglin's Orn. Nord-Ost-Afr. pl. 1 (male, immature).
> *Polihierax semitorquatus*, Gray's Hand-list of Birds, No. 197.
> *Falco semitorquatus*, Finsch & Hartlaub's Vögel Ost-Afrika's, p. 77.

This exquisite little Falcon may be regarded as very

rare in both Great Namaqua and Damara Land, especially in the latter, where I have only seen it once or twice; altogether I have probably not come across above eight individuals, three-fourths of which I have secured.

It is always met with in pairs, and usually perches on bushes or on the lower or middle branches of small trees, though I have seen it on the topmost boughs of lofty trees. I never saw it soar like other Falcons; it is not shy, and when disturbed it never moves further than to the next conspicuous tree or bush.

It feeds on small birds, mice, lizards, and coleopterous insects, the latter being, I apprehend, its chief food.

28. Elanus cæruleus (Desfont.). Sonnini's Kite.

Elanus melanopterus, Gould's Birds of Europe, pl. 31.
    ,,   ,,   Layard's Cat. No. 38.
Falco melanopterus, Chapman's Travels in S. Afr., App. p. 392.

This is a rare bird in Damara Land, and is only found about the estuaries of some of the large periodical watercourses; but it is not uncommon at Lake Ngami and its watersheds. It is invariably seen either singly or in pairs, and is usually difficult to approach on account of its watchful habits and from its perching on the tops of trees or lofty bushes. A pair observed in Ondonga were heard to whistle to each other as they flew from tree to tree. This species hunts at a moderate height, but now and then ascends to a considerable elevation, and at such times may be heard to utter a succession of piercing cries.

The irides are deep bright orange; the bill yellow at the base, the remainder being bluish black; the legs are yellow.

Measurements of a male and a female:—

	Male. in. lin.	Female. in. lin.
Entire length	11 9	12 5
Length of folded wing	9 8	10 9
„ tarsus	1 0	1 3
„ middle toe	1 0	1 0
„ tail	4 6	4 9
„ bill	1 0	1 1

29. Milvus migrans (Bodd.). Black Kite.

Milvus ater, Gould's Birds of Europe, pl. 29.
  „   „   Layard's Cat. No. 36.
  „   „   Chapman's Travels in S. Afr., App. p. 392.
Milvus migrans, Newton's Edition of Yarrell's Br. Birds, vol. i. p. 98.

The Black Kite appears in Damara and Great Namaqua Land with the first rains or even before; the earliest arrival that I have noticed was on the 24th August, or about two months before any rain would fall. Usually it arrives in October and November: at first only a few individuals make their appearance; but in a few days their name is legion; indeed this Kite and its congener, the Yellow-billed Kite, are then more abundant than almost any other species of bird. In 1866 the first Kites were unusually late, and did not appear till the 23rd December. The Black Kite is a very bold and fearless bird; it is by no means an uncommon occurrence for it to swoop down under your very nose and carry off the meat set before you, and I have even known it to snatch a piece of flesh out of a person's hand. It attacks birds much its superior in size and strength, either with the view of depriving them of their prey or from sheer pugnacity. The chief food of this species consists of carrion and offals; but it devours with equal relish fish, mice, lizards, snakes, insects of all kinds, especially locusts, and not unfrequently it also proves destructive to young poultry.

The irides are brown, the bill black, the tarsi and feet lemon-yellow.

Measurements of a male:—

	in.	lin.
Entire length	20	3
Length of folded wing	15	10
,, tarsus	2	0
,, middle toe	1	5
,, tail	0	1
,, bill	1	7

[Mr. Andersson's last collection contained specimens of this Kite from Ondonga, in both adult and immature plumage; the specimens in apparently adult dress did not, however, exhibit the grey tints on the head which distinguish the adult Black Kites of Europe and of Northern Africa, but which I have not yet met with in any South-African specimen.—ED.]

30. Milvus Forskahli (Gmel.).   Yellow-billed Kite.

Le Parasite, Levaillant's Ois. d'Afr. pl. 22.
Milvus parasiticus, Strickland & Sclater, Birds Damar., Cont. Orn. 1852, p. 142.
Milvus Forskahli, Strickland's Ornitholog. Synonyms, No. 225.
Milvus parasiticus, Layard's Cat. No. 37.
Milvus ægyptius, Chapman's Travels in S. Afr., App. p. 392.
Milvus Forskali, Finsch & Hartlaub's Vögel Ost-Afrika's, p. 63.

What has been said of the Black Kite will equally apply to this species, which, however, is probably the more common of the two in Damara and Great Namaqua Land.

Measurements of a male:—

	in.	lin.
Entire length	21	0
Length of folded wing	16	6
,, tarsus	2	0
,, middle toe	1	6
,, tail	10	4
,, bill	1	7

[A specimen of this Kite contained in Mr. Andersson's last collection was obtained in Ondonga, Ovampo Land, which must therefore be recorded as an additional locality for this species.

Mr. Campbell, a gentleman formerly residing at Abeokouta, in West Africa, informed me that this Kite is sometimes extremely common in that locality, but always leaves in the rainy season, which appears to be the time when it makes its migratory appearance further south in Damara Land.—Ed.]

31. Machærhamphus Anderssoni (Gurney). Andersson's Pern.

Stringonyx Anderssoni, Gurney, in Proc. of Zool. Soc. 1865, p. 618.
Machaerhamphus alcinus, Bartlett, in Proc. of Zool. Soc. 1866, p. 324.
Macheirhamphus alcinus, Gurney, in Trans. of Zool. Soc. vol. vi. p. 117, pl. 29.
Machærhamphus Anderssoni, Sharpe in Proc. of Zool. Soc. 1871, p. 502.

On the 10th March, 1865, I obtained one specimen, a female, of this singular bird at Objimbinque, Damara Land; it was shot by my servant, who observed another, probably the male. I imagine that I have myself observed it once or twice in the neighbourhood of Objimbinque just before dusk. When brought to me, I instinctively suspected the bird to be a feeder at dusk or at night, and called out " Why, that fellow is likely to feed on bats!" And truly enough, so it turned out; for, on dissection, an undigested bat was found in the stomach; and in another specimen, subsequently killed by Axel, there were several bats in the stomach.

In this species the irides are bright lemon-yellow, extremities of mandibles black, basal parts and gape bluish lead-colour, tarsi and toes bluish white.

Measurements of a male and a female:—

	Male.	Female.
	in. lin.	in. lin.
Entire length	17 8	18 2
Length of folded wing	13 9	13 11
„ tarsus	2 2	2 5
„ middle toe	2 0	2 1
„ tail	7 3	7 6
„ bill	1 10	1 9

[The female specimen above referred to, which was obtained at Objimbinque on the 10th March, 1865, was presented to me by Mr. Andersson, and was added to the collection of Raptorial birds in the Norwich Museum. Amongst the skins of birds left by Mr. Andersson at his decease was the male specimen which he subsequently obtained; and this, on the sale of his collection, was secured for the British Museum, in the ornithological gallery of which it is now exhibited. These two specimens only differ from each other in the somewhat smaller dimensions of the male bird. The female example was described by me in the 'Proceedings of the Zoological Society' for 1865, under the belief that it was both generically and specifically new, and I suggested for it the name of *Stringonyx Anderssoni*.

Mr. A. D. Bartlett very obligingly undertook to mount this specimen for the Norwich Museum; and whilst the bird was thus passing through his hands, it occurred to him that it certainly belonged to the same genus, and probably to the same species, as a bird in the Museum at Leyden, which had been figured and described by Mr. G. F. Westerman, under the name of *Machaerhamphus* alcinus*, in the first volume of a scientific work published at Amsterdam under the title of 'Bijdragen tot de Dierkunde, &c.' This specimen had been purchased for the Leyden

---

* Equivalent to *Machærhamphus* in the type used by English printers.

Museum from Mr. Frank, the well-known dealer in birds and animals, and was stated to have been brought from Malacca; but Mr. Bartlett was of opinion that this locality had been erroneously assigned to it, and that the bird was really a native of Damara Land, especially as many birds collected by Mr. Andersson in Damara Land had passed through the hands of Mr. Frank about the same date as that at which the *Machærhamphus* was acquired for the Leyden Museum; and it was therefore presumed that some accidental confusion of tickets might have caused a mistaken habitat to be asssigned in error to this specimen. Mr. Bartlett's views on this subject were recorded in the 'Proceedings of the Zoological Society' for 1866, p. 324; and as I concurred in his opinion, the female specimen from Damara Land, now in the Norwich Museum, was figured and described under the name of *Macheirhamphus alcinus* in the 'Transactions of the Zoological Society,' vol. vi. pl. 29.

This conclusion, however, has proved incorrect—two specimens of the true *Machærhamphus alcinus*, agreeing with that at Leyden, having subsequently occurred, an examination of which has proved that the Damara-Land bird, though a nearly allied, is yet a distinct species, and therefore entitled to retain the specific name of *Anderssoni* which I originally proposed for it.

Of the two additional specimens of *M. alcinus* above referred to, one is in the possession of Count Turati, of Milan (as I am informed by my friend M. Jules Verreaux); but the locality whence it was obtained has not been recorded.

The second additional specimen, which is now in the collection of Viscount Walden, was obtained by the late Dr. Maingay at Malacca, thus confirming the correctness of the locality originally assigned to the Leyden specimen.

Mr. R. B. Sharpe in an able paper on this subject, published in the 'Proceedings of the Zoological Society' for 1871, thus sums up the distinctions between these two nearly allied species:—
"The Malaccan species coincides with the Damara bird in the form and style of plumage, having the white ring round the eye and the stripe down the throat, but differs in its larger bill, darker colours, brown abdomen, and long occipital crest: there seems, however, to be a difference in the white feathers round the eye: *M. Anderssoni* has a white superciliary line and a white spot

below the eye; *M. alcinus* has the latter plainly mottled, but has no distinct supercilium, though the feathers round the rim of the eye are whitish."—ED.]

32. Kaupifalco monogrammicus, Temm. One-streaked Hawk.

Aster monogrammicus, Swainson's Birds of West Africa, vol. i. pl. 4.
Astur monogrammicus, Hartlaub's Birds of West Africa, No. 30.
Melierax monogrammicus, Gurney, Birds Damar., Proc. Zool. Soc. 1864, p. 2.
Kaupifalco monogrammicus, Gray's Hand-list of Birds, No. 352.

[The Norwich Museum contains a male of this species, obtained by Mr. Andersson at Elephant's Vley*, on October 26th, 1859, which is the most southern example of this species that has come under my notice.—ED.]

33. Melierax musicus (Daud.). Chanting Hawk.

Le Faucon chanteur, Levaillant's Ois. d'Afr. pl. 27.
Melierax musicus, Layard's Cat. No. 46.
Falco musicus, Chapman's Travels in S. Afr., App. 391.

This handsome species is one of the most common Hawks in Damara and Great Namaqua Land, where it is found throughout the year; it is particularly abundant about Walvisch Bay, and is usually found in open country. It perches on a dry branch on the top of some low acacia, whence it will watch with the greatest assiduity for hours together. It has a skimming flight, occasionally moving its large wings with a slow and heavy undulation.

From the contents of the stomachs of those I have

* [Mr. Andersson appears to have given this name to a locality about six days' journey to the south of the Okavango River, where he remained encamped from about July 20th to November 23rd, 1859, and made a large collection of birds' skins. *Vide* Andersson's ' Okavango River,' pp. 220, 234, 230, 244.—ED.]

dissected I am inclined to think that the chief food of this species consists of rats, mice, small reptiles, and many kinds of insects; it also occasionally devours young birds.

The irides in adult specimens are of a deep reddish brown; the cere and base of the bill bright brick-red, the rest of the bill dark horn-colour; the legs and feet vermilion.

Measurements of a male and a female:—

	Male. in. lin.	Female. in. lin.
Entire length	20 6	21 5
Length of folded wing	14 0	14 6
,, tarsus	4 0	4 5
,, middle toe	1 7	2 0
,, tail	9 4	10 1
,, bill	1 6	1 7

34. Melierax polyzonus, Rüpp. Rüppell's many-zoned Hawk.

Nisus polyzonus, Rüppell's Fauna Abyssin. pl. 15.
Melierax polyzonus, Gurney, Birds Damar., Proc. Zool. Soc. 1864, p. 2.
,, ,, Gray's Hand-list of Birds, No. 354.

[An adult male of this species was obtained by Mr. Andersson at Elephant's Vley on November 10th, 1859, and is preserved in the Norwich Museum. I know no other instance of this species occurring so far south; and it is remarkable that it should have been obtained at the same spot, and within fifteen days of the same date, as the specimen of *Kaupifalco monogrammicus* previously referred to.

Mr. Andersson, in his notes on some of the Birds of Damara Land, published in the Zoological Society's 'Proceedings' for 1864, p. 4, erroneously gave the name of this species to *Accipiter polyzonoides*, Smith, the bird to which he then applied the latter name being in reality *A. tachiro* (Daud.).—ED.]

35. Melierax gabar (Daud.). Gabar Hawk.

Le Gabar, Levaillant's Ois. d'Afr. pl. 33.
Accipiter gabar, Strickland & Sclater, Birds Damar., Cont. Orn. 1852, p. 142.
,,   ,,   Andersson in Proc. of Zool. Soc. 1864, p. 4.
Melierax gabar, Layard's Cat. No. 44.
Accipiter gabar, Chapman's Travels in S. Afr., App. p. 391.
Nisus gabar, Finsch & Hartlaub's Vögel Ost-Afrika's, p. 86.

This is one of the commonest Hawks in Damara and Great Namaqua Land, as well as in the Lake-regions.

In Damara Land it is partially migratory, arriving with the rain and retiring on the return of the dry season; nevertheless stragglers remain throughout the year.

Its favourite resorts are thickly wooded districts, where it hides amongst the foliage, never being seen to perch on the topmost boughs of trees.

It feeds chiefly on mice, lizards, white ants, locusts, and many coleopterous insects.

In the adult birds the anterior parts of the bill, together with the legs and toes, are a bright reddish orange, the irides are a brilliant purple.

Average dimensions of seventeen males and of eleven females:—

	Males. in. lin.	Females. in. lin.
Entire length	11 10	13 8
Length of folded wing	7 3	7 11
,, tarsus	2 0	2 0
,, middle toe	1 3	1 5
,, tail	6 1	6 8
,, bill	0 9	0 10

36. Melierax niger (Vieill.). African Black Hawk.

Sparvius niger, Vieillot's Gal. des Ois. pl. 22.
Accipiter niger, Strickland & Sclater, Birds Damar., Contr. to Orn. 1852, p. 142.
Melierax niger, Layard's Cat. No. 45.
Nisus niger, Finsch & Hartlaub's Vögel Ost-Afrika's, p. 88.

This rather scarce Hawk is usually met with either singly or in pairs. I do not recollect having seen it in Great Namaqua Land; but it is found to the north as far as the Okavango and eastward to the lake, though nowhere numerous. It is not particularly shy; its food consists of small birds, mice, &c.

The irides are cherry-colour; the base of the bill yellowish red, the tips of the mandibles black; the legs and toes vermilion, with the scutellated parts yellowish.

Measurements of a male and a female:—

	Male. in. lin.	Female. in. lin.
Entire length	12 9	13 6
Length of folded wing	7 6	8 0
„ tarsus	1 10	1 11
„ middle toe	1 3	1 4
„ tail	6 7	7 0
„ bill	0 10	0 10

37. Accipiter tachiro (Daud.). Tachiro Sparrow-Hawk.

Falco tachiro, Temminck's Pl. Col. pl. 377 & 420.
Accipiter tachiro, Gurney, Birds Damar., Proc. Zool. Soc. 1864, p. 2.
Accipiter polyzonoides, Andersson in Proc. Zool. Soc. 1864, p. 4.
Accipiter tachiro, Layard's Cat. No. 40.
Nisus tachiro, Finsch & Hartlaub, Vögel Ost-Afrika's, p. 78.

I only obtained two or three specimens of this Hawk, which is very rare in Damara Land.

38. Accipiter polyzonoides, Smith. Many-banded Sparrow-Hawk.

Accipiter polyzonoides, Smith's Zool. of S. Afr. pl. 11.
    ,,   ,,   Gurney, Birds Damar., Proc. Zool. Soc. 1864, p. 2.
Melierax polyzonus, Andersson, ibid. p. 4.
Accipiter polyzonoides, Layard's Cat. No. 41.
Falco polyzonoides, Chapman's Travels in S. Afr., App. p. 391.
Nisus badius (part.), Finsch & Hartlaub's Vögel Ost-Afrika's, p. 81.

I think the adult birds of this species are rather rare both in Damara and Great Namaqua Land, as I succeeded in obtaining comparatively few of them; the young birds, however, are in some years very abundant. It is a migratory species, arriving in Damara Land after the first rains have fallen and retiring again on the approach of the dry season, though a few individuals probably remain throughout the year. The females are the first to arrive, and are followed by the males after a considerable interval. Both sexes are much emaciated on their first appearance; during their temporary stay in Damara Land they chiefly feed on white ants; but their food also includes grasshoppers, lizards, and mice. They are partial to well and deeply wooded districts, where they seek shelter amongst the foliage and are never seen to perch in any very conspicuous situations.

The base of the upper mandible is yellow, that of the lower mandible bluish black, and the remainder of the bill darkish; the edges of the gape are dusky yellow, the irides bright orange, the legs and toes yellow.

[Drs. Finsch and Hartlaub, *loc. cit.*, blend in one species *Accipiter badius*, Gmel. (= *A. dussumieri*, Temm.), of South-eastern Asia and Ceylon, *Accipiter sphenurus*, Rüpp. (= *A. brachydactylus*, Swains.), of intertropical Africa, and *Accipiter polyzonoides*, Smith, of Southern Africa; but it appears to me that these three local races, though unquestionably very nearly allied, are constantly distinguishable, and may therefore be correctly treated as specifically distinct; as regards the two former, Mr. Blanford's observations in his 'Geology and Zoology of Abyssinia,' p. 294, may be consulted with advantage*.

In Mr. Andersson's notes in the 'Proceedings of the Zoological Society' for 1864, there is an accidental error at page 4 as to the nomenclature of this species, which I have already explained under the head of *Melierax polyzonus*.—ED.]

39. Accipiter minullus (Daud.). Minulle Sparrow-Hawk.

Le Minulle, Levaillant's Ois. d'Afr. pl. 34 (immature).
Accipiter minullus, Gurney, Birds Damar., Pr. Zool. Soc. 1864, p. 2.
,,   ,,   Andersson, ibid. p. 4.
,,   ,,   Layard's Cat. No. 42.
,,   ,,   Chapman's Travels in S. Afr., App. p. 392.

To the best of my recollection I have never met with this bird either in Damara or Great Namaqua Land, though I have observed it sparingly in the Lake-regions and near the river Okavango. Notwithstanding its diminutive size, it is a bold and fearless bird, more especially during the time of incubation, when it will unhesitatingly face enemies more than twice or three times its own size, and is much assisted in its attacks by the swiftness of its flight. A curious incident occurred to myself which will serve to illustrate its daring and recklessness; I was crouching on the ground near a water-hole in the desert for the purpose of securing such birds as might come there to quench their thirst, when a Hawk of this species perched on a dry tree in my neighbourhood, but out of range of my gun; suddenly it left its perch and

* Some valuable remarks on the differences between these three species will be found in Messrs. Sharpe and Dresser's article on *Accipiter brevipes* in their excellent work on the Birds of Europe.

flew straight at me, almost sweeping the ground, and with such rapidity that before I could raise myself and point the fowling-piece, I felt the Hawk's wings fanning my face; and, to save my head from its claws, I had, actually to throw myself on my back, at the same time making a sweep at the bird with my gun; even after this it hovered over me for a while almost within reach of the gun's muzzle, and evidently only abstained with great reluctance from further attempts at molestation.

[I am not aware that any figure of this species in its adult plumage has yet been published.—ED.]

40. Accipiter rufiventris, Smith. Rufous-bellied Sparrow-Hawk.

Accipiter rufiventris, Smith's Zool. of S. Africa, pl. 93.
        „   „   Layard's Cat. No. 43.

This must be a scarce bird in Damara and Great Namaqua Land, as it has but very rarely come under my notice; it is, however, common to the south of the Orange River and in various parts of the Cape Colony; Mr. Layard also found it amongst the birds collected by the Messrs. Chapman in the Lake-regions.

41. Circus Swainsoni, Smith. Swainson's Harrier.

Circus Swainsonii, Smith's Zool. of S. Africa, pls. 43 & 44.
Circus Swainsoni, Gurney, Birds Damar., Proc. Zool. Soc. 1864, p. 2.
    „   „   Andersson, ibid. p. 4.
Circus Swainsonii, Layard's Cat. No. 49.
    „   „   Chapman's Travels in S. Afr., App. p. 392.

This bird is migratory, appearing towards the return of the rainy season in Damara and Great Namaqua Land; I have observed very few adult specimens; but young and middle-aged birds are pretty numerous.

This Harrier chiefly haunts the sides of marshes, the banks of rivers, and other humid places, in search of lizards, mice, moths, white ants, &c. ; it usually flies low and straight, and only occasionally in circles.

[This Harrier also occurs in Ovampo Land, a specimen obtained in Ondonga on January 22nd, 1867, having been contained in Mr. Andersson's last collection. In addition to this species, Mr. Andersson states that *Circus cyaneus* " occurs very sparingly in Damara Land;" but I suspect that in this remark there has been an error of identification, and that the specimens thus referred to in reality belonged either to the present or to the succeeding species, as I have never seen an example of *Circus cyaneus* from any locality south of the equator.—ED.]

42. Circus cinerarius (Mont.). Montagu's Harrier.

Circus cineraceus, Gould's Birds of Europe, pl. 35.
Circus ater, Vieillot's Nouv. Dict. d'Histoire Nat. vol. iv. p. 459 (melanism).
Circus cinerarius, Strickland's Orn. Synonyms, No. 247.
„ „ Gurney, Birds Damar., Proc. Zool. Soc. 1864, p. 2.
Circus cinerascens, Layard's Cat. No. 50.

[Mr. Andersson's record of this species is limited to the following measurements of a male :—

		in.	lin.
Entire length		17	0
Length of folded wing		13	2
„	tarsus	2	4
„	middle toe	1	5
„	tail	8	3
„	bill	1	2

I have seen specimens of this Harrier which were obtained by Mr. Andersson at Objimbinque in Damara Land, and in Ondonga, Ovampo Land.—ED.]

43. Circus maurus (Temm.). Fuliginous Harrier.

Falco maurus, Temminck's Pl. Col. pl. 461 (adult).
Circus maurus, Smith's Zool. of S. Africa, pl. 58 (immature).
„ „ Layard's Cat. No. 52.

I have observed this bird in Little Namaqua Land, but am not aware that it is an inhabitant of Great Namaqua or Damara Land; it frequents the banks of lakes and rivers and marshy places in general, along which it leisurely hunts for birds, mice, lizards, frogs, &c.

44. Circus ranivorus (Daud.).   Levaillant's Harrier.

Le Grenouillard, Levaillant's Ois. d'Afr. pl. 23.
Circus ranivorus, Layard's Cat. No. 51.

I do not recollect to have met with this Harrier in Damara or Great Namaqua Land; but Mr. Layard informs me that he received specimens collected in the Lake-country by Mr. James Chapman.

SERPENTARIIDÆ.

45. Sagittarius secretarius (Scop.).   Secretary bird.

Le Mangeur de Serpents, Levaillant's Ois. d'Afr. pl. 25.
Sagittarius secretarius, Strickland's Orn. Syn. No. 242.
Serpentarius reptilivorus, Layard's Cat. No. 48.
Secretarius reptilivorus, Chapman's Travels in S. Afr., App. p. 392.

The Secretary bird is found sparingly in Great Namaqua and Damara Land and on the plains of Ondonga in the Ovampo country; it also occurs about Lake Ngami. It spends most of its time upon the ground, rarely, if ever, taking to the wing; and if compelled to do so, it is only for very short flights, as it seems to prefer seeking its safety by means of its long legs, which are admirably adapted for running. Its swiftness is wonderful, and it actually seems to skim the ground when briskly pursued; sometimes, however, this confidence in its legs costs the bird its life, when the well-mounted horseman, aware of its terrestrial propensities, steadily pursues it until it becomes too much exhausted to avail itself of its wings, and ultimately falls a prey to its enemy. When undisturbed, it usually stalks about with considerable ease, grace, and dignity; but it is difficult to approach, as its long legs and neck, and its habit of frequenting open and exposed localities, enable it to espy an enemy at a great distance, and thus to guard against any sudden surprise. When seen making steadily for a particular point, it may sometimes be successfully cut off by pressing forward rapidly across its path, as on such occasions, instead of deviating from its straight course, it trusts to its legs for outstripping its pursuer by holding on at all risks, in this respect resembling the Ostrich. The food of the Secretary bird is very various, consisting of snakes, lizards, tortoises, mice, rats, insects of almost every kind, and even young birds; but these latter, I believe, it only devours when distressed by hunger; for amongst the old Dutch colonists it was frequently kept in captivity as an excellent mediator in the poultry-yard, as well as a protector to the young fowls from the attacks of snakes, rats, &c.

Many snakes show fight when attacked by the Secretary bird; and it is a most amusing and ludicrous sight to witness a combat between such different opponents; the bird, however, invariably comes off victorious after a short but desperate resistance: the reptile hisses and darts at the Secretary, which not only skilfully wards off the attack, but, by a rapid succession of violent blows from its formidably armed wings, generally succeeds, in a short time, in prostrating its wily enemy; and sometimes a well-directed blow on the vertebræ of the snake at once ends the combat. As soon as this is accomplished the bird dexterously seizes its fallen enemy in its bill, and, after having well tossed it backwards and forwards, finally puts an end to the death-struggle by transfixing the brain with its powerful beak.

STRIGIDÆ.

46. Strix poensis, Fraser. South-African Screech-Owl.

Strix poensis, Fraser, in Proc. Zool. Soc. 1842, p. 189.
Strix affinis, Layard's Cat. No. 65.
Strix flammea, Chapman's Travels in S. Afr., App. p. 393.

South of the Orange River this Owl is exceedingly common; but north of that river it is a very scarce bird, though widely distributed over all the countries of which these notes treat.

Measurements of a male and a female:—

	Male.		Female.	
	in.	lin.	in.	lin.
Entire length	13	5	13	8
Length of folded wing	11	4	11	3
„ tarsus	2	8	2	6
„ middle toe	1	3	1	3
„ tail	5	0	5	0
„ bill	1	7	1	7

[I believe that no figure of this Owl has yet been published. It is very closely allied to *Strix flammea* of Europe and Northern Africa, from which it appears only to differ in its slightly larger average measurements, in the somewhat deeper colouring of the upper surface generally, and in the under surface being more profusely sprinkled with small dark spots.—ED.]

47. Athene perlata (Vieill.). African Pearl-spotted Owl.

La Chevêchette Perlée, Levaillant's Ois. d'Afr. pl. 284.
Strix perlata, Vieillot's Nouv. Dict. vol. vii. p. 26.
Athene licua, Strickland & Sclater, Birds Damar., Contr. Orn. 1852, p. 142.
„ „ Layard's Cat. No. 56.
Strix senegalensis, Chapman's Travels in S. Afr., App. p. 393.
Athene perlata, Finsch, in Trans. of Zool. Soc. vol. vii. p. 209.

This is the smallest as well as the commonest Owl in Damara Land, Great Namaqua Land, and Ovampo Land. It is met with singly or in pairs, and though nocturnal in its habits it seems nearly as much at home in the daylight as by night, and can see its way perfectly well even in dazzling sunshine. It possesses quite an intelligent look, and hops about on its perch in the liveliest and briskest manner. It is quite tame, allowing a person to approach within a very few paces; and when at last compelled to retreat it only moves to the next convenient perch. At night it utters a shrill cry very unlike the usual hooting of Owls. It appears to feed largely on insects.

The sexes in this species are of about the same size. The iris is bright lemon-yellow, the bill and legs greenish yellow.

[Specimens of this Owl from Damara Land and also from Trans Vaal appear to be identical with those obtained on the Gambia; and I therefore do not here use the specific name of "*licua*" applied by Lichtenstein to South-African examples of this species under the idea that they could be specifically distinguished from those which occur to the north of the equator.—ED.]

48. Tænioglaux capensis (Smith). African Barred-tail Owl.

Athene capensis, Smith's Zool. of S. Africa, pl. 33.
" " Layard's Cat. No. 55.
Tænioglaux capensis, Gurney, in Ibis, 1868, p. 147.

This, I apprehend, is a very scarce bird in Damara Land, as I saw but very few specimens in all my travels. I have never observed it in Great Namaqua Land.

[Mr. Andersson's collection contained a pair of these Owls obtained at the river Cunéne, and also a specimen from Ovampo Land.—ED.]

49. Scops capensis, Smith. Cape Scops Owl.

Scops capensis, Smith in South-African Quarterly Journal, 1834, p. 314 (sub *Scops europæus*).
Scops senegalensis, Strickland & Sclater, Birds Damar., Contr. Orn. 1852, p. 142.
Scops latipennis, Kaup, in Trans. Zool. Soc. vol. iv. p. 223.
Ephialtes capensis, Gurney in Ibis, 1859, p. 242.
Ephialtes senegalensis, Layard's Cat. No. 60.
Strix scops, Chapman's Travels in S. Afr., App. p. 393.
Ephialtes senegalensis, Finsch, in Trans. Zool. Soc. vol. vii. p. 210.

To the best of my belief this is a very rare species in Damara Land; and I have never met with it elsewhere; indeed I had been many years in the Damara country before I became aware of its existence, and at last made the discovery in the following rather singular manner:—
One day, whilst reloading my gun in a wood, my attention was attracted by a noise like that of the gentle and measured tappings of a Woodpecker against the stem of a tree. I approached cautiously the spot from which the sound appeared to proceed; but, after making several turns round the tree, I could find nothing, and began to doubt whether the sound could have been caused by a

Woodpecker, as it was scarcely jarring or hollow enough; still there the sound was. At last, when I was on the point of giving up the search, I espied in a small cavity, caused by the breaking off of a branch close to the stem, a small, dark, and all but immovable mass; I fired at once, and down came one of the prettiest and most perfect little Owls I had ever seen. Undoubtedly this was the bird which had attracted my attention in so peculiar a manner; and the noise was probably caused by the opening and shutting of its bill. At long intervals I obtained three more specimens in the same locality; and as the four consisted of an adult male and female and two young, I inferred that they all belonged to the same family.

The iris in this species is bright yellow, darkest on the outer side of the ring; the bill and toes are a bluish horn-colour.

Measurements of two specimens, both males:—

		in. lin.		in. lin.
Entire length	of one	7 0	of the other	6 10
Length of folded wing	,,	5 4	,,	5 3
,, tarsus	,,	1 1	,,	1 0
,, middle toe	,,	0 8	,,	0 8
,, tail	,,	2 7	,,	2 4
,, bill	,,	0 8	,,	0 8

[I am not aware that any figure of this species has yet been published; a specimen brought from Abyssinia, by Mr. Jesse, and now in the collection of Viscount Walden, though referred by Dr. Finsch (*loc. cit.*) to *S. senegalensis*, appears to me to belong to this species, as it exactly agrees with those obtained in Damara Land by Mr. Andersson. I have seen several Damara specimens of this bird, all of which, as well as the Abyssinian example above referred to, exhibit a much darker tint of grey over the entire plumage than is to be found in any other Old-

World species of the genus *Scops* with which I am acquainted. The only other individual of *Scops capensis* which has come under my notice was obtained in Natal by Mr. Ayres, and is now preserved in the Norwich Museum; this is a more rufous specimen, but appears to me to be nevertheless referable to this species. From a recent comparison of specimens I believe *Scops capensis* to be quite distinct from the West-African *S. senegalensis*, which I have only seen from Senegal, Bissao, and the Gaboon, and which I consider is also distinct from the more northern *S. zorca*. This last-named species occurs in Morocco, Algeria, Egypt, and Nubia, which seems to be its most southern limit.—ED.]

50. Scops leucotis (Temm.). White-faced Scops Owl.

Strix leucotis, Temminck's Pl. Col. pl. 16.
Scops leucotis, Strickland & Sclater, Birds Damar., Contr. Orn. 1852, p. 142.
Tufted Owl, Baines's South-west Africa, woodcut at p. 213.
Ephialtes leucotis, Layard's Cat. No 61.
Strix leucotis, Chapman's Travels in S. Afr., App. p. 393.
Bubo leucotis, Finsch & Hartlaub's Vögel Ost-Afrika's, p. 106.

Next to *Athene perlata*, this is about the most common Owl in Damara Land and the parts adjacent to the northward; it is also pretty frequent in the Lake-regions, but is less frequently observed in Great Namaqua Land.

It is always seen in pairs; and though strictly a night Owl, its vision by day is by no means bad.

This Owl lays two or three pure-white eggs, rather smaller than those of *Otus vulgaris*, very broad, and equally rounded at both ends.

I found nests of this species on the respective dates of September 18th, October 15th, October 31st, and November 10th.

The first of these was in the hollow of a tree; the

female bird and one egg were brought to me; and she laid a second egg during the ensuing night. The second was in one of the hollows or nests in a mass of nests constructed by *Textor erythrorhynchus*; but it appeared as if it had been enlarged by the Owl. The third was a small open stick nest, evidently constructed by the bird itself and composed of a few sticks so loosely put together that the single egg the nest contained could be discerned between the interstices; the Owl remained upon this nest till the tree began to give way under the strokes of the axe, and did not appear to care for our presence or our shouting. The fourth nest was about twelve feet from the ground, and situated on a branch; it was composed of only a few straggling sticks, and was probably an abandoned pigeon's nest.

The iris in this species is bright orange-yellow, and its eyes are exquisitely beautiful; the bill and feet are of a light bluish white. There does not seem to be any marked difference of size between the sexes.

51. Huhua Verreauxi (Bon.). Verreaux's Eagle-Owl.

Bubo verreauxi, Bonaparte's Consp. Avium, p. 49.
Bubo lacteus, Gurney, Birds Damar., Proc. Zool. Soc. 1864, p. 2.
Bubo verreauxi, Layard's Cat. No. 57.
Huhua verreauxi, Gurney, in Ibis, 1868, p. 147.
Nyctaëtus verreauxi, Gray's Hand-list of Birds, No. 450.
Strix lactea, Chapman's Travel's in S. Afr., App. p. 392.

This is the largest Owl found in Damara Land. It is of not unfrequent occurrence from the Okavango River northward, to the Cape Colony in the south, and it is also met with in the Lake-regions; it is generally found in pairs, and perches on large trees; it utters at night, and sometimes during the day, a most formidable, hollow, and sepulchral cry or hoot. Its food consists of mice and other small quadrupeds, birds, lizards, and large beetles. The iris is very dark hazel.

Measurements of a male:—

		in.	lin.
Entire length		23	10
Length of folded wing		16	7
,,	tarsus	3	0
,,	middle toe	2	3
,,	tail	9	8
,,	bill	2	0

[I believe that no figure of this South-African species has yet been published, though the nearly allied but smaller and more northern race, *Huhua lacteus* (Temm.) is figured in the ' Planches Coloriées,' pl. 4.—ED.]

52. Bubo maculosus (Vieill.). Spotted Eagle-Owl.

Strix africana, Temminck's Pl. Col. pl. 50.
Bubo maculosus, Layard's Cat. No. 59.
,, ,, (part.), Finsch & Hartlaub's Vögel Ost-Afrika's, p. 103.

This Owl is sparingly met with in Damara and Great Namaqua Land; its food comprises rats and large beetles. The irides are bright yellow, and the bill black.

[Some naturalists, including Drs. Finsch and Hartlaub,*loc. cit.,* have supposed this species to be identical with *Bubo cinerascens,* of Guérin, from Equatorial Africa, a very distinct species belonging to the genus *Huhua* (or *Nyctaëtus*), and having, like the other species of *Huhua,* dark brown irides, very different from the golden-yellow iris which is universal in the genus *Bubo.*—ED.]

53. Phasmaptynx capensis (Smith). African Short-eared Owl.

Otus capensis, Smith's Zool. of S. Africa, pl. 67.
   ,,   ,,   Layard's Cat. No. 64.
Phasmaptynx capensis, Gray's Hand-list of Birds, No. 553.

[Mr. Andersson's last collection contained examples of this species, one of which was marked as having been obtained in Ondonga, Ovampo Land.—ED.]

PASSERES.

FISSIROSTRES.

CAPRIMULGIDÆ.

54. Caprimulgus rufigena, Smith.   Rufous-cheeked Goatsucker.

Caprimulgus rufigena, Smith's Zool. of S. Africa, pl. 100.
Caprimulgus damarensis, Strickland & Sclater, Birds Damar., Contr. Orn. 1852, p. 143.
Caprimulgus rufigena, Layard's Cat. No. 68.
Caprimulgus damarensis, Chapman's Travels in S. Afr., App. p. 410.

I have found this bird tolerably common in the southern portion of Damara Land; and it is also an inhabitant of Great Namaqua Land. It is met with singly or in pairs, and frequents moderately dense brushwood.

I have, to the best of my recollection, always found these Goatsuckers settled on the ground during the day, and not on the lower branches of small trees as observed by Sir A. Smith.

They are fond of settling in open spaces, and more especially in roads and footpaths; they make their appearance a little before dusk, and appear to be partially migratory, as they are much more common in the rainy than during the dry season.

The irides are very dark brown, the legs and toes bright flesh-colour.

[I have ascertained the identity of *Caprimulgus damarensis* of Strickland with this species by examination of the type specimen preserved in the Museum of Zoology at Cambridge.—ED.]

55. Caprimulgus pectoralis, Cuv. Pectoral Goatsucker.

L'Engoule-vent à collier, Levaillant's Ois. d'Afr. pl. 49.
Caprimulgus pectoralis and *C. rufigena*, Strickland & Sclater, Birds of Damar., Contr. Orn. 1852, p. 143.
Caprimulgus atrovarius, Layard's Cat. No. 72.

This species is tolerably common in the north of Damara Land. It is usually found singly, and is partial to open roads and paths about dusk. Its food consists of beetles and other insects, their eggs, and small seeds.

The irides are deep dark brown, the bill black, the feet grey.

Measurements of a male and a female:—

	Male. in. lin.	Female. in. lin.
Entire length	9 5	9 5
Length of folded wing	6 6	6 6
„ tarsus	0 10	0 9
„ middle toe	0 8	0 9
„ tail	4 9	4 11
„ bill	1 3	1 3

56. Caprimulgus lentiginosus, Smith. Freckled Goatsucker.

Caprimulgus lentiginosus, Smith's Zool. of S. Africa, pl. 101.
„   „   Strickland & Sclater, Birds Damar., Contr. Orn. 1852, p. 143.
„   „   Layard's Cat. No. 70.
„   „   Sharpe's Cat. No. 4.

I am inclined to think that this is a scarce bird in South Africa, though, at the same time, somewhat widely diffused, as I have obtained specimens in every part traversed by myself.

57. Cosmetornis vexillarius (Gould). Standard-wing Goatsucker.

Semeiophorus vexillarius, Gould's Icones Av. pl. 3.
Cosmetornis vexillarius, Hartlaub in Ibis, 1862, p. 143.

Cosmetornis vexillarius, Sclater in Ibis, 1864, pl. 2.
   „   „   Finsch and Hartlaub's Vögel Ost-Afrika's, pp. 129, 856.
   „   „   Sharpe's Cat. No. 11.

I only observed this singular Nightjar in the large forests about one degree south of the river Okavango; I never saw many of them; and the few that were observed were all found within from twenty to one hundred yards of each other.

This bird, when seen on the wing at dusk, presents a most singular appearance, giving the idea of a huge double-winged bat.

[Mr. R. B. Sharpe possesses a specimen of this Goatsucker which was obtained by Mr. Andersson at Elephant's Vley, on October 13th, 1859.—ED.]

CYPSELIDÆ.

58. Cypselus gutturalis, Vieill. White-throated Swift.

Le Martinet à gorge blanche, Levaillant's Ois. d'Afr. pl. 243.
Cypselus gutturalis, Tristram in Proc. Zool. Soc. 1867, p. 887.
Cypselus melba, Layard's Cat. No. 74.
   „   „   Sharpe's Cat. No. 16.

On July 8th, 1866, I observed at Objimbinque a large flight of these Swifts, which, to the best of my recollection, were the first I ever saw in Damara Land.

Subsequently I observed immense numbers in various places, and particularly noted them as very numerous on the Omaruru River on November 2nd.

The food of this Swift consists of flies and beetles. The iris is brown, the bill black, the legs flesh-coloured, and the toes brownish.

[The Rev. H. B. Tristram, in the 'Proceedings of the Zoological Society of London' for 1867, p. 887, has recorded his reasons for considering this Swift to be specifically distinct from the more northern but closely allied *C. melba*. One of the differences which he mentions, the greater breadth of the brown gorget in the South-African bird, is, however, by no means constant, and the whole question as to the specific distinctness of the two races can probably only be satisfactorily decided by a comparison of a larger series of northern and southern examples than has hitherto been instituted. Mr. R..B. Sharpe has given it as his opinion, in 'The Ibis,' for 1870, p. 427, that the two supposed species are not in reality distinct; and it is quite possible that further investigations may confirm this view.—ED.]

59. Cypselus barbatus, Temm. MS.   South-African Black Swift.

Cypselus barbatus, Sclater in Proc. Zool. Soc. 1865, p. 599 (sub *Cypselus apus*).
  „ „ Tristram in Proc. Zool. Soc. 1867, p. 887.
Cypselus apus, Layard's Cat. No. 75.
  „ „ Finsch in Trans. Zool. Soc. vol. vii. p. 213.
  „ „ Sharpe's Cat. No. 17.

I am far from certain that the Black Swift of Damara Land is identical with *Cypselus apus*, of Europe. The bird found in Damara and Great Namaqua Land is common during the rainy season. Great numbers are often found in the neighbourhood of the sea, near the mouths of periodical watercourses that have a moist bed.

Measurements of two specimens, both males:—

		in. lin.		in. lin.
Entire length	of one	7 6	of the other	7 4
Length of folded wing	„	6 10½	„	6 7
„ tarsus	„	0 6	„	0 5
„ middle toe	„	0 3	„	0 3
„ tail	„	3 0	„	3 0
„ bill	„	0 10½	„	0 9

[Dr. Tristram, in his paper in the Zoological Society's 'Proceedings,' already alluded to under the head of *Cypselus gutturalis*, also describes the distinguishing differences between the Black Swift of Europe and that of South Africa. These, though slight, appear to be constant; but at the same time the specific value of the distinction must probably remain in abeyance until a larger number of specimens have been obtained for comparison than have as yet been made available for that purpose. I am not aware that the South-African Black Swift has yet been figured; and in the opinion of Dr. Finsch, *loc. cit.*, the South-African bird is merely the young of *C. apus*; but if so, it seems singular that only the immature birds should migrate so far south.—ED.]

60. Cypselus parvus, Licht. Little African Swift.

Cypselus ambrosiacus, Temm. in Pl. Col. pl. 460. fig. 2.
Cypselus parvus, Sclater, in Proc. Zool. Soc. 1865, p. 601.
Cotyle ambrosiacus, Layard's Cat. No. 92.
Cypselus parvus, Sharpe's Cat. No. 13.

I observed this species at Ondonga, where I found it pretty common. At the end of February these Swifts appeared to be nesting, as they were seen in pairs, and a male and female were both shot with feathers in their bills. The flight of this species is generally lofty.

The iris is dark brown, the legs and feet brown, and the bill black.

Measurements of two males :—

		in.	lin.		in.	lin.
Entire length	of one	6	4	of the other	6	3
Length of folded wing	,,	5	4	,,	5	4
,, tarsus	,,	0	5	,,	0	4
,, middle toe	,,	0	3	,,	0	3
,, tail	,,	3	6	,,	3	6
,, bill	,,	0	7	,,	0	7

Measurements of two females:—

		in. lin.		in. lin.
Entire length	of one	6 3	of the other	5 11
Length of folded wing	,,	5 4	,,	5 2
,, tarsus	,,	0 4	,,	0 5
,, middle toe	,,	0 3	,,	0 3
,, tail	,,	3 6	,,	3 4
,, bill	,,	0 7	,,	0 7

HIRUNDINIDÆ.

61. Hirundo Monteiri, Hartl. Monteiro's Swallow.

Hirundo Monteiri, Hartlaub, in Ibis, 1862, pl. 11.
,, ,, Gurney, Birds Damar., Proc. Zool. Soc. 1864, p. 2.
,, ,, Finsch & Hartlaub's Vögel Ost-Afrika's, p. 139.
,, ,, Sharpe, in Proc. Zool. Soc. 1870, p. 316.
,, ,, Sharpe's Cat. No. 437.

To the best of my knowledge this fine Swallow (of which I first obtained a few individuals on the river Okavango in 1859) never extends its migration so far south as Damara Land proper; and, indeed, very few individuals come much further south than the Okavango. Those that came under my notice were always found in large open forests, flying high above the tree tops in pursuit of their insect prey, or occasionally perching on lofty, isolated, and aged trees, and they were in consequence by no means easy to procure.

Measurements of a male and a female:—

	Male. in. lin.	Female. in. lin.
Entire length	9 5	9 3
Length of folded wing	5 9	5 9
,, tarsus	0 8	0 8
,, middle toe	0 5	0 6
,, tail	4 11	4 10
,, bill	0 9	0 9

[A specimen obtained by Mr. Andersson at Elephant's Vley,

and two others from Ondonga, are in the collection of Mr. R. B. Sharpe.—ED.]

62. Hirundo rustica, Linn. Chimney-Swallow.

Hirundo rustica, Gould's Birds of Europe, pl. 54.
„ „ Strickland & Sclater, Birds Damar., Contr. to Orn. 1852, p. 144.
„ „ Layard's Cat. No. 79.
„ „ Chapman's Travels in S. Afr., App. p. 410.
„ „ (part.), Finsch & Hartlaub's Vögel Ost-Afrika's, p. 134.
„ „ Sharpe & Dresser, in Proc. Zool. Soc. 1870, p. 244.

This well-known species is pretty common in Damara and Great Namaqua Land during the rainy season, and I have found it very numerous at Walvisch Bay and in other localities near the coast.

In uncivilized parts of Africa these Swallows affix their nests to some projection of a rock or trunk of a tree, or occupy cavities in rocks or banks.

The iris is dark brown, the bill black, the legs brownish. Measurements of a male and a female:—

	Male.		Female.	
	in.	lin.	in.	lin.
Entire length	7	3	6	8
Length of folded wing	5	0	4	10
„ tarsus	0	6	0	5
„ middle toe	0	5	0	5
„ tail	3	3	3	0
„ bill	0	8	0	8

[Messrs. Sharpe and Dresser, in their paper in the Zoological Society's 'Proceedings' above referred to, have recorded some interesting facts relative to the changes of plumage which this species undergoes during its migration to Southern Africa.—ED.]

63. Hirundo cucullata, Bodd. Rouselline Swallow.

L'Hirondelle rouselline, Levaillant's Ois. d'Afr. pl. 245. fig. 1.

Hirundo capensis, Layard's Cat. No. 81.
" " Chapman's Travels in S. Afr., App. p. 410.
Hirundo cucullata, Gray's Hand-list of Birds, No. 795.
" " Sharpe, in Proc. Zool. Soc. 1870, p. 318.
" " Sharpe's Cat. No. 440.

This is not a very common Swallow in Damara Land, where it usually arrives later than *Hirundo dimidiata*; it courts the society and neighbourhood of man, and, where permitted, will unhesitatingly enter his dwellings and construct its nest and rear its young in the midst of the household duties of the family. The nest is built of clay, and at first resembles in shape that of *Hirundo rustica*; but gradually the hollow bowl is narrowed into a tube of some extent. If the nest be destroyed at this stage, the poor bird at once sets about repairing the damage, but generally contents itself with rebuilding the dome, to which a narrow entrance is added.

I have known a pair of these Swallows reconstruct their nest three times in one season, the female depositing a nearly full complement of eggs on each occasion. At the Cape this species commences its incubation towards the latter end of September or early in October, but in Damara Land it is somewhat later. The eggs are four or five in number, of a pure white, dotted over with minute brown spots; the irides are brown.

Measurements of a male and a female:—

	Male.	Female.
	in. lin.	in. lin.
Entire length	7 6	7 0
Length of folded wing	4 10	4 6
" tarsus	0 7	0 6
" middle toe	0 7	0 6
" tail	3 11	3 8
" bill	0 7½	0 7

64. Hirundo dimidiata, Sundev. Pearly-breasted Swallow.

Hirundo dimidiata, Sundevall, Öfvers. 1850, p. 107.
Hirundo scapularis, Cassin, in Pr. Ac. Phil. 1850, pl. 12. fig. 3.
Hirundo dimidiata, Layard's Cat. No. 87.
,,   ,,   Chapman's Travels in S. Afr., App. p. 410.
,,   ,,   Finsch & Hartlaub's Vögel Ost-Afrika's, p. 133.
,,   ,,   Sharpe, in Proc. Zool. Soc. 1870, p. 310.

These Swallows are tolerably common in Damara Land, where they arrive about November; but on the Okavango River I have seen them as early as the 1st of September. They do not stay any great length of time in Damara Land, in fact barely long enough to rear their young.

In December 1863 a pair of these birds took up their abode in my dining-room at Objimbinque, where they half completed a nest and then abandoned it; another pair (at least I conjectured that they were not the same) after a time continued the labour; but finally they also abandoned the nest whilst still incomplete; the next season, however, it was finished, probably by the original projectors, and the parent birds safely brought up their young.

The nest of this Swallow is cup-shaped, and the eggs pure white. The irides in this species are dark brown; the bill, legs, and toes are black.

65. Cotile fuligula, Licht. Fawn-breasted Martin.

L' Hirondelle fauve, Levaillant's Ois. d'Afr. pl. 246. fig. 1.
Cotyle fuligula, Layard's Cat. No. 89.
Hirundo rupestris, Chapman's Travels in S. Afr., App. p. 409.
Cotyle fuligula, Sharpe, in Proc. Zool. Soc. 1870, p. 299.
,,   ,,   Sharpe's Cat. No. 425.

This Martin is common in Damara and Great Namaqua

Land, and is the only species of Swallow which remains throughout the year, a few couples being always to be found in suitable localities. I once saw a very large number at Hykomkap on the 20th of May.

This species breeds in the holes of low rocks and clay-banks. The nest is cup-shaped, and built of the usual clay materials; the eggs are five or six in number, white, tinged with fawn, and spotted with brown.

The iris in this species is of a very dark brown; the bill is brown, the upper mandible being darker than the lower; the legs and toes are brown.

Measurements of a male:—

		in.	lin.
Entire length		5	8
Length of folded wing		5	0
,,	tarsus	0	6
,,	middle toe	0	6
,,	tail	2	3
,,	bill	0	7¼

CORACIADÆ.

66. Coracias caudata, Linn. Green-necked Roller.

Coracias caudata, Des Murs, Icon. Orn. pl. 28.
   ,,   ,,   Strickland & Sclater, Birds Damar., Cont. to Orn. 1852, p. 154.
   ,,   ,,   Layard's Cat. No. 96.
Coracias abyssinica, Chapman's Travels in S. Afr., App. p. 408.
Coracias caudata, Finsch & Hartlaub's Vögel Ost-Afrika's, p. 154.
   ,,   ,,   Sharpe's Cat. No. 36.
   ,,   ,,   Sharpe, in Ibis, 1871, p. 194.

This species is common in the Lake-regions, and is also pretty common in Damara Land, where, however, I imagine that it must be partially migratory, as during the dry season comparatively few individuals are seen. It is more shy and difficult of approach than *Coracias pilosa*, which it otherwise resembles as to food and habits.

The iris is yellowish brown, the ring round the eyes greenish yellow, as are also the legs and toes; the bill is black.

[It may be right here to mention that Mr. Layard, in his 'Catalogue of the Birds of South Africa,' p. 60, states that he has received *Coracias abyssinica* " from the neighbourhood of Springbok Fontein, in Namaqua Land, and from Damara Land;" but it is not referred to in Mr. Andersson's MS. notes, and I have not met with it in any of his collections.

Possibly some confusion may have arisen between specimens of *C. caudata* and supposed examples of *C. abyssinica*, the two species being very nearly related.

It appears clear from Mr. Andersson's MS. notes that the species referred to in Mr. Chapman's appendix, *loc. cit.*, under the name of *C. abyssinica*, is in reality *C. caudata.*—ED.]

67. Coracias nævia, Daud. White-naped Roller.

Le Rollier varié d'Afrique (jeune âge), Levaillant's Ois. de Paradis et Rolliers, pl. 29.
Coracias nævia, Daudin's Traité d'Orn. vol. ii. p. 258.
Coracias pilosus, Gurney, Birds Damar., Proc. Zool. Soc. 1864, p. 2.
Coracias nuchalis, Layard's Cat. No. 93.
Coracias pilosa et nuchalis, Chapman's Travels in S. Afr., App. p. 408.
Coracias pilosa, Gray's Hand-list of Birds, No. 898.
Coracias nævia, Sharpe's Cat. No. 34.
„   „   Sharpe, in Ibis, 1871, p. 190.

This richly coloured and exceedingly handsome though somewhat coarse-looking Roller is not uncommon throughout Damara Land, and it is also found in the Lake-regions; it is usually met with in pairs, and is not particularly shy. It seldom extends its flight far, but occasionally rises suddenly to a considerable height, rocking violently to and fro, and descending in a similar manner, with a motion resembling that of a boy's kite when falling to the ground on the guiding force being withdrawn. When on the wing it makes a great noise, rapidly uttering harsh and discordant sounds; its notes are at times not unlike the sound produced by a broad-bladed knife passing through a tough piece of cork, but are in a louder key. This species seeks much of its food on the ground; but sometimes it watches from some elevated position, and, the moment its prey comes within sight, darts upon it with unerring certainty, its habit in this respect being very like that of the Butcher-bird Shrikes. It is a most useful bird, feeding largely on centipedes, scorpions, tarantulas, and other insects, as well as on small snakes and lizards.

This Roller is one of the earliest breeders in Damara Land, and makes its nest in the hollows of trees, usually such as have been previously occupied by some Woodpecker: the stems of these trees are mostly very tall and straight; and in consequence of this, and of the smallness of the apertures, the nests are very inaccessible. I have frequently seen such breeding-places without being able to reach them; but I believe the eggs are white and two in number, and that both parents assist in their incubation.

The iris in this species is dark brown, the bill black, the legs and toes greenish brown.

[Mr. Chapman, who refers to this Roller in his 'Travels in South Africa,' vol. i. p. 282, and Appendix, p. 386, confirms

Mr. Andersson's belief that the egg of this species is white, and adds the following information respecting its habits :—" These birds before they are fledged, as well as the hen while breeding, are fed by the male bird; the hen never leaves the nest until the brood are fledged. The birds cannot fly well, and if seen in an open field, where an occasional resting-place is not to be found, are easily run down and eaten by the Bushmen."—ED.]

68. Coracias garrula, Linn. European Roller.

Coracias garrulus, Gould's Birds of Europe, pl. 60.
Coracias garrula, Layard's Cat. No. 94.
    ,,   ,,   Sharpe's Cat. No. 33.
    ,,   ,,   Sharpe & Dresser's Birds of Europe, pl. 5.

This species is common in Ondonga, but is less so in Damara Land proper, than either *C. pilosa* or *C. caudata*. As far as I recollect, it is only seen during the rainy season.

The iris is dirty brown, the bill black, the legs and toes brownish yellow.

ALCEDINIDÆ.

69. Halcyon cyanoleuca (Vieill.). Angola Kingfisher.

Halcyon senegalensis, Gurney, in Ibis, 1865, p. 265.
    ,,   ,,   Layard's Cat. No. 98.
Halcyon cyanoleuca, Sharpe's Alcedinidæ, pl. 69.
    ,,   ,,   Sharpe's Cat. No. 65.

This species is very abundant in Ondonga. It generally perches on or near the summit of lofty trees, from whence it sends forth a succession of rather pleasant, thrilling or whirring notes.

The iris is dark brown; the upper mandible red, the lower black.

Measurements of a male and a female:—

	Male. in. lin.	Female. in. lin.
Entire length	9 6	9 9
Length of folded wing	4 7	4 7
,, tarsus	0 7	0 7
,, middle toe	0 8	0 8
,, tail	2 10	2 10
,, bill	2 4	2 4

70. Halcyon semicærulea (Forsk.). African White-headed Kingfisher.

Halcyon Swainsonii, Layard's Cat. No. 100.
Halcyon semicærulea, Sharpe's Alcedinidæ, pl. 63.
,, ,, Sharpe's Cat. No. 59.

[This species is not mentioned in Mr. Andersson's MS. notes; but a single immature specimen obtained in Ondonga, Ovampo Land, formed part of his last collection, and is now in the possession of Mr. R. B. Sharpe, who refers to it, in his excellent 'Monograph of the Kingfishers,' as the most southern example of this species with which he is acquainted.—ED.]

71. Halcyon chelicutensis (Stanl.). Striped Kingfisher.

Halcyon damarensis, Strickland & Sclater, Birds Damar., Contr. Orn. 1852, p. 153.
Halcyon striolata, Layard's Cat. No. 102.
Halcyon damarensis, Chapman's Travels in S. Afr., App. p. 409.
Halcyon chelicutensis, Sharpe's Alcedinidæ, pl. 67.
,, ,, Sharpe's Cat. No. 62.

This Kingfisher is very sparingly met with in Damara Land and the parts adjacent to the northward; it is partial to localities where the vegetation has been destroyed or partially injured by fire; and it selects, if possible, a low dry branch on an isolated tree, where it watches by the hour for its prey; this, as far as I could ascertain, consists of insects, which it generally seizes on the wing.

It utters loud, sharp, and shrill cries, and is always found either singly or in pairs.

The irides in this species are claret-coloured, the bill reddish brown on the upper mandible and orange-red on the lower, the lores are dusky, the legs and toes yellowish.

[Mr. Sharpe, in his ' Monograph of the Alcedinidæ,' *loc. cit.*, makes the following remarks on this species:—

" All the examples of the present species from South Africa are much larger than those from West Africa and Abyssinia. Strickland separated them under the name of *Halcyon damarensis* on receipt of some specimens collected by Andersson. It is, however, impossible to separate them specifically, as, taking the Abyssinian bird as the type of the species, a regular series of gradations is reached according as specimens from the different parts of Western Africa are examined, those from Angola being intermediate in size and nearly attaining the large form of *H. damarensis*.

" I therefore regard these races in a subspecific light only, as in the case of the species of *Corythornis*."

I have thought it best to follow the view taken by Mr. Sharpe on this subject, and have accordingly used for this Kingfisher the specific name which he has adopted.—ED.]

72. Alcedo semitorquata, Swains. Half-collared Kingfisher.

Alcedo semitorquata, Layard's Cat. No. 105.
,, ,, Sharpe's Alcedinidæ, pl. 7.
,, ,, Sharpe's Cat. No. 48.

[This Kingfisher is not mentioned in Mr. Andersson's MS. notes, neither did I meet with any specimen of it in those of his collections which I had the opportunity of examining; but Mr. R. B. Sharpe possesses a pair obtained on the Orange River, which probably entitles this species to be included among the birds of Namaqua Land.—ED.]

73. Ceryle maxima, Pall. Great African Kingfisher.

Ceryle maxima, Gurney, in Ibis, 1859, p. 243.
„   „   Layard's Cat. No. 109.
Alcedo giguntea, Chapman's Travels in S. Afr., App. p. 409.
Ceryle maxima, Sharpe's Alcedinidæ, pl. 20.
„   „   Sharpe's Cat. No. 47.

The Great Kingfisher is occasionally found on the Teoughe River and also on the Okavango, but is everywhere very shy.

74. Ceryle rudis (Linn.). Black-and-White Kingfisher.

Ceryle rudis, Layard's Cat. No. 110.
Ceryle bicincta, Chapman's Travels in S. Afr., App. p. 409.
Ceryle rudis, Sharpe's Alcedinidæ, pl. 19.
„   „   Sharpe's Cat. No. 46.

I do not remember to have seen this bird in Damara Land proper, but have occasionally met with it along the periodical watercourses and temporary rain-pools of Great Namaqua Land; and I have reason to think that it may be found permanently on the banks of the Great Fish River, where large pools of water, containing fish exist at all times of the year. It frequents both salt and fresh water, and feeds on fish, shrimps, small crabs, &c., for which it watches from some dry branch of a tree immediately overhanging the water. It breeds in the sandy and clayey banks of rivers, and is comparatively tame. The iris is dark brown.

Measurements of a male and a female:—

	Male. in. lin.	Female. in. lin.
Entire length	10 6	10 0
Length of folded wing	5 7	5 6
„ tarsus	0 5	0 5
„ middle toe	0 6	0 6
„ tail	3 0	2 11
„ bill	2 10	2 7

75. Corythornis cyanostigma (Rüpp.). African Malachite-crested Kingfisher.

Alcedo cristata, Layard's Cat. No. 106.
Alcedo cyanostigma, Chapman's Travels in S. Afr., App. p. 409.
Corythornis cristata, Sharpe's Alcedinidæ, pl. 11.
Corythornis cyanostigma, ibid. (Introduction), p. 6.
„   „   Sharpe's Cat. No. 51.

Probably from want of permanently running rivers, this exquisite little species is not found in Damara or Great Namaqua Land; but it is common on all the waters north of those countries; it invariably perches on some low twig or bough almost on a level with the water.

MEROPIDÆ.

76. Merops apiaster, Linn. European Bee-eater.

Merops apiaster, Gould's Birds of Europe, pl. 50.
,,   ,,   Layard's Cat. No. 111.
„   „   Sharpe's Cat. No. 18.

This species is very common in Ondonga during the rainy season, when it is also not uncommon in Damara Land proper; but I do not think that it is abundant in Great Namaqua Land.

These Bee-eaters are observed during their annual migrations in small flocks; but having arrived at their temporary destination they scatter somewhat over the country, though several may still be seen in close proximity. They seem to live chiefly on a species of red wasp, and sometimes seize their food on the wing like Swallows, though they more frequently watch for it from some elevated perch, whence they suddenly pounce upon any prey which may chance to come within their ken, returning invariably to the same spot whether successful or not. When their capture proves a bee or other stinging insect, it is always seized across the body, when the bird, after giving it a sharp squeeze or two between the mandibles of the bill, quickly swallows it. I have seen lizards pursue exactly the same plan when catching hymenopterous insects.

When on the wing, this Bee-eater utters a pleasant but rather subdued warbling chirp.

The iris in this species is red, the legs and toes reddish brown, and the bill almost black.

The female bird is not quite so large, nor so brightly coloured, as the male.

77. Merops superciliosus, Linn. Blue-cheeked Bee-eater.

Merops Savigni, Swainson's Birds of W. Africa, vol. ii. pl. 7.
    ,,   ,,   Layard's Cat. No. 112.
Merops ægyptius, ibid. No. 113.
Merops superciliosus, Finsch & Hartlaub's Vögel Ost-Afrika's, No. 79.
    ,,   ,,   Sharpe, in Proc. Zool. Soc. 1870, p. 145.
    ,, ,   ,,   J. H. Gurney, Jun., Orn. of Algeria, Ibis, 1871, p. 75.
Merops ægyptius, Sharpe's Cat. 20.

I have only once observed this species, which I then met with near the river Okavango.

The iris is reddish brown, the legs and toes brown, and the bill black.

[Mr. Andersson's last collection contained several specimens of this Bee-eater obtained in Ondonga in November 1866, no doubt subsequently to the date of the note above recorded. —ED.]

78. Merops nubicoides, Des Murs.   Carmine-throated Bee-eater.

Merops nubicoides, Des Murs, Icon. Orn. pl. 35.
Merops natalensis, Schlegel's Mus. des Pays-Bas, *Merops,* p. 7.

I have only once observed this species, when a specimen occurred a few days' journey south of the river Okavango; its appearance on the wing was beautiful. I understand from the hunters that at certain seasons this Bee-eater is common on the Okavango, and breeds in the banks of that river.

[I have not seen the specimen mentioned in the above note by Mr. Andersson; but another memorandum left by him describes its plumage in detail, and leaves no doubt of its having been an example of *Merops nubicoides,* though supposed by Mr. Andersson to be a specimen of the more northern *M. nubicus,* from which *M. nubicoides* is chiefly distinguishable by the carmine colouring of the throat, that part being green in *M. nubicus.*—ED.]

79. Melittophagus pusillus, Müll.   Rufous-winged Bee-eater.

Le Guépier à collier bleu, Levaillant's Hist. Nat. des Promerops et des Guépiers, pl. 7.
Le Guépier minulle, ibid. pl. 17.
Merops erythropterus, Layard's Cat. No. 115.
   „   „   Chapman's Travels in S. Afr., App. p. 409.
Melittophagus pusillus, Gray's Hand-list of Birds, No. 1222.

This exquisite and diminutive species is common on the banks of the rivers Okavango, Teoughe, and Botletlé, as well as on the Lake-watersheds in general, and also about Lake Ngami itself; but I have never observed it so far south as Damara Land proper. It seems to be partial to the immediate neighbourhood of the reedy banks of rivers and of swamps and morasses; and I have never found it at any distance from water.

80. Dicrocercus hirundinaceus (Vieill.). Swallow-tailed Bee-eater.

Merops hirundinaceus, Swainson's Birds of W. Africa, vol. ii. pl. 10.
Melittophagus hirundineus, Strickland & Sclater, Birds Damar., Contr. Orn. 1852, p. 154.
Merops hirundinaceus, Layard's Cat. No. 117.
„   „   Chapman's Travels in S. Afr., App. p. 409.
Dicrocercus hirundinaceus, Gray's Hand-list of Birds, No. 1221.
Merops hirundineus, Finsch & Hartlaub's Vögel Ost-Afrika's, p. 193.
Merops hirundinaceus, Sharpe's Cat. No. 28.

This is the commonest species of Bee-eater in Damara Land, and it is also found in Great Namaqua Land and in the Lake-country: it chiefly visits Damara Land during the wet season; but a few may be found throughout the year.

I took a nest of this Bee-eater on the Omaruru River on 31st October. It was situated in a soft sandy bank, some three feet deep horizontally: the entrance was not above two fingers wide; but the hole was slightly enlarged where the nest was found. The nest, which had no lining, contained three beautifully white eggs.

The iris in this species is carmine-red, the bill black, the tarsi and toes brownish.

Measurements of a male and a female :—

	Male. in. lin.	Female. in. lin.
Entire length	8 10	8 5
Length of folded wing	3 9	3 9
„ tarsus	0 5	0 4
„ middle toe	0 5	0 5
„ tail	4 2	4 1
„ bill	1 6	1 3

TENUIROSTRES.

UPUPIDÆ.

81. Upupa minor, Shaw. South-African Hoopoe.

Upupa cristatella, Vieillot's Gal. des Ois. pl. 184.
Upupa minor, Strickland & Sclater, Birds Damar., Contr. Orn. 1852, p. 155.
„ „ Layard's Cat. No. 118.
Upupa africana, Finsch & Hartlaub's Vögel Ost-Afrika's, p. 200.
Upupa minor, Sharpe's Cat. No. 80.

This species is very abundant in Damara Land during the wet season, but gradually disappears with the return of the hot weather, though a few individuals remain throughout the year; it is also common at Lake Ngami. When it first arrives in Damara Land it is seen in straggling flocks, which soon, more or less, disperse; yet a number of individuals are often found in close proximity, leading a person unacquainted with the habits of the bird to believe that it is really gregarious; it is, however, most frequently found singly. This species, to some extent, seeks its food (which consists of insects) upon the ground; but, like the Bee-eater, it will also watch for and pounce upon its prey from some commanding position. Besides frequenting the ground in search of food, it also loves to dust itself in the sand. When in a state of rest the crest of this bird is generally recumbent; but on the least excitement it is alternately elevated and depressed, not rapidly, but in a graceful manner, with deliberation and ease.

In Damara Land the Hoopoe is not very difficult to approach within range; yet there is some difficulty in obtaining specimens, as the bird, the moment it finds itself observed, flits about incessantly amongst the foliage, or is lost to view by gliding rapidly to the opposite side of a tree. Its flight is short, rising and dipping alternately.

The irides in this species are intensely dark brown, almost black, the legs and toes bluish brown.

Average dimensions of four males:—

	in.	lin.
Entire length	9	8
Length of folded wing	5	4
,, tarsus	0	10
,, middle toe	0	8
,, tail	3	8½
,, bill	2	1

82. Irrisor erythrorhynchos (Lath.). Red-billed Irrisor.

Le Promerops moqueur, Levaillant's Hist. Nat. des Promerops et Guépiers, pls. 1, 2, 3.
Irrisor erythrorhyncus, Strickland & Sclater, Birds Damar., Contr. Orn. 1852, p. 154.
Irrisor senegalensis, Gurney, Birds Damar., Proc. Zool. Soc. 1864, p. 2.
Irrisor erythrorynchos, Layard's Cat. No. 119.
Irrisor erythrorhynchus, Chapman's Travels in S. Afr., App. pp. 255 and 406.
,, ,, Finsch & Hartlaub's Vögel Ost-Afrika's p. 202.
,, ,, Finsch, in Trans. of Zool. Soc. vol. vii. p. 226.
,, ,, Sharpe's Cat. No. 81.

This species is not uncommon in Damara Land and the parts adjacent to the north and east, extending to Lake Ngami.

It lives in small flocks, probably consisting of entire families—which frequent trees, chiefly of the larger kinds, and examine them most assiduously in search of insects and their larvæ, which they extract from crevices in the wood and from beneath the bark. These birds climb like Woodpeckers; and their long tails come into constant contact with the rough surface of the trees, by which the tail-feathers are much injured. When they have finished their examination of one tree they move to the next convenient one, but not altogether, as a short interval generally elapses after the departure of each individual. The moment flight is decided on, they utter harsh discordant cries or chatterings, which are continued until they are all safely lodged in their new quarters. These harsh notes are also heard when they conceive themselves in danger from either man, beast, or bird; and they thus often betray their presence.

The bodies of these birds emit a peculiarly powerful and disagreeable odour.

The bill in this species is sometimes of a black or of a semidark colour instead of red; the size and curve of the bill also vary somewhat in different individuals; and the white spots on the wings are more developed in some than in others. The following note was jotted down by me respecting four individuals which were killed out of a flock of five, and which formed a most interesting series. One of the four had the bill almost black; two had it half black and half red; and the fourth, which I believe to be an adult, had the bill quite red: all had the legs and toes and insides of both mandibles red. In two the oval white spots which often exist on the first three secondaries were present; in the others (including the red-billed specimen) only one such spot was discernible; and among the four were distinctly exhibited the two kinds of curve shown in the outlines of the bill, as depicted in 'The Naturalist's Library,' vol. xii. p. 119.

The irides in this species are a deep dark brown.

Average dimensions of seven males and five females:—

	Males. in. lin.	Females. in. lin.
Entire length	16 11	15 0
Length of folded wing	6 0	5 7½
„ tarsus	0 11	0 11
„ middle toe	0 11	0 10
„ tail	9 7	8 1
„ bill	2 3	1 11

83. Irrisor cyanomelas (Vieill.). Namaqua Irrisor.

Rhinopomastus Smithii, Jardine, in Zool. Journ. vol. iv. pl. 1.
Rhinopomastus cyanomelas, Strickland & Sclater, Birds Damar., Contr. Orn. 1852, p. 155.
Irrisor cyanomelas, Layard's Cat. No. 120.
Promerops niger, Chapman's Travels in S. Afr., App. p. 407.
Irrisor cyanomelas, Finsch & Hartlaub, Vögel Ost-Afrika's, p. 207.
„ „ Sharpe's Cat. No. 83.

This species is sparingly found throughout Damara and Great Namaqua Land, but more rarely in the latter than in the former country; it is also met with on the Okavango and Teoughe Rivers and about Lake Ngami. It partakes much of the habits and manners of the true Creepers, attaching itself to trees and examining them in a similar manner, but sometimes with this difference, that, after settling on a tree or stump (which it generally does about halfway up), it carefully examines it in a downward direction, and with its head downwards, thus seeking for its usual food, which consists of ants and other insects with their larvæ; having reached the base of the tree or stump, it moves onwards to another, in a similar manner to that adopted by the preceding species.

These birds are usually observed in pairs; but occasionally a solitary individual may be seen perched on the topmost bough of a lofty tree, uttering peculiar and plaintive notes.

The iris in this species is very dark brown; the bill is yellowish towards the gape, the remainder being of a dark-brown colour; the legs and feet are dusky black, with a brownish tint on the tarsi anteriorly, and with the soles of the feet olive.

84. Irrisor aterrimus (Steph.). Black Irrisor.

Scoptelus aterrimus, Gray's Hand-list of Birds, No. 1267.
Irrisor aterrimus, Finsch & Hartlaub's Vögel Ost-Afrika's, p. 209.

[Dr. Finsch and Dr. Hartlaub, in their valuable work above cited, on the birds of East Africa, mention a specimen of the Black Irrisor obtained in Damara Land by the late Mr. Andersson; but I have no further information respecting its occurrence in that country. It is a species which I believe has not as yet been figured.—ED.]

PROMEROPIDÆ.

85. Nectarinia famosa (Linn.). Malachite Sun-bird.

Le Sucrier malachitte, Levaillant's Ois. d'Afr. pls. 289 & 290.
Nectarinia famosa, Layard's Cat. No. 127.

This splendid species is exceedingly abundant in Little Namaqua Land, and also occurs, though but rarely, in the southernmost parts of Great Namaqua Land. It is usually found permanently established where it has once taken up its abode. Its food consists of insects and the saccharine juices of flowers, in search of which it flits incessantly from one flowering tree or plant to another, now settling and now hovering, but glittering all the while in the sunshine like some brilliant insect or some precious gem.

The male bird, in addition to the beauty of its plumage, possesses a very pleasant warble.

The iris in this species is very dark, and the bill, legs, and feet black.

Measurements of a male and a female:—

	Male. in. lin.	Female. in. lin.
Entire length	9 0	5 9
Length of folded wing	2 10½	2 8
,, tarsus	0 7	0 7
,, middle toe	0 5	0 5
,, tail	5 1	2 0
,, bill	1 5	1 3

86. Cinnyris chalybea (Linn.). Lesser Double-collared Sun-bird.

Nectarinia chalybea, Jardine's Sun-birds, pl. 1.
,, ,, Layard's Cat. No. 122.
Cinnyris chalybea, Gray's Hand-list of Birds, No. 1282.

I do not recollect having observed this species north of the Orange River; but I have not unfrequently met with it in Little Namaqua Land, and I am informed by Mr. Layard that it was brought by Mr. Chapman from the Lake-regions. It is found both in open and wooded localities, and generally singly.

Measurements of a male and a female:—

	Male. in. lin.	Female. in. lin.
Entire length	4 8	4 3
Length of folded wing	2 0	1 11
,, tarsus	0 8	0 8
,, middle toe	0 5	0 5
,, tail	1 9	1 6
,, bill	0 10	0 10

87. Cinnyris afra (Linn.). Greater Double-collared Sun-bird.

Nectarinia afra, Jardine's Sun-birds, pl. 2.
„   „   Layard's Cat. No. 123.
Cinnyris afra, Gray's Hand-list of Birds, No. 1281.

This is another species brought by Mr. Chapman from the Lake-country, as I have been assured by Mr. Layard; it never came under my personal observation, except in the south-eastern districts of the Cape Colony.

It frequents forests, but may occasionally be observed in the more open parts during the flowering-season.

The iris is dark brown, the bill and legs are black.

Measurements of a male and a female:—

	Male.		Female.	
	in.	lin.	in.	lin.
Entire length	5	6	5	3
Length of folded wing	2	6	2	3
„ tarsus	0	8	0	8
„ middle toe	0	5	0	5
„ tail	1	0	1	0
„ bill	1	0	1	0

88. Cinnyris bifasciata, Shaw. Bifasciated Sun-bird.

Nectarinia bifasciata, Jardine's Sun-birds, pl. 4.
„   „   Strickland & Sclater, Birds Damar., Contr. Orn. 1852, p. 153.
„   „   Layard's Cat. No. 126.
„   „   Chapman's Travels in S. Afr., App. p. 407.
Cinnyris bifasciata, Gray's Hand-list of Birds, No. 1283.
Nectarinia bifasciata, Sharpe's Cat. No. 359.

This species is very common in Ondonga, and is not uncommon in Damara Land; it is also found at Lake Ngami. It is usually seen in pairs, and frequents the banks of periodical streams; I never saw it far away from such localities.

The iris is dark brown.

Measurements of a male and a female:—

	Male. in. lin.	Female. in. lin.
Entire length	5 2	4 9
Length of folded wing	2 7	2 4
,, tarsus	0 8	0 8
,, middle toe	0 4	0 4
,, tail	1 11	1 9
,, bill	0 11	0 10

89. Cinnyris fusca (Vieill.). White-vented Sun-bird.

Le Sucrier namaqois, Levaillant's Ois. d'Afr. pl. 296.
Nectarinia fusca, Strickland & Sclater, Birds Damar., Contr. Orn. 1852, p. 153.
,, ,, Layard's Cat. No. 131.
,, ,, Chapman's Travels in S. Afr., App. p. 407.
Cinnyris fusca, Gray's Hand-list of Birds, No. 1284.
Nectarinia fusca, Sharpe's Cat. No. 371.

This is the commonest Sun-bird in Damara and Great Namaqua Land, where it is really abundant, especially towards the sea-coast. The scantier and more dreary the vegetation the more common is this bird; and though unattractive in dress, it helps to enliven the monotonous solitudes which it frequents, by its activity and pleasant subdued warbling chirp.

The male assumes a somewhat more attractive garb during the breeding-season than at other times of the year, when it resembles the female, whose colouring is of the most sombre description.

Levaillant tells us that this bird nests in the hollows of trees; but this differs from my experience, as I have always found its nest suspended from the branch of some low acacia. The nest is chiefly composed of soft grasses and the fine inner bark of trees, and is lined with a quantity of feathers. I found the young just fledged on April 3rd.

The iris in this species is very dark brown.

[Mr. Sharpe possesses a specimen of this Sun-bird, obtained by Mr. Andersson in Little Namaqua Land.—ED.]

90. Cinnyris talatala (Smith). Andersson's Sun-bird.

Nectarinia talatala, Smith's Report of South Afr. Exp. 1836, p. 58.
Nectarinia Anderssoni, Strickland & Sclater, Birds Damar., Contr. Orn. 1852, p. 153.
Cinnyris Anderssonii, Gray's Hand-list of Birds, No. 1285.
Nectarinia talatala, Sharpe's Cat. No. 375.

I only met with this exquisite little species in my journeyings to the Okavango, in the neighbourhood of which river it was very abundant during the rainy season, being a migratory species, and arriving a little before the commencement of the rains.

I also found it very common, though exceedingly shy, on the edge of the bush in Ondonga, where I obtained its nest on February 19th: the nest was very large and strongly built, and resembled in form and material that of *C. fusca*; it contained five small, oblong, and pure-white eggs. Another nest, taken on March 27th, also contained five eggs.

This Sun-bird is exceedingly lively in its habits, and at the approach of the pairing-season it becomes inspired with the most lovely and exquisite melodies; in fact its voice is then enchanting beyond description, being a concentration of the softest thrilling and melodious notes. I always found it either singly or in pairs.

Measurements of a male:—

		in.	lin.
Entire length		4	6
Length of folded wing		2	3
,,	tarsus	0	8
,,	middle toe	0	4
,,	tail	1	6
,,	bill	0	10

[I am not aware that any figure of this beautiful Sun-bird has yet been published; and as the descriptions which have been given of it are not very generally accessible, I transcribe that drawn up by the late Mr. Strickland from a male in full plumage, and inserted in the 'Contributions to Ornithology,' *loc. cit.*:—

" Head, back, and lesser wing-covers metallic green, the crown with a coppery gloss; upper tail-covers bluish green; greater wing-covers and remiges deep fuscous, margined externally with greyish brown; chin bluish green; cheeks and throat bright coppery green; a broad zone on the breast of violet-purple, followed by a narrow one of dull greyish brown; axillary tufts gamboge-yellow; abdomen, sides, and lower tail-covers dirty white; beak and legs black."—ED.]

91. Chalcomitra gutturalis (Linn.). Proteus Sun-bird.

Le Sucrier Protée, Levaillant's Ois. d'Afr. pl. 295. fig. 2.
Nectarinia senegalensis, Strickland & Sclater, Birds Damar., Contr. Orn. 1852, p. 153.
,, ,, Layard's Cat. No. 133.
Nectarinia natalensis, Layard's Cat. No. 134.
Nectarinia senegalensis, Chapman's Travels in S. Afr., App. p. 407.
Chalcomitra gutturalis, Gray's Hand-list of Birds, No. 1307.
Nectarinia gutturalis, Sharpe's Cat. No. 364.

This species occurs at Lake Ngami, and is pretty abundant in the neighbourhood of the Okavango. In Damara Land proper it is not common; but in June 1866 I obtained several specimens at Objimbinque, where they seemed chiefly to seek their food amongst the "tobacco" trees now growing so abundantly in the bed and on the banks of the Swakop. Can the increase of this tree of late years have brought more of these birds? I hardly remember to have seen them at Objimbinque previously.

Measurements of a male and a female:—

	Male.		Female.	
	in.	lin.	in.	lin.
Entire length	6	0	5	0
Length of folded wing	2	11	2	7
,, tarsus	0	8	0	8
,, middle toe	0	5	0	5
,, tail	2	1	1	9
,, bill	1	1	1	1

92. Anthobaphes violacea (Linn.). Orange-breasted Sun-bird.

Le Sucrier orangé, Levaillant's Ois. d'Afr. pl. 292.
Nectarinia violacea, Layard's Cat. No. 130.
Anthobaphes violacea, Gray's Hand-list of Birds, No. 1353.

I have found this species pretty abundant in Little Namaqua Land; but to the best of my knowledge it is not an inhabitant of Great Namaqua or Damara Land, though Mr. Layard informs me that Mr. Chapman brought specimens from the Lake-country. It is found singly and in pairs, often also in flocks, frequenting the slopes of hills and mountains, whence it descends to the lower grounds, but only during the flowering season of the garden plants and trees, amongst which it is especially fond of the sweet-scented orange-blossom. With the exception of such excursions, it is not migratory. The male bird has a brisk, pleasant song.

The iris is dark brown, the bill and legs are black.

Measurements of a male and a female :—

	Male.		Female.	
	in.	lin.	in.	lin.
Entire length	6	1	6	0
Length of folded wing	2	1	2	0
,, tarsus	0	7	0	7
,, middle toe	0	5	0	4
,, tail	3	0	3	0
,, bill	1	0	1	0

MELIPHAGIDÆ.

93. Zosterops capensis, Sund. Cape White-eye.

Le Tcheric, Levaillant's Ois. d'Afr. pl. 132.
Zosterops capensis, Layard's Cat. No. 215.

I have only once or twice observed this species in the southernmost parts of Great Namaqua Land, along the periodical watercourses bordered by Mimosas; but from thence southwards it becomes more numerous, and at the Cape and in many parts of the Colony it is abundant: a pair or two may be seen any day in most of the gardens in the immediate environs of the Cape. It is met with in small families, probably the entire broods of the season. It feeds on small insects and larvæ, for which it searches diligently amongst low bushes and trees. It is quite tame; and it is not very difficult to approach it near enough to distinguish the colour of its eye, beak, &c. It forms its nest on the extremity of some branch of a low tree; the nest is very prettily shaped, and is composed of loose tendrils interlaced, covered with moss outside, and lined internally with hair &c. The eggs are four or five in number, and are said to be incubated by both parents.

The irides are yellowish brown; the bill bluish black, but lighter on the under mandible; the legs and feet are lead-colour, with sometimes a tinge of brown.

The sexes in this species do not differ in their dimensions.

Measurements of a male:—

		in.	lin.
Entire length		4	6
Length of folded wing		2	5
,, tarsus		0	9
,, middle toe		0	5
,, tail		1	10
,, bill		0	7

94. Zosterops senegalensis, Bon. Yellow White-eye.

Zosterops flava, Swainson's Birds of West Africa, vol. ii. pl. 3.
Zosterops senegalensis, Hartlaub's Birds of West Africa, No. 215.
,, ,, Sharpe's Cat. No. 336.

I never met with this exquisite little bird in either Great Namaqua or Damara Land; and it was only as I approached the Okavango that I became aware of its existence. In the thornless forests bordering upon this stream it is not uncommon, but it migrates northwards during the dry season. It is found in small flocks, and diligently explores in search of insects the branches of the smaller trees, and especially the buds and flowers, suspending itself in a variety of positions while it is thus employed.

[Two specimens, obtained by Mr. Andersson at Elephant's Vley, are in the collection of Mr. R. B. Sharpe.—ED.]

TROGLODYTIDÆ.

95. Sylvietta rufescens (Vieill.). South-African Crombec.

Le Crombec, Levaillant's Ois. d'Afr. pl. 135.
Sylvietta brachyura, Strickland & Sclater, Birds Demar., Contr. Orn. 1852, p. 148.
Dicæum rufescens, Layard's Cat. No. 144.
Sylvietta rufescens, Gray's Hand-list of Birds, No. 2870.
 „   „   Sharpe's Cat. No. 324.

I have found this species widely distributed in all parts which I have traversed, from the Okavango to Table Mountain, but nowhere very common.

It frequents dwarf vegetation, which it examines carefully as it hops and glides quickly onwards. The irides are yellowish brown, the upper mandible dusky, the under mandible a purplish flesh-colour, as also are the legs and toes.

Measurements of a female:—

		in.	lin.
Entire length		4	0
Length of folded wing		2	6
„	tarsus	0	9
„	middle toe	0	4½
„	tail	1	1
„	bill	0	8

DENTIROSTRES.

PARIDÆ.

96. Parisoma subcæruleum (Vieill.). Rufous-vented Grignet.

Le Grignet, Levaillant's Ois. d'Afr. pl. 126.
Parisoma subcæruleum, Strickland & Sclater, Birds Damar., Contr. Orn. 1852, p. 149.

Parisoma rufiventer, Layard's Cat. No. 213.
" " Chapman's Travels in S. Afr., App. p. 397.
Parisoma subcæruleum, Sharpe's Cat. No. 390.

This species is common in Damara and Great Namaqua Land, but, from its small size and secluded habits, often escapes notice; it is rather a pretty songster, and utters, at times, varied and singular notes, and occasionally also a clear ringing call rapidly repeated. It is very familiar, active but not rapid in its movements, and careful in its examination of the branches of trees and bushes in search of insects; it is found singly or in pairs.

A nest of these birds, taken on the 21st of September, was situated in a hedge and composed outside of grass, fine twigs, and tendrils; internally it was lined with hair, and contained two eggs. A second nest, obtained on the 1st of October, was similarly composed externally, but was lined with the softer tendrils of flexible roots; it contained two eggs, hard sat upon. A third nest, taken on the 29th of November, also contained two eggs.

The iris in this species is yellowish white.

97. Parisoma Layardi, Hartl. Layard's Grignet.

Parisoma Layardi, Hartlaub, in Ibis, 1862, p. 147.
" " Layard's Cat. No. 214.

This species greatly resembles the preceding in its habits, but is not so common; I have observed it, though very sparingly, in Damara and Great Namaqua Land and near the west coast of the Cape Colony. I have also obtained specimens from the Okavango, which are of a darker and richer hue than those from Damara and

Great Namaqua Land; this is also the case with specimens from the western parts of the Colony.

98. Anthoscopus minutus (Shaw). Dwarf Blossom-pecker.

Le Becque-fleur, Levaillant's Ois. d'Afr. pl. 134.
Ægithalus Smithi, Strickland & Sclater, Contr. Orn. 1852, p. 149.
Paroides capensis, Layard's Cat. No. 212.
Anthoscopus minutus, Gray's Hand-list of Birds, No. 3417.
Ægithalus capensis, Sharpe's Cat. No. 331.

This diminutive species is sparingly found from the Okavango River to the neighbourhood of Cape Town, following the line of the coast and occurring in small flocks amongst brushwood, low trees, or flowers; in such situations it hunts with great assiduity for minute insects, in which occupation it strongly reminded me of some of the titmice, which it much resembles in its mode of climbing and feeding. It utters a low and almost inaudible chirp or whistle.

I once found, in Ondonga, on March 27th, a nest of this species containing one young bird and one egg, the latter being of so extraordinary a size that, had I not shot the old bird at the nest, and had not the identity of the egg been corroborated by the young bird, I should not have believed that it belonged to this species. The nest had been rudely displaced from its original site and was hanging down several inches, very much the worse for its misfortune; I only wonder the birds stuck so bravely to it. The rim of the nest was composed of very fine twigs of small tender bushes interlaced with decomposed silky grasses; the rest of the nest was chiefly of the latter material, but lined within with fine tendrils.

When I first observed the nest there was no bird present; but after waiting awhile, one appeared and was on the point of entering the nest, when it perceived me and moved slowly out of view. Not feeling quite certain of its identity, I waited for nearly another hour to get a second sight of it, when both parent birds came and settled quite close to the nest, and I killed the male.

The irides in this species are brownish yellow.

Measurements of a male and a female:—

	Male. in. lin.	Female. in. lin.
Entire length	3 7	3 3
Length of folded wing	2 0	1 11
„ tarsus	0 7	0 7
„ middle toe	0 4	0 4
„ tail	1 6	1 4
„ bill	0 5	0 4

[The disturbed state of the nest above referred to, and the presence in it of only one egg, and that of comparatively large size, suggest the probability of a parasitic intrusion by one of the smaller Cuckoos.

The male specimen from which Mr. Andersson took the measurements above recorded is now in the collection of Mr. R. B. Sharpe.—ED.]

99. Anthoscopus Caroli (Sharpe). Andersson's Blossom-pecker.

Ægithalus minutus, Sharpe's Cat. No. 330.
Ægithalus Caroli, Sharpe, in Ibis, 1871, p. 415.

This species, which, as far as my experience goes, is very rare, resembles the preceding one in its habits and also in its general appearance; but the lower part of the belly, instead of being light yellow, is yellowish brown.

[Mr. R. B. Sharpe, who possesses three specimens of this little bird, which were all procured by Mr. Andersson at Ovaquenyama in May and June 1867, has favoured me with the following description of this species. "The entire throat and upper breast white, the abdomen yellowish rufous, the head and back grey throughout, the lores white, the forehead without spots." In size this species closely resembles *A. minutus.* Mr. Sharpe, considering this to be an undescribed species, has assigned to it the specific name of *Caroli,* in remembrance of its discoverer, Charles James Andersson. It has not been figured.—ED.]

100. Parus afer, Gmel. Grisette Tit.

La Mésange grisette, Levaillant's Ois. d'Afr. pl. 138.
Parus cinerascens, Andersson, iu Proc. Zool. Soc. 1864, p. 6.
„ „ Layard's Cat. No. 210.
„ „ Chapman's Travels in S. Afr., App. p. 398.
Parus afer, Gray's Hand-list of Birds, No. 3331.
„ „ Sharpe's Cat. No. 332.

This species is found sparingly in all the regions between the Okavango towards the north, Lake Ngami towards the east, and the Orange River towards the south, and it also occurs in some parts of the Cape Colony.

The irides are dark brown.

Measurements of a male and a female:—

	Male. in. lin.	Female. in. lin.
Entire length	5 6	5 3
Length of folded wing	3 0	2 11
„ tarsus	0 9	0 9
„ middle toe	0 6	0 5
„ tail	2 3	2 3
„ bill	0 7	0 7

101. Parus niger, Vieill. Southern Black-and-White Tit.

La Mésange noire, Levaillant's Ois. d'Afr. pl. 137.
Parus niger, Gurney, Birds Damar., in Proc. Zool. Soc. 1864, p. 2.
„ „ Andersson, ibid. p. 6.

Parus leucopterus, Layard's Cat. No. 211.
Parus niger, Chapman's Travels in S. Afr., App. p. 398.
„ „ Sharpe's Cat. No. 333.

This Tit is to be met with, though more sparingly than the preceding species, in Damara Land and in the neighbourhood of the Okavango River and of Lake Ngami; it is, however, more frequent in the last two districts than in Damara Land proper, and in Great Namaqua Land I have never observed it. It is generally found in pairs, searching amongst the larger trees for insects and their larvæ; it also feeds on seed. The female is distinguished from the male by her inferior size and duller plumage. The irides are dark brown, the bill dark horn-colour, the legs and toes greenish lead-colour.

Measurements of a male and a female :—

	Male. in. lin.	Female. in. lin.
Entire length	5 9	5 6
Length of folded wing	3 2	2 10
„ tarsus	0 9	0 9
„ middle toe	0 6½	0 6
„ tail	2 9	2 8
„ bill	0 6½	0 6

LUSCINIDÆ.

102. Drymoica maculosa (Bodd.). Cape Drymoica.

Drymoica capensis, Smith's Zool. of S. Africa, pl. 76. fig. 1.
„ „ Andersson, in Proc. Zool. Soc. 1864, p. 7.
„ „ Layard's Cat. No. 161.
Drymoica maculosa, Gray's Hand-list of Birds, No. 2731.
Drymœca maculosa, Sharpe's Cat. No. 277.

I have reason to think that this bird is common in some of the southern parts of Great Namaqua Land;

further south, on the west coast and within the Cape Colony, I have frequently met with it; and in the neighbourhood of Cape Town it is exceedingly common, a pair or two inhabiting almost every garden.

It is found singly or in pairs; and its whereabouts is easily discovered by the harsh querulous notes that it is in the habit of uttering almost incessantly. It builds in low bushes; and the nest is composed of moss, wool, and other soft material, which is artistically and strongly put together. This species feeds on insects, searching for them either on the ground or amongst the low bushes which form its favourite resort; it runs with great rapidity along the ground, and steals through tangled foliage with equal celerity. The iris is yellow, bill brown, and legs pale flesh-colour.

Measurements of a female:—

	in.	lin.
Entire length	5	0
Length of folded wing	1	10
,, tarsus	0	10
,, middle toe	0	5
,, tail	2	8
,, bill	0	8

[Mr. Sharpe possesses a specimen of this *Drymoica*, obtained by Mr. Andersson in Little Namaqua Land.—ED.]

103. Drymoica affinis, Smith. Allied Drymoica.

Drymoica affinis, Smith's Zool. of S. Africa, pl. 77. fig. 1.
    ,,   ,,   Layard's Cat. No. 154.
Drymœca affinis, Sharpe's Cat. No. 276.

[Specimens of this *Drymoica*, obtained at Ovaquenyama by Mr. Andersson, are in the possession of Mr. R. B. Sharpe. —ED.]

104. Drymoica flavicans (Vieill.). Pectoral Drymoica.

Le Citrin, Levaillant's Ois. d'Afr. pl. 127.
Sylvia flavicans, Vieillot's Nouv. Dict. d'Hist. Nat. vol. ii. p. 175.
Drymoica pallida, Smith's Zool. of S. Africa, pl. 72. fig. 2.
Drymoica pectoralis, ibid. pl. 75. fig. 2.
Drymœca flavicans, Strickland & Sclater, Birds Damar., Contr. Orn. 1852, p. 148.
Drymoica pectoralis, Layard's Cat. No. 146.
Drymoica pallida, ibid. No. 147.
Drymoica subflava, ibid. No. 169.
Drymœca Ortleppi, Tristram, in Ibis, 1869, p. 207.
Drymœca flavicans, Sharpe's Cat. No. 275.

I have obtained specimens of *Drymoica pectoralis* both in Damara Land and in the neighbourhood of the Okavango. I am sadly puzzled about specimens of *D. flavicans* (as well as about some others) called *D. pectoralis*; and sometimes I fancy they are all identical, the difference being merely in age, sex, and colour. The male is evidently both larger and more robust than the female, and the latter has the black on the breast less well defined than the male. I have found the nests of these birds (usually containing three, but sometimes four, eggs) at various dates, extending from December 20th to April 1st. The nest is very light and graceful, composed of fine grass both externally and internally, and built on a low bush a few feet above the ground.

The iris is brownish yellow, the bill jet-black, the legs and toes flesh-colour.

Measurements of a female:—

	in.	lin.
Entire length	6	0
Length of folded wing	2	1
„ tarsus	0	9
„ middle toe	0	4
„ tail	3	0
„ bill	0	$6\frac{1}{2}$

[The female specimen whose measurements are given above by Mr. Andersson is now in the collection of Mr. R. B. Sharpe, who has a fine series of specimens of this *Drymoica* and who is of opinion that Mr. Andersson was correct in his views respecting this species, and that *Drymoica flavicans, pallida, pectoralis,* and *Ortleppi* must be regarded as identical; this opinion is shared by Mr. Jules P. Verreaux, who has recently examined Mr. Sharpe's collection; and a similar examination has led me to concur in the same view. Some specimens show no trace of a black gorget, as in Levaillant's plate of *Le Citrin*, fig. 2; others show it partially, as in fig. 1 of the same plate; and others, again, fully, as portrayed in Sir A. Smith's plate of *D. pectoralis* above referred to. The absence or presence of the black gorget is proved by Mr. Sharpe's specimens not to be a sexual difference; but the gorget may not improbably be a nuptial dress assumed by both sexes: Sir A. Smith, however, gives it as his opinion that the absence of the gorget is a mark of immaturity, which may perhaps be the correct explanation.—ED.]

105. Drymoica ocularius, Smith.   Rufous-cheeked Drymoica.

Drymoica ocularius, Smith's Zool. of S. Africa, pl. 75. fig. 1.
    „   „   Layard's Cat. No. 150.
Drymœca ocularius, Chapman's Travels in S. Afr., App. p. 398.
    „   „   Sharpe's Cat. No. 278.

This species is very sparingly found in Damara Land, but is more common in some parts of Great Namaqua Land. It is generally met with singly or in pairs; and it is usual to find it amongst the most arid scenes, hopping slowly about amongst the branches of low bushes in search of insects.

The iris is ochry brown, the bill bluish black, the legs and toes flesh-coloured.

Measurements of a male:—

	in.	lin.
Entire length	5	6
Length of folded wing	1	10
,, tarsus	0	9
,, middle toe	0	6
,, tail	2	8
,, bill	0	7

106. Drymoica Smithii, Bon. Smith's Drymoica.

Drymoica ruficapilla, Smith's Zool. of S. Africa, pl. 73. fig. 1.
Drymoica Smithii, Bonaparte's Conspectus, vol. i. p. 283.
Drymoica ruficapilla, Gurney, Birds Damar., Proc. Zool. Soc. 1864, p. 2.
 ,, ,, Layard's Cat. No. 156.
Drymœca Smithii, Sharpe's Cat. No. 294.

I am not aware that this compact little bird is an inhabitant of either Great Namaqua or Damara Land. I first became acquainted with it on penetrating to the Okavango, but even there I do not remember to have seen much of it.

[A specimen collected by Mr. Andersson, and now in the possession of the Rev. H. B. Tristram, is ticketed "Elephant Vley, October 12th, 1859;" two others in the collection of Mr. Sharpe are from the same locality.—ED.]

107. Drymoica chiniana, Smith. Kurichane Drymoica.

Drymoica chiniana, Smith's Zool. of S. Africa, pl. 79.
Drymoica subruficapilla, Gurney, Birds Damar., in Proc. Zool. Soc. 1864, p. 2.
Drymoica chiniana, Layard's Cat. No. 158.
Drymœca chiniana, Sharpe's Cat. No. 281.

[Mr. Andersson's collections contained examples of this species obtained at Objimbinque, and also two obtained in Ondonga; but the latter were both of them somewhat smaller than the usual dimensions of this species, of which they are considered to be " a small variety " by the Rev. H. B. Tristram, who possesses one of these specimens, the other being in the collection of Mr. R. B. Sharpe.—ED.]

108. Drymoica subruficapilla, Smith. Tawny-headed Drymoica.

Drymoica subruficapilla, Smith's Zool. of S. Africa, pl. 76. fig. 2.
" " Layard's Cat. No. 160.
Drymœca subruficapilla, Sharpe's Cat. No. 280.

[Mr. R. B. Sharpe possesses a specimen of this bird obtained by Mr. Andersson in Little Namaqua Land.—ED.]

109. Drymoica rufilata, Hartl. Andersson's Drymoica.

Drymoica chiniana, Gurney, Birds Damar., Proc. Zool. Soc. 1864, p. 2.
Drymoica rufilata, Finsch & Hartlaub, Vögel Ost.-Afrika's, p. 238.

I have a couple of specimens somewhat resembling the figure of *Drymoica chiniana* as given by Smith (Ill. of Zool. of S. Afr. *Aves*, pl. 79); they were obtained in the neighbourhood of the Okavango.

[These remarks of Mr. Andersson's were probably intended to apply to the present species, which is nearly allied to *D. chiniana*, Smith, but has been recently described as distinct by Drs. Finsch and Hartlaub (*loc. cit.*) from a pair which were sent to me by Mr. Andersson in 1859, and which are now preserved in the Bremen Museum. This species has not been figured.—ED.]

110. Drymoica Levaillantii, Smith. Levaillant's Drymoica.

Drymoica Levaillantii, Smith's Zool. of S. Africa, pl. 73. fig. 2.
Drymœca Levaillantii, Strickland & Sclater, Birds Damar., Contr. Orn. 1852, p. 147.
Drymoica Levaillantii, Layard's Cat. No. 157.
Drymœca Levaillantii, Sharpe's Cat. No. 279.

I found this bird by no means uncommon in the neighbourhood of the Okavango; its favourite haunts seem to be along the sedgy streams and amongst the rank vegetation of marshy localities. It flits quickly from reed to reed in quest of insects, and is a comparatively tame species.

The iris is brown; the upper mandible black; the lower reddish, but black at the extremity; the legs are a pale flesh-colour.

111. Cisticola terrestris (Smith). Ground-Cisticole.

Drymoica terrestris, Smith's Zool. of S. Africa, pl. 74. fig. 2.
„   „   Layard's Cat. No. 159.
Cisticola terrestris, Sharpe's Cat. No. 273.
„   „   Gurney, in Ibis, 1871, p. 151.

This species came under my notice in Great Namaqua Land in about 24° or 25° S. lat.; I have also met with it abundantly in southern Damara Land, and have obtained it in Ondonga. Specimens from Damara Land are of a lighter tint than those from Ondonga, but I have no doubt they are identical. It is common at some large waters on the Omaruru River, but is most difficult to shoot; it can generally only be shot on the wing as it rises; and when shot it invariably falls in the reeds, where its diminutive size easily eludes the eye. It is, however, found in many other situations besides reedy localities, but chiefly amongst tall, coarse grasses growing about small periodical watercourses. When disturbed, it rises almost perpendicularly, descending nearly as abruptly, and either burying itself at once in the rank vegetation or first perching on a grass-stalk and gradually creeping out of view, and also out of reach; for it is difficult to flush it again.

The food of this little bird consists of small insects. Its eggs, which are four or, rarely, five in number, are sometimes white, or more frequently white freely sprinkled with minute brown spots; but occasionally they are tinged with green, whilst others are of a reddish colour. The nests also vary in form, material, and construction; some are airy and fragile, like the home of a spider, whilst others are pretty compact and more or less pasted on the outside with decomposed grasses; and it is a remarkable fact that the eggs in the spider-like nests are always whitish, spotted with brown, whilst those in the more complete nests are of a greenish tint but with the same spotting. The nests, which are sometimes globular, are suspended to the stalks of long grasses about a foot above the ground. I have found them with eggs from the 18th of February to the 31st of March.

The irides in this species are brownish yellow; the upper mandible is light brown, except the edges, which, with the lower mandible, are of a light flesh-colour, as are also the legs and toes.

Measurements of a male and a female:—

	Male. in. lin.	Female. in. lin.
Entire length	4 3	4 2
Length of folded wing	2 0	1 10
,, tarsus	0 9	0 9
,, middle toe	0 5	0 5
,, tail	1 7	1 6
,, bill	0 8	0 7

[This species is a true *Cisticola*, and possesses twelve rectrices, being the number characteristic of that genus.—ED.]

112. Aëdon subcinnamomea (Smith). Kamiesberg Aëdon.

Drymoica subcinnamomea, Smith's Zool. of S. Africa, pl. 111. fig. 1.
     „   „   Layard's Cat. No. 151.
Drymœca subcinnamomea, Sharpe's Cat. No. 295.

[Sir A. Smith, in his 'Illustrations of the Zoology of South Africa' (*loc. cit.*), figures a single individual of this species, which he obtained " on the top of one of the mountains of the Kamiesberg in Little Namaqua Land," a locality which brings the species within the geographical range of the present volume. I believe that this specimen is the same that is preserved in the British Museum; and I only know one other example of this species, which was obtained at Colesburg, and is now in the collection of Mr. R. B. Sharpe, to whom it was sent by Mr. Layard.

Sir A. Smith has placed this species and the succeeding one, which is very closely allied to it, in the genus *Drymoica*, but with the following remark in his article on the present species:—" In its general aspect this bird has much of what characterizes birds of the group *Drymoica*; but when its individual characters are closely surveyed it is found to want that which would entitle it to be classed among typical species; or, in other words, it exhibits what requires it to be viewed as an aberrant form."

Both this and the succeeding species appear to me to be more nearly allied to the genus *Aëdon* than to *Drymoica*; and I have therefore so arranged them, though not without some doubt as to whether they ought not rather to be placed by themselves in a separate and distinct genus.—ED.]

113. Aëdon fasciolata (Smith). Fasciolated Aëdon.

Drymoica fasciolata, Smith's Zool. of S. Africa, pl. 111. fig. 2.
     „   „   Layard's Cat. No. 148.
Drymœca fasciolata, Chapman's Travels in S. Afr., App. p. 398.
     „   „   Sharpe's Cat. No. 296.
Aëdon fasciolata, Gurney, in Ibis, 1871, p. 152.

This species is common in the neighbourhood of Objimbinque; and I have found a few individuals between that place and Rehoboth; it greatly reminds me of our Swedish Gärdsmygg (the common Wren of England) in its habits, which are somewhat secluded. It frequents dense bush and occasionally trees, searching diligently amongst the branches for insects; it carries its tail erect when moving about. When disturbed it flies but a short distance at a time, and is easily distinguished by the bright brown on the rump.

The iris is yellowish brown; legs pale flesh-colour; toes the same, but a trifle darker; upper mandible dark liver-brown, the lower edge and the under mandible dark bluish or purple.

Measurements of a male and a female:—

	Male. in. lin.	Female. in. lin.
Entire length	5 3	5 0
Length of folded wing	2 4	2 2
,, tarsus	0 11	0 10
,, middle toe	0 5	0 5
,, tail	2 3	2 0
,, bill	0 8	0 8

[Under the head of *Aëdon subcinnamomea* I have explained my reason for assigning that species and the present to the genus *Aëdon*; and I may remark that the typical species of that genus, the North-African *Aëdon galactodes* (Temm.), has the same habit of carrying its tail raised which Mr. Andersson noticed in *Aëdon fasciolata* (vide J. H. Gurney, Jun., "On the Ornithology of Algeria," in ' The Ibis ' for 1871, p. 83).—ED.]

114. Aëdon leucophrys (Vieill.). White-browed Aëdon.

Erythropygia pectoralis, Smith's Zool. of S. Africa, pl. 49.
Aëdon leucophrys, Gurney, Birds Damar., Proc. Zool. Soc. 1864, p. 2.
Ædon leucophrys, Layard's Cat. No. 182.
Erythropygia pectoralis, Chapman's Travels in S. Afr., App. p. 398.
Aëdon leucophrys, Sharpe's Cat. No. 269.

This is a pretty common species in the middle and northern parts of Damara Land, and also further to the northward, being very common near Ombongo. During the pairing- and breeding-season it occasionally sings most exquisitely; and it, moreover, has the power of imitating almost every other bird to be found in its vicinity. It forms its nest, in November and December, on the lower branches of small bushes, rarely more than one or two feet from the ground; the nest is composed of rough grasses, and is lined with material of the same kind, but of a softer texture. The eggs are two in number.

115. Aëdon paena (Smith). Smith's Aëdon.

Erythropygia paena, Smith's Zool. of S. Africa, pl. 50.
Ædon pæna, Layard's Cat. No. 181.
Aëdon paena, Gray's Hand-list of Birds, No. 2982.
Erythropygia pæna, Chapman's Travels in S. Afr., App. p. 398.
Aëdon pæna, Sharpe's Cat. No. 270.

This bird is pretty generally distributed over Damara and Great Namaqua Land and the parts adjacent. Its habits are exactly like those of the preceding species; and I know no birds which they both resemble so greatly in manner, habits, food, &c. as those of the genus *Saxicola*. The present species spends much of its time on the ground and amongst the roots of bushes; it runs with great swiftness, and raises and droops its tail in quick succession, but does not expand it; at times it stands quite vertically, with its whole body vibrating with excitement, whilst it rapidly utters a succession of harsh, jarring chirps.

Two nests of this species, taken on the 5th and 6th of January, contained two eggs each: the nests were composed of grass, and lined with fine, soft tendrils; they were built in thorn-bushes, and placed from twelve to eighteen inches above the ground.

Measurements of a male:—

	in.	lin.
Entire length	6	3
Length of folded wing	2	10
„ tarsus	1	0
„ middle toe	0	7
„ tail	2	8
„ bill	0	10

116. Thamnobia coryphæus (Vieill.). Coriphée Bush-creeper.

Le Coriphée, Levaillant's Ois. d'Afr. pl. 120.
Brachypterus coriphæus, Layard's Cat. No. 183.
Thamnobia coryphæus, Gray's Hand-list of Birds, No. 2990.

I do not remember to have seen this species in Damara Land; but I first observed it in the central part of Great Namaqua Land, and from thence southward it became more abundant. It seems partial to low bushes, and may be seen running along the ground from one bush to another with wonderful rapidity; it also usually adopts this mode of endeavouring to make its escape when pursued.

The male has a very agreeable song during the

breeding-season; it is very brusque in its movements, and frequently utters a chirping noise, at the same time elevating and spreading its tail Peacock-fashion over its back.

The food of this species consists of insects and berries. The irides are brown, the bill, legs, and toes blackish.

My specimens are not so dark, especially in the upper parts, as those from the Cape Colony.

[This species was not contained in Mr. Andersson's last collection; but it was identified in a collection sent by him to England some years previously, as I learn from a memorandum left amongst his papers.—ED.]

117. Camaroptera olivacea, Sund. Olivaceous Camaroptera.

Camaroptera olivacea, Sundevall, Öfvers. 1850, p 103.
    ,,   ,, Gurney, Birds Damar., Proc. Zool. Soc. 1864, p. 2.
Calamodyta olivacea, Layard's Cat. No. 180.
Camaroptera olivacea, Von Heuglin, in Ibis, 1869, p. 140.
Camaroptera brevicaudata, Sharpe's Cat. No. 309.

This species is pretty common in the neighbourhood of the Okavango River, and is also not uncommon in Damara Land proper; but I did not meet with it in Great Namaqua Land. It is a very tame bird, and hunts slowly and with great care, examining alike, in quest of insects, both thickets and the largest trees; it generally carries its wings slightly drooping when thus engaged, and its tail raised at an angle of 45°.

The iris is light brown, the eyelids flesh-coloured, the bill livid horn-colour, and the legs and toes brownish flesh-colour.

Measurements of a male and a female :—

	Male. in. lin.	Female. in. lin.
Entire length	4 9	4 6
Length of folded wing	2 3	2 1
,, tarsus	0 11	0 10
,, middle toe	0 6	0 6
,, tail	1 10	1 9
,, bill	0 9	0 7½

[Mr. G. R. Gray in his 'Hand-list of Birds,' No. 2864, and Drs. Finsch and Hartlaub in their work on the 'Birds of East Africa,' p. 241, unite this species with its more northern congener *C. brevicaudata* (Rüpp.). Notwithstanding the view taken by these high authorities, I am disposed to follow Von Heuglin (*loc. cit.*) and Blanford (' Geology and Zoology of Abyssinia,' p. 377) in regarding these two nearly allied races as perceptibly and specifically distinct. I am not aware that the southern race has been figured.—ED.]

118. Dryodromas damarensis (Wahl.). Damara Dryodrome.

Eremomela damarensis, Wahlberg, Öfvers. 1855, p. 213.
,, ,, Wahlberg, in Journal für Orn. 1857, p. 2.
Dryodromas damarensis, Gray's Hand-list of Birds, No. 2861.
,, ,, Finsch & Hartlaub's Vögel Ost-Afrika's, p. 240.
,, ,, Sharpe's Cat. No. 318.

I only observed this diminutive species in one locality, a place called Oosoop, on the lower course of the Swakop River; and even there it was very scarce, so that, though very tame, I have hunted for it whole days unsuccessfully. It is found in small families of from two to six individuals amongst the widely scattered dwarf vegetation; it hops slowly and systematically amongst the branches, searching diligently for small insects, which constitute its sole food, and uttering all the while a low but distinct chirp.

The iris is bright yellow; legs liver-brown; upper mandible and tip of lower bluish black, remainder of lower mandible bluish brown.

Measurements of a male and a female:—

	Male. in. lin.	Female. in. lin.
Entire length	4 2	4 0
Length of folded wing	1 10	1 10
„ tarsus	0 8	0 8
„ middle toe	0 5	0 5
„ tail	1 11	1 9
„ bill	0 6	0 6

[Professor Wahlberg, *loc. cit.*, gives no other locality for this bird than the river Swakop. This species, I believe, has not yet been figured. The following is a copy of Wahlberg's description of it:—" Superne cinereo-olivacea, auchenio capiteque superne et in lateribus cinereis, vertice olivaceo tincto, stria supraoculari vix ulla. Subtus alba, crisso dilute sulphureo. Cauda rotundata (rectr. ext. 9 mm. intermediâ brevior) fusca, olivaceo tincta. Tectrices alarum inferiores albæ, remiges fuscæ, extus olivaceo limbatæ; 1^a brevis ($=\frac{1}{2}$ 2^{dæ}); 2=7; 3, 4, et 5 æquales, longissimæ ... Femina similis mari sed major et magis olivaceo tincta."—Ed.]

119. Dryodromas flavida (Strick.). Black-breasted Dryodrome.

Drymœca flavida, Strickland & Sclater, Birds Damar., Contr. Orn. 1852, p. 148 (female).
Eremomela flavida, Gurney, Birds Damar., Proc. Zool. Soc. 1864, p. 2.
Dryodromas flavida, Gray's Hand-list of Birds, No. 2862.
Dryodromas flavidus, Finsch & Hartlaub's Vögel Ost-Afrika's, p. 240.
Dryodromas flavida, Sharpe's Cat. No. 320.

It is only in the neighbourhood of the Okavango that I have found this pretty little species at all numerous, and I never saw it either in Damara or in Great Namaqua Land. I generally found it in small flocks, probably consisting of entire families. It resorts to the denser parts of the forests, and lives entirely on small insects.

The iris is yellowish brown, the bill almost black, the tarsus flesh-coloured.

Measurements of a male:—

		in.	lin.
Entire length	. . . . . .	4	10
Length of folded wing	. . . . .	2	0
,, tarsus	. . . . . .	0	9
,, middle toe	. . . . .	0	5
,, tail	. . . .	2	0
,, bill	. . . .	0	7

[The females of this species are destitute of the black gorget which distinguishes the males; the type specimen, which is preserved in the Museum of Zoology, at Cambridge, is a female. Neither sex has as yet been figured.—ED.]

120. Eremomela flaviventris (Burch.). Yellow-bellied Bush-chirper.

Sylvia flaviventris, Burchell's Travels, vol. i. p. 335 (note).
Eremomela flaviventris, Sundevall, Öfvers. 1850, p. 102.
Drymoica brachyura, Layard's Cat. No. 171 (part.).
Eremomela flaviventris, Gray's Hand-list of Birds, No. 2849.
   ,,   ,,   Sharpe's Cat. No. 317.

This little bird is sparingly met with from the Okavango River on the north to the Orange River on the south, but, on account of its diminutive size and secluded habits, is probably often overlooked. It is found in small families amongst dwarf vegetation, where it diligently searches for small insects and their larvæ. It is very tame, and is often found in company with othe small birds.

The iris is dark brown; the bill dark horn-colour, but yellow at the angle of the mouth, and livid flesh-colour on the basal part of the lower mandible.

Measurements of a male and a female:—

	Male. in. lin.	Female. in. lin.
Entire length	4 1	3 9
Length of folded wing	2 3	2 1
,, tarsus	0 9	0 0
,, middle toe	0 5	0 4
,, tail	1 8	1 6
,, bill	0 6	0 6

[This species has not been figured.—ED.]

121. Eremomela usticollis, Sund. Brown-throated Bush-chirper.

Eremomela usticollis, Sundevall, Öfvers. 1850, p. 102.
,,   ,,   Gray's Hand-list of Birds, No. 2853.
,,   ,,   Sharpe's Cat. No. 316.

I first obtained an example of this species at Objimbinque, on June 7th, 1866; it was shot whilst hopping amongst the branches of a lofty tree; and I have since obtained several other specimens.

The iris is lemon-colour, the legs yellowish flesh-colour, the upper mandible brown, the lower brownish flesh-colour.

Measurements of a male and a female:—

	Male. in. lin.	Female. in. lin.
Entire length	4 7	4 4
Length of folded wing	2 2	2 2
,, tarsus	0 10	0 10
,, middle toe	0 5	0 5
,, tail	0 9	0 9
,, bill	0 7	0 7

[As this species has not been figured, and as Professor Sundevall's description of it (*loc. cit.*) is not readily accessible in England, I annex a copy of it: "Superne cinerea, subtus tota pallide fulvescens lunulâ juguli fusco-rufescente. Ala 55, tars. 21. Rostrum pallidum culmine fusco, pedes pallidi, alæ et cauda cinereofusca, tenuissime pallescenti marginatæ. Macula juguli, colore obscuro, maculam adustam refert. Adultæ quoque in genis eodem colore tinctæ. Penna 1. brevis, 2. = 7.—*Hab.* in Caffraria Superiori (25°)."—ED.]

122. Calamodyta arundinacea (Linn.). Thrush-like Reed-Warbler.

Salicaria turdoides, Gould's Birds of Europe, pl. 106.
Calamodyta arundinacea, Layard's Cat. No. 178.

This bird bears some resemblance to the British species, but seems much larger. I found it plentiful in the reedy marshes at Omanbondé. It was always singing, but on the approach of danger immediately retired to the thickest parts of its reedy resorts.

[The above note is merely headed by Mr. Andersson " Reed-Warbler;" but I believe it is intended to apply to this species, as, although I did not meet with any examples of it in Mr. Andersson's last collection, Mr. Layard writes (*loc. cit.*) :—" Mr. Andersson brought specimens from Damara Land apparently identical with a European bird in the South-African Museum."—ED.]

123. Calamodyta bæticata (Vieill.). Isabelle Reed-Warbler.

L'Isabelle, Levaillant's Ois. d'Afr. pl. 121. fig. 2.
Sylvia bæticata, Vieillot's Nouv. Dict. vol. xi. p. 195.
Calamodyta bæticula, Layard's Cat. No. 175.
Calamodyta rufescens, Layard's Cat. No. 176.
Calamoherpe rufescens, Gurney in Ibis, 1869, p. 292.
Calamodyta bæticula, Layard in Ibis, 1869, pp. 365 & 366.
Calamodyta bæticata, Ibis, 1869, p. 365 (note).
Calamoherpe bæticata, Sharpe's Cat. No. 305.

I have obtained this bird on a few occasions in Damara Land; it feeds on small worms, aquatic insects, &c.

124. Calamodus schœnobænus (Linn.). British Sedge-Warbler.

Salicaria phragmitis, Gould's Birds of Europe, pl. 110.
Calamodus schœnobænus, Gray's Hand-list of Birds, No. 2964.
Calamodyta schœnobænus, Sharpe's Cat. No. 308.

I obtained two specimens on December 22nd, 1866, at Objimbinque in the bed of the river, amongst some "cotton-plants," on which, and on the ground, they were hopping about in search of insects; they were new to me.

Iris dark brown; upper mandible dark horn-colour, lower brownish, but both yellowish at the edge; gape bright orange-yellow; legs and toes livid brown.

[Mr. Andersson's last collection contained the two above-named specimens, which are now in the possession of Mr. R. B. Sharpe. They appear to be identical with examples obtained in England.—ED.]

125. Sylvia hortensis, Gmel. British Garden-Warbler.

Curruca hortensis, Gould's Birds of Europe, pl. 121.
Sylvia hortensis, Hartlaub's Birds of West Africa, No. 744.
   „   „ Malmgren in Ibis, 1869, p. 230.
Curruca hortensis, Sharpe's Cat. No. 325.

[Mr. R. B. Sharpe possesses two specimens of this Warbler, which were obtained in Damara Land by Mr. Andersson.—ED.]

126. Phyllopseuste hypolais (Linn.). European Melodious Warbler.

Sylvia hypolais, Gould's Birds of Europe, pl. 133.
Sylvia obscura, Smith's Zool of S. Africa, pl. 112. fig. 1.
  „   „ Layard's Cat. No. 190.
  „   „ Chapman's Travels in S. Afr., App. p. 397.
Ficedula hypolais, Malmgren in Ibis, 1869, p. 230.
Phyllopseuste hypolais, Gray's Hand-list of Birds, No. 3042.
Hypolais salicaria, Sharpe's Cat. No. 318.

I observed this *Sylvia* sparingly in the neighbourhood of the Okavango and in Damara Land; it is migratory and amongst the earlier arrivals. It sings deliciously, and is found in small flocks hopping about industriously amongst the branches of the smaller trees, preferring such as have a thick tangled foliage, which it slowly examines for small insects and larvæ.

The iris is brown, the legs and toes slate-coloured; the gape melon-yellow; the upper mandible brownish with yellowish tip and edges, the under mandible yellow tinged with flesh-colour; the ring round the eyes light yellow.

[Mr. Andersson's last collection contained specimens of this Warbler, which appeared to be identical with European examples. —Ed.]

127. Phyllopseuste trochilus (Linn.). British Willow-Warbler.

Sylvia trochilus, Gould's Birds of Europe, pl. 131. fig. 1.
Phyllopneuste trochilus, Gurney, Birds Damar., Proc. Zool. Soc. 1864, p. 2.
Sylvia trochilus, Layard's Cat. No. 189.
Phylloneuste trochilus, Sharpe's Cat. No. 314.

I have only observed this species in the neighbourhood of the Okavango; it is rather lively in its habits, hopping incessantly amongst the foliage, and even suspending itself head downwards from the buds and flowerets of the trees, almost every part of which it explores in search of the small insects which constitute its food.

The irides are very dark brown, almost black; the bill brownish, lighter beneath and tinged with yellow; the legs and toes are yellowish brown, but a narrow band at the back of the legs and the soles of the feet are orange-yellow.

Measurements of a male and a female:—

	Male. in. lin.	Female. in. lin.
Entire length	4 11	4 6
Length of folded wing	2 8	2 5
,, tarsus	0 10	0 9
,, middle toe	0 5	0 5
,, tail	2 1	1 10
,, bill	0 7	0 7

[Mr. Sharpe's collection contains specimens of this species obtained by Mr. Andersson both in Damara Land and in Ovampo Land.—ED.]

128. Pratincola torquata (Linn.). South-African Stonechat.

Le Traquet pâtre, Levaillant's Ois. d'Afr. pl. 180.
Pratincola pastor, Layard's Cat. No. 207.
Saxicola rubicola, Chapman's Travels in S. Afr., App. p. 399.
Pratincola torquata, Gray's Hand-list of Birds, No. 3276.
       ,,   ,,   Sharpe's Cat. No. 257.

This bird is common in Little Namaqua Land; but, to the best of my recollection, I have never seen it either in Great Namaqua or in Damara Land, though it is not infrequent at Lake Ngami. Its habits are solitary, and it is not often that two individuals are seen together. It frequents low bush and marshy ground, perching on the tops of bushes, tall grasses, and reeds, whence it descends in search of insects, which constitute its food, and which it sometimes also captures on the wing, like a Flycatcher.

Measurements of a male and a female:—

	Male. in. lin.	Female. in. lin.
Entire length	5 3	4 10
Length of folded wing	2 7	2 7
,, tarsus	0 10	0 11
,, middle toe	0 6	0 6
,, tail	2 0	1 10
,, bill	0 9	0 9

[The South-African Stonechat appears only to differ from the European *Pratincola rubecola* in the greater clearness of its general colouring when adult, especially in the males, and in the somewhat greater extent of the white patch on the rump and upper tail-coverts, together with the absence of any rufous tint from the white feathers in the centre of the abdomen and from the under tail-coverts.—ED.]

129. Saxicola familiaris, Steph. Familiar Wheatear.

Le Traquet familier, Levaillant's Ois. d'Afr. pl. 183.
Saxicola familiaris, Stephen's Gen. Zool. vol. xiii. p. 241.
Erythropygia Galtoni, Strickland & Sclater, Birds Damar., Cont. Orn. 1852, p. 147.
Saxicola sperata, Chapman's Travels in S. Afr., App. p. 398.
Aëdon familiaris, Gray's Hand-list of Birds, No. 2980.
Aëdon Galtoni, ibid., No. 2986.
Saxicola familiaris, Sharpe's Cat. No. 249.

This is the most common *Saxicola* with which I am acquainted both in Damara and Namaqua Land, from whence it ranges southward along the west coast as far as Table Mountain. It is very familiar in its manners, and will fearlessly approach human habitations, which it not unfrequently enters by the doors and windows.

It invariably perches on low bushes, whence it watches for passing insects, which it usually seizes on the wing, though it occasionally descends to the ground for a similar purpose.

Like all the birds of this family it is very restless, now flapping its wings, then raising and expanding the tail or alternately raising and depressing its whole body. It makes its nest on the ground, laying three or four eggs, which are either greenish grey spotted with brown, or nearly white spotted with brown and grey.

The iris is very dark brown, and the bill, legs, and feet are black.

Measurements of a male and a female:—

	Male. in. lin.	Female. in. lin.
Entire length	6 2	5 8
Length of folded wing	3 3	3 2
,, tarsus	1 0½	1 1
,, middle toe	0 7	0 7
,, tail	2 7½	2 6
,, bill	0 10	0 8

[I have examined the type specimen of *Erythropygia Galtoni*, Strickland, which is now preserved in the Museum of Zoology at Cambridge, and which appears to belong to the present species.

The true *Saxicola sperata* of Gmelin is quite a distinct species, and may be readily recognized by the whitish colour of the rump, which is rufous in *S. familiaris*.—ED.]

130. Saxicola Schlegelii (Wahl.). Schlegel's Wheatear.

Erithacus Schlegelii, Wahlberg, Öfvers. 1855, p. 213.
,, ,, Wahlberg, Journal für Orn. 1857, p. 3.
Saxicola modesta, Tristram, in Ibis, 1869, p. 206.
Saxicola Schlegeli, Sharpe's Cat. No. 254.

This species is very common in Great Namaqua and Damara Land, and frequents alike broken ground, low bush, and old abandoned "werfts;" it seems to prefer low bushes for perching on, whence it descends to the ground in search of insects, and runs with great swiftness in pursuit of its prey. It is tolerably easy of approach.

Specimens are frequent in Damara Land of a smaller size and paler colour than the ordinary type, which, however, they exactly resemble in habits and manners.

Measurements of two specimens from Objimbinque, male and female:—

	Male. in. lin.	Female. in. lin.
Entire length	6 8	6 0
Length of folded wing	3 8	3 3
,, tarsus	1 1	1 0
,, middle toe	0 6½	0 6
,, tail	2 6	2 6
,, bill	0 9	0 9

[This species has not been figured.—ED.]

131. Saxicola Stricklandii, Bon. Strickland's Wheatear.

Saxicola albicans, Wahlberg, Öfvers. 1855, p. 213.
Saxicola Stricklandii, Gray's Hand-list of Birds, No. 3214.
      ,,   ,,   Sharpe's Cat. No. 258.

I have only met with this bird in Damara Land, and that at no very great distance from the sea-coast; it is not uncommon on the extensive plains bordering on Walvisch Bay, and seems to be chiefly confined to such localities. It is of a very friendly, fearless disposition, almost rivalling *Saxicola sperata* in its familiarity with man; it will approach a person to within a very short distance, and it is no uncommon thing to see it hopping about amongst the cooking-utensils that may chance to be scattered about a temporary encampment. It spends nearly all its time on the ground, along which it runs with great swiftness; but now and then it may be seen perched on a stone raised a few feet above the level of the plain. Its flight is a kind of alternate dip and rise, and never extends far at a time.

The food of this species consists of small insects; its eggs are laid on the ground in a small excavation sheltered by a stone or bush.

Measurements of a male :—

	in.	lin.
Entire length	5	9
Length of folded wing	3	8
,, tarsus	1	2
,, middle toe	0	7
,, tail	2	2
,, bill	0	11

In the adult of this Wheatear the head, auriculars, neck, and back are drab tinted with fawn, the lower part of the back being more strongly tinted than the upper; a streak over the eye, the chin, throat, breast, belly, under wing-coverts, sides, vent, under and upper tail-coverts, and part of the rump are all more or less white tinted with pale fawn; the wings are grey, with the outer vanes, shafts, and extremities of the primaries brown, and the upper parts narrowly margined with fawn-white, the wing-coverts, secondaries, and tertials being more broadly margined and also tipped with this colour; the tail is white, with the posterior third brown margined with fawn-white and tipped with grey; the bill is brown, the legs, toes, and claws black.

The young bird resembles the adult, but the upper parts and sides of the head are mottled with brown; the breast is also mottled, the rest of the underparts are white; the tail is fawn-white for two-thirds of its length, the remainder being less dark and more liberally margined and tipped with fawn-white than in the adult; the bill, legs, and toes are lighter-coloured than in the adult.

[As this species has not been figured, and as the descriptions that have been published of it are not very accessible, I have here included that contained in Mr. Andersson's notes.—ED.]

132. Saxicola infuscata, Smith. Great Fuscous Wheatear.

Saxicola infuscata, Smith's Zool. of S. Africa, pl. 28.
Saxicola No. 29, Strickland & Sclater, Birds Damar., Contr. Orn. 1852, p. 146.
Saxicola infuscata, Layard's Cat. No. 198.
    ,,   ,,   Layard, in Ibis, 1869, p. 367.
    ,,   ,,   Sharpe's Cat. No. 252.

This species is found abundantly in Great Namaqua Land, and also occurs in a few localities in southern Damara Land; it is found singly or in pairs in open localities interspersed with low bush. It is extremely wary and difficult to approach; perched on the top of a conspicuous bush it quickly espies the hunter, and immediately takes its departure; it does not fly far at a time, but always takes care to be beyond the range of the gun. It feeds on insects, which it catches on the wing or on the ground, but it never stays on the ground to search for them there.

The iris is brown, and the bill, legs, and feet are a very dark brown.

Measurements of a male :—

	in.	lin.
Entire length	7	7
Length of folded wing	4	4
,, tarsus	1	1
,, middle toe	0	7
,, tail	3	5
,, bill	1	0

133. Saxicola pileata (Gmel.). Imitative Wheatear.

Le Traquet imitateur, Levaillant's Ois. d'Afr. pls. 181 & 182.
Saxicola hottentotta, Strickland & Sclater, Birds Damar., Contr. Orn. 1852, p. 146.
Saxicola pileata, Layard's Cat. No. 192.
Saxicola hottentotta, Chapman's Travels in S. Afr., App. p. 399.
Saxicola pileata, Sharpe's Cat. No. 248.

I have found this species common from Table Mountain in the south to the Okavango in the north, in the neighbourhood of which river it may be seen at all times of the year, though in Damara Land proper it only appears during the wet season, and again gradually retreats to more favoured regions as the dry season returns. In the Cape Colony it is one of the best-known birds, and, from its familiar habits and its being frequently seen near cattle and sheep, the Dutch boors have given it the name of " Schaap Wagter " or Shepherd; it has also the more local name of " Nagtgaal " and " Rossignol," from a habit it is said to have of singing by night. It is a very tame bird, of a most inquisitive nature, and seems to seek the society of man.

The male has a very pleasant and varied song during the breeding-season, and is especially remarkable for its strange power of imitating sounds, such as the notes of other birds, the barking of a dog, the bleating of a goat, &c.

The iris in this species is dark brown.

[Specimens of this bird from Damara Land appear to belong to a race somewhat smaller than that which is found in more southern localities.—ED.]

134. Saxicola leucomelæna, Burch. Mountain-Wheatear.

Saxicola leucomelæna, Burchell's Travels, vol. i. p. 335, note.
" " Strickland & Sclater, Birds Damar., Contr. Orn. 1852, p. 146.
Saxicola cursoria, Layard's Cat. No. 203.
Saxicola alpina, Chapman's Travels in S. Afr., App. p. 399.
Saxicola leucomelæna, Sharpe's Cat. No. 241.

This fine *Saxicola* is abundantly met with throughout Great Namaqua Land, and also in the south of Damara Land, especially in the valley of the Swakop; it is partial to localities which abound in rocks, and is found throughout the year in the dreariest and most arid spots, but never at any great distance from the hills, to which it immediately resorts on the least approach of danger.

Like the rest of its family, this Wheatear is constantly moving about, now fluttering its wings, then rapidly elevating and depressing its tail, and next all at once vibrating in every part of its body with frolic and excitement; it perches on a dry branch, a bush, a stone, or any other spot from which it can obtain a clear view of surrounding objects. It usually seeks its food upon the ground, but will also seize insects on the wing as they happen to pass within its ken and reach; if not disturbed it will return times without number, after such excursions, to the same perch.

The iris in this species is dark brown.

[This species has not been figured.—ED.]

135. Saxicola Atmorii, Tristr. Atmore's Wheatear.

Saxicola Atmorii, Tristram, in Ibis, 1869, p. 206.
„ „ Gray's Hand-list of Birds, No. 3235.
Saxicola Atmorei, Sharpe's Cat. No. 244.

[Mr. Andersson's last collection contained several specimens of this new Wheatear, obtained at Objimbinque and Hykomkap; and from these it was described by Mr. Tristram in the 'Ibis' as cited above, but has not yet been figured.

Mr. Andersson's notes furnish no account of this species, except the following measurements of a female:—

	in. lin.
Entire length	7 3
Length of folded wing	4 1
„ tarsus	1 2
„ middle toe	0 8
„ tail	2 8½
„ bill	0 10½

—ED.]

136. Myrmecocichla formicivora (Vieill.). Southern Ant-eating Wheatear.

Le Traquet fourmilier, Levaillant's Ois. d'Afr. pls. 186 & 187.
Myrmecocichla formicivora, Layard's Cat. No. 205.
Saxicola formicivora, Chapman's Travels in S. Afr., App. p. 399.
Myrmecocichla formicivora, Sharpe's Cat. No. 239.

I have met with this species, though only sparingly, in Damara Land, and in the parts adjacent to the north and east, but I do not recollect having seen it in Great Namaqua Land. It always occurs in pairs in open localities interspersed with bush, on which, or on anthills, it usually perches. It seeks its food on the ground, watching for insects from its elevated perch; and when these are observed, descending at once upon them and quickly returning to its post of observation. Its flight is straight, and it moves its short wings with extraordinary rapidity; but it does not go far at a time, and I never found any difficulty in approaching it.

[Specimens of this bird from Damara Land were found, on comparison, to be a little smaller than an example obtained in Natal.—ED.]

MOTACILLIDÆ.

137. Motacilla capensis, Linn. Cape-Wagtail.

La Lavandière brune, Levaillant's Ois. d'Afr. pl. 177.
Motacilla capensis, Layard's Cat. No. 219.
„ „ Sharpe's Cat. No. 697.

This is rather a local bird in Damara and Great Namaqua Land, but is found somewhat numerously in moist and humid localities, and is also at times pretty freely met with on the sea-shore. It occurs sometimes in pairs, and sometimes in small flocks. It captures its prey both on the wing and by running along the ground, frequently following in the wake of cattle and picking up such small insects as may chance to be thus disturbed.

The nest of this Wagtail is found in a variety of situations, and is composed of tendrils and soft pliable plants. The eggs are three to four in number, and are generally of a yellow-drab tint, profusely speckled with obscure spots of pale brown, especially towards the larger end; but the eggs, even when from the same nest, are subject to some differences, both as regards colour and size.

The iris in this species is brown.

138. Motacilla Vaillantii, Cab. Levaillant's Wagtail.

L'Aguimp, Levaillant's Ois. d'Afr. pl. 178.
Motacilla aguimp, Layard's Cat. No. 221.
Motacilla Vaillantii, Gray's Hand-list of Birds, No. 3572.

I have only observed this Wagtail on the borders of the Orange River, where it is not uncommon. It is generally to be seen either singly or in pairs, and usually settles on stones or on the ground, along which it runs with great celerity in pursuit of small insects, which constitute its chief food; and it also skims the surface of the water for the same purpose.

139. Budytes flava (Linn.). Blue-headed Yellow Wagtail.

Motacilla neglecta, Gould's Birds of Europe, pl. 140.
Motacilla flava, Layard, in Ibis, 1869, p. 73.
Budytes flavus, Malmgren, in Ibis, 1869, p. 230.
Budytes flava, Gray's Hand-list of Birds, No. 3578.
Motacilla flava, Finsch & Hartlaub's Vögel Ost-Afrika's, p. 268.
  „   „   Finsch, in Trans. of Zool. Soc. vol. vii. p. 239.
Budytes flava, Sharpe's Cat. No. 702.
  „   „   Ayres, in Ibis, 1871, p. 154.

I had been fifteen years in Damara Land before I became aware of the existence of this Wagtail, which I first observed at Objimbinque in 1865, when I obtained a few specimens, nearly all of which were immature. It is a migratory bird, and appears only in or about the rainy season.

The irides in this species are an intensely dark brown; the legs and feet dusky black, as are also the upper mandible and the point of the lower; the remainder of the lower mandible is a light horn-colour; in young birds the bill is rather paler.

140. Anthus Raalteni, Temm. Raalten's Pipit.

Anthus Raaltenii, Layard's Cat. No. 229.
Anthus Raalteni, Gray's Hand-list of Birds, No. 3616.
„   „   Finsch & Hartlaub's Vögel Ost-Afrika's, p. 274.

[The Rev. H. B. Tristram possesses a specimen of this Pipit obtained in Damara Land by the late Mr. Andersson; and the same locality is also quoted for this species by Drs. Finsch and Hartlaub, *loc. cit.*
This Pipit has not been figured.—ED.]

141. Anthus caffer, Sund. Caffre Pipit.

Anthus caffer, Sundevall, Öfvers. 1850, p. 100.
„   „   Layard's Cat. No. 230.
„   „   Sharpe's Cat. No. 691.

I have found these Pipits common at Objimbinque. Their favourite resorts are open places near moist situations; a great number are sometimes found together, yet not in flocks; they mix much with the Wagtails. These birds offer considerable variety of plumage; sometimes they are very light-coloured, and at others their tints are very deep.

[This species has not been figured.—ED.]

142. Anthus pyrrhonotus (Vieill.). Cinnamon-backed Pipit.

L'Alouette à dos roux, Levaillant's Ois. d'Afr. pl. 197.
Alauda pyrrhonota, Vieillot's Nouv. Dict. d'Histoire Nat. vol. i. p. 361.

Alauda erythronota, Stephens's Gen. Zool. vol. xiv. p. 24.
Anthus leucophrys, Vieillot's Gal. des Oiseaux, vol. i. p. 262.
Anthus sordidus, Rüppell's Neue Wirbelth. p. 103, pl. 39. fig. 1.
Anthus cinnamomeus, Rüppell's Neue Wirbelth. p. 103.
Anthus Gouldii, Fraser, Proc. Zool. Soc. 1843, p. 27.
 „ „ Gurney, in Ibis, 1860, p. 208.
Anthus sordidus, Layard's Cat. No. 226.
Anthus leucophrys, Layard's Cat. No. 228.
Anthus erythronotus, Sharpe's Cat. No. 693.
Anthus pyrrhonotus, Gurney, in Ibis, 1871, p. 156.
 „ „ Layard, in Ibis, 1871, p. 228.

This Pipit resembles the foregoing species, but is much larger; it is widely dispersed over both Damara and Great Namaqua Land.

143. Anthus campestris, Bechst. Tawny Pipit.

Anthus rufescens, Gould's Birds of Europe, pl. 137.
Anthus campestris, Layard's Cat. No. 232.

[Mr. R. B. Sharpe possesses a single specimen of this Pipit collected in Damara Land by the late Mr. Andersson.—ED.]

TURDIDÆ.

144. Turdus letsitsirupa, Smith. Ground-scraper Thrush.

Turdus letsitsirupa, Smith's Append. to Report of Exp. p. 45.
Turdus strepitans, Smith's Zool. of S. Africa, pl. 37.
 „ „ Strickland & Sclater, Birds Damar., Contr. Orn. 1852, p. 145.
 „ „ Layard's Cat. No. 237.
 „ „ Chapman's Travels in S. Afr., App. p. 396.
Turdus letsitsirupa, Sharpe's Cat. No. 183.

This Thrush is pretty abundant in Damara and Great Namaqua Land, especially the former; it also occurs in the Lake-regions. It is partially migratory, only a few remaining in Damara Land throughout the year.

It lives chiefly on insects, for which it searches at the roots of trees and amongst low bushes, old leaves, and decayed wood. It scratches somewhat after the manner of a Fowl, and is thence called by the Bechuanas the "Ground-scraper;" it also runs with great celerity. It lives singly or in pairs, and occasionally perches on the topmost branch of some lofty tree. It utters a plaintive half-song, half-call, just as if it were troubled with a bad cold.

This species breeds about Objimbinque; and I took a nest containing three young on the 29th of October: it was built on a branch about ten feet from the ground, and was composed exteriorly of grass, the interior being lined with down and feathers.

The legs in this Thrush are of a light greenish yellow; the upper mandible of the bill is horn-colour, the lower mandible yellowish.

There does not appear to be any particular difference in the size of the sexes.

145. Turdus libonyanus, Smith. Kurichane Thrush.

Turdus libonyana, Smith's Zool. of S. Africa, pl. 38.
„ „ Layard's Cat. No. 236.
Turdus libonyanus, Sharpe's Cat. No. 186.

In all my wanderings north of the Orange River I have but once met with this Thrush, which I then fell in with in the neighbourhood of the river Okavango.

[Mr. R. B. Sharpe possesses a specimen of this Thrush obtained by Mr. Andersson at Ombongo, Damara Land.—ED.]

146. Turdus olivaceus, Linn. Olivaceous Thrush.

Le Grivron, Levaillant's Ois. d'Afr. pls. 98 & 99.
Turdus olivaceus, Layard's Cat. No. 240.
" " Sharpe's Cat. No. 188.

Messrs. J. and H. Chapman brought away specimens of this Thrush from the Lake-regions; but I have never met with it to the west of the Lake-country. It is one of the commonest species in the Cape Colony, where it is partially migratory, being found most abundantly at the grape- and fruit-season; it is fond of almost all kinds of fruit, but when these fail it subsists on beetles and other insects. Its flesh is well-tasted.

Measurements of a male and a female:—

	Male.		Female.	
	in.	lin.	in.	lin.
Entire length	8	2	8	10
Length of folded wing	4	7	4	6
" tarsus	1	0	1	3
" middle toe	0	8	0	9
" tail	4	0	3	5
" bill	0	10	1	3

[Mr. Sharpe possesses a specimen of this Thrush from Lake Ngami, and another from the Orange River.—ED.]

147. Monticola brevipes, Waterh. Short-footed Rock-Thrush.

Monticola brevipes, Strickland & Sclater, Birds Damar., Contr. Orn. 1852, p. 147.
Petrocincla explorator, Chapman's Travels in S. Afr., App. p. 399.
Monticola brevipes, Sharpe's Cat. No. 229.

This species is not uncommon throughout Great Namaqua Land and the southern parts of Damara Land; and in one year I found it particularly abundant at Objimbinque. In its habits and manners it resembles *Saxicola monticola,* and, like it, is partial to localities of a rocky nature, as also to abandoned "werfts" and villages. It is rather a voracious feeder, preying on all kinds of insects, from the minutest beetle to the scorpion, of which I have found specimens in its stomach; it also occasionally eats soft seeds.

The iris is reddish brown; the legs and toes bluish black, as also is the bill, but with the base of the gape yellow.

[This species has not yet been figured; seven specimens obtained at Objimbinque by Mr. Andersson are now in the collection of Mr. R. B. Sharpe.—ED.]

148. Chætops pycnopygius (Sclat.). Damara Chætops.

Sphenœacus pycnopygius, Strickland & Sclater, Birds Damar., Contr. Orn. 1852, p. 148, pl. 102.
Drymœca anchietæ, Bocage, in Jorn. Acad. Lisb. 1868, p. 41.
Chætops anchietæ, Bocage, in Jorn. Acad. Lisb. 1868, p. 351.
Chætops Grayi, Sharpe, in Proc. Zool. Soc. 1869, pl. 14.
Chætops pycnopygius, Sharpe's Cat. No. 226.

The only places where I can recollect having caught a glimpse of this rarely seen Rock-Thrush are the Kaaru River, Ongari Ombo, near Jacongana, Okamaluté, and, lastly, the Omaruru River, where I obtained one specimen on October 30th, 1866, which I found hopping about amongst some stones thickly overgrown with bush and coarse grass, and strewn with decaying wood. As soon as the bird perceived that it was observed, it immediately slipped into the thickest part of this tangled bush, and for some time I thought I had lost it; but after pelting it with stones it flew out and settled on the lower branch of a small acacia. I found, when shot, that its bill was very dirty, from which I infer that it seeks much of its food on the ground; its stomach contained only small whitish ants (not termites); its flesh was light-coloured, and excellently flavoured. The iris was dark brown; the upper mandible dark horn-colour, the lower livid lead-colour; the tarsi pale brown, the toes a shade or two darker; there was a space under the eye quite naked; the tail when closed was of equal breadth throughout, but was cuneated when expanded.

Measurements of this specimen:—

	in.	lin.
Entire length	7	1
Length of folded wing	2	9
„ tarsus	1	0
„ middle toe	0	8
„ tail	3	3
„ bill	0	10½

[The specimen above referred to is in the collection of Mr. Sharpe, by whom it was figured, *loc. cit.*, under the name of *Chætops Grayi*; but another example sent home from Mr. Andersson's first journey, and now in the Museum of Zoology at Cambridge, had been previously described by Mr. Sclater, *loc. cit.*, under the name of *Sphenœacus pycnopygius*.—ED.]

149. Cossypha caffra (Vieill.). Jan-frédric Chat-Thrush.

Le Jan frédric, Levaillant's Ois. d'Afr. pl. 111.
Bessonornis phœnicurus, Layard's Cat. No. 248.
Bessonornis caffra, Gray's Hand-list of Birds, No. 3868.

This species is sparingly met with on the borders of the Orange River, whence it extends southwards to the Cape, where it is very numerous. It is of a most inquisitive nature, and seems to court the neighbourhood of man. It is very lively in its movements, either hopping and gliding amongst bushes and plants, or running along the ground with astonishing swiftness, generally accompanying all such movements by rapid expansions and depressions of its tail and wings. The male sings very pleasantly; and his notes have been likened to the following differently intoned syallables, jan-fredric-dric-dric fredric, whence its colonial name of Jan frédric.

The irides in this species are dark brown, the bill black, the legs and feet livid brown.

150. Cossypha bicolor (Sparr.). Vociferous Chat-Thrush.

Le Réclameur, Levaillant's Ois. d'Afr. pl. 104.
Bessonornis vociferans, Layard's Cat. No. 245.
Cossypha bicolor, Sharpe's Cat. No. 236.

[Mr. R. B. Sharpe possesses a bird of this species, which was obtained by Mr. Andersson at Objimbinque.—ED.]

PYCNONOTIDÆ.

151. Pycnonotus nigricans (Vieill.). Brunoir Bulbul.

Le Brunoir, Levaillant's Ois. d'Afr. pl. 106. fig. 1.
Pycnonotus nigricans, Layard's Cat. No. 261.
    „   „   Finsch & Hartlaub's Vögel Ost-Afrika's, p. 297.
    „   „   Sharpe's Cat. No. 204.

This bird is found abundantly as far north as the Okavango River, as also in the Lake-regions. It is gregarious in its habits, often congregating in considerable numbers, and is never found far away from water; it is active, lively, and noisy, but chatters rather than sings; its food consists of berries, insects, &c. A nest taken in

Damara Land on the 11th of October contained two eggs rather hard sat on; the nest was situated in a dabbe bush, and was composed of twigs and grass externally, lined internally with finer grass.

The iris in this species is a rather pale yellowish red, and the skin round the eye bright orange; the bill black, the legs and toes dark shiny brown.

[Mr. R. B. Sharpe possesses three specimens of this Bulbul, obtained by Mr. Andersson at Objimbinque, in Damara Land, in the months of June, July, and September.—ED.]

152. Pycnonotus tricolor, Hartl. Angola Bulbul.

Pycnonotus capensis, Strickland & Sclater, Birds Damar., Contr. Orn. 1852, p. 145.
Ixos tricolor, Hartlaub, in Ibis, 1862, p. 341.
Pycnonotus tricolor, Sharpe's Cat. No. 211.
    „   „ Sharpe, in Proc. Zool. Soc. 1871, p. 130, pl. 7. fig. 2.

[Mr. R. B. Sharpe possesses two specimens of this species obtained by Mr. Andersson in January 1867, one at Ovaquenyama, the other in Oudonga. A previous specimen transmitted by Mr. Andersson to this country is now in the late Mr. Strickland's collection at Cambridge, and appears to have been the bird included in Messrs. Strickland and Sclater's Damara list, under the name of *Pycnonotus capensis*.—ED.]

153. Phyllastrephus capensis, Swains. Cape Jaboteur.

Le Jaboteur, Levaillant's Ois. d'Afr. pl. 112. fig. 1.
Phyllastrephus capensis, Layard's Cat. No. 206.
    „   „ Sharpe's Cat. No. 207.

This species feeds on seeds. The iris is yellow; the upper mandible horn-colour, the lower bluish; tarsus bluish.

Measurements of a male and a female:—

	Male. in. lin.	Female. in. lin.
Entire length	8 2	8 2
Length of folded wing	3 6	3 5
" tarsus	0 11	0 11
" middle toe	0 8	0 6
" tail	3 8	3 7
" bill	1 0	0 10

[Mr. Andersson gives the above particulars from specimens obtained at the Knysna; but as his last collection contained one example from Lake Ngami, I, on that account, include the species in the present volume.—ED.]

154. Criniger flaviventris (Smith). Yellow-bellied Criniger.

Trichophorus flaviventris, Smith's Zool. of S. Africa, pl. 59.
Criniger flaviventris, Finsch, in Journ. für Orn. 1869, p. 22.
    "   "   Layard's Cat. No. 259.
    "   "   Sharpe's Cat. No. 196.

[Mr. Sharpe possesses two specimens of this bird, which were obtained by Mr. Andersson at Ovaquenyama.—ED.]

155. Crateropus bicolor, Jard. Southern Black-and-White Babbler.

Crateropus bicolor, Jardine, in Edinburgh Journal, pl. 3.
    "   "   Strickland & Sclater, Birds Damar., Contr. Orn. 1852, p. 145.
    "   "   Andersson, in Proc. Zool. Soc. 1864, p. 7.
    "   "   Layard's Cat. No. 252.
    "   "   Chapman's Travels in S. Afr., App. p. 396.
    "   "   Sharpe's Cat. No. 216.

This species is common throughout Damara and Great Namaqua Land, and is also found in the Lake-regions; it always occurs in flocks of many individuals, and creeps and glides through the mazes of tangled wood and dense thickets with amazing ease and celerity. When alarmed, it flies slowly from tree to tree, its flight being feeble. It climbs excellently, and also seems equally at home upon the ground.

I was fortunate enough to fall in with a nest of this species on the 15th of October, 1866; and, considering how common the bird is, I wonder that I have not met with more nests; this one contained three eggs, in every way very similar to those of *Crateropus melanops*. The nest was situated in a fork on the very top of a small anna tree, some ten or twelve feet from the ground; it was composed externally of fine twigs and coarse grasses, and was lined with somewhat finer grass; it was circular, deep, and very compact. I could both see and hear the parent bird whilst we were robbing the nest; but it did not come near or appear very solicitous.

On the 11th of December, 1866, I observed a family of these birds, consisting of an old pair and their young, hopping about in an anna wood almost as carelessly and fearlessly as Robins. One of them, evidently the female, led the way, followed by the young, which uttered a querulous, subdued note. In the young birds the tail and wings are of somewhat the same colouring as those of their parents; but the body differs much from the colour of the adult bird, being grey or brownish grey, instead of white.

The irides in this species are light reddish brown, the bill dark horn-colour, the legs brownish black.

156. Crateropus melanops, Hartl. Dark-faced Babbler.

Crateropus melanops, Hartlaub in Proc. Zool. Soc. 1866, pl. 37.
      „   „   Sharpe's Cat. No. 219.

I have only met with this bird in the northern district of Damara Land, and in the parts adjacent towards the north and east; there it is common, and, when not too much disturbed, becomes quite familiar with man. During my encampment in the desert, on my return from the Okavango, there was scarcely a day that a family of these birds did not pay me a visit, coming quite close to my tent, searching for insects amongst the débris, and especially attaching themselves to my cook's establishment.

The favourite resorts of this species are tangled brakes, where it restlessly hops about amongst the bushes, gradually descending to the ground, exploring on the way, and searching about the roots and amongst the fallen dry leaves for insects and their larvæ. It is gregarious in its habits; and several of its nests are frequently found on the same bush or in the immediate neighbourhood of each other. The nest is firmly constructed of fine twigs lined with some softer materials. The eggs are of a greenish-blue colour, smooth at the extremities, but quite rough on the central parts, with numerous little tubercles.

157. Crateropus Jardinii, Smith. Jardine's Babbler.

Crateropus Jardinii, Smith's Zool. of S. Africa, pl. 6.
      „   „   Layard's Cat. No. 251.
      „   „   Sharpe's Cat. No. 217.

[Mr. Andersson's last collection contained specimens of both this and the preceding species, alike obtained at the river Cunéné on the 25th of June, 1867.—ED.]

158. Crateropus Hartlaubi, Boc. Hartlaub's Babbler.

Crateropus Hartlaubi, Bocage, Jorn. Acad. Lisboa, 1868, p. 48.
Crateropus senex, Finsch & Hartlaub's Vögel Ost-Afrika's, p. 290.
Crateropus Hartlaubi, Sharpe's Cat. No. 220.

[Mr. Andersson's last collection also contained two specimens of this species from the river Cunéné, obtained on June 25th, 1867, one of which is now in the possession of Mr. R. B. Sharpe. This species has not been figured.—ED.]

ORIOLIDÆ.

159. Oriolus galbula (Linn.). Golden Oriole.

Oriolus galbula, Gould's Birds of Europe, pl. 71.
„   „   Layard's Cat. No. 255.
„   „   Chapman's Travels in S. Afr., App. p. 397.
„   „   Sharpe's Cat. No. 506.

The European Golden Oriole arrives in Damara Land with the return of the rainy season; but it is comparatively rare, and very few adult birds are seen; it is excessively shy and difficult to approach, both when perched on lofty trees and also when gliding rapidly through the underwood. Its food consists of insects and fruits.

[Mr. Sharpe possesses an example of this species obtained by Mr. Andersson in Ondonga.—ED.]

160. Oriolus notatus, Peters. Andersson's Oriole.

Oriolus auratus, Gurney, Birds Damar., Proc. Zool. Soc. 1864, p. 2.
„   „   Andersson, ibid. p. 6.
„   „   Chapman's Travels in S. Afr., App. p. 397.
Oriolus notatus, Peters in Journ. für Orn. 1868, p. 132.

Oriolus Anderssoni, Bocage in Jorn. Acad. Lisb. 1869, p. 342.
Oriolus notatus, Sharpe in Ibis, 1870, pl. 7. fig. 2.
" " Finsch & Hartlaub's Vögel Ost-Afrika's, p. 291.
" " Sharpe's Cat. No. 508.

I have only obtained the adult of this splendid Oriole in Damara Land on a few occasions, and that always during the rainy season; the young, however, are frequently met with; and at the Okavango River the species is more common than in Damara Land proper. The young birds are easily obtained; but the old are excessively shy and difficult to procure, as they always perch on the most elevated and conspicuous trees and retire into the densest parts of tangled brakes and thickets on the least approach of danger.

The food of this Oriole consists of seeds, berries, and insects. The irides are brown in the young birds and bright red in the adult, the bill is reddish brown, the legs are lead-coloured.

DICRURIDÆ.

161..Dicrurus musicus, Vieill. Musical Drongo.

Le Drongear, Levaillant's Ois. d'Afr. pl. 167.
Dicrurus divaricatus, Strickland & Sclater, Birds Damar., Contr. Orn. 1852, No. 14.
Dicrurus musicus, Layard's Cat. No. 301.
" " Sharpe's Cat. No. 446.

This bird is common in almost all parts of Great Namaqua and Damara Land, and I also found it plentiful in Ondonga; but it is partially migratory. It is an exceedingly fierce bird, and will fearlessly attack any other, the most powerful Falcon not excepted, on which occasions its very swift flight and its habit of always keeping above its enemy more than compensates for its inferiority in strength. It is very particular about keeping its own locality to itself, and will not rest till it has fairly expelled an intruder, thus becoming a useful ally to the naturalist, to whom it rarely fails to give notice of the presence of a bird of prey, if at all near to its own domain. This species is generally found in pairs; and at times, especially during the season of incubation, one of these birds may be heard late at night, perched on the top of a tree and uttering the most melodious notes, perhaps rather plaintive, but very soothing and varied; and again early in the morning, an hour or so before daybreak, he may also be heard carolling to his mate. The song of the Damara *Dicrurus* is, however, inferior to that of the *Dicrurus musicus* of the Cape, and its call-note is also not so loud. It breeds in Damara Land, and builds in a fork or branch of a tree. The nest is composed of twigs, and is lined with tendrils or roots of tiny plants, but with no softer lining; it is so loosely constructed that a person standing underneath it may see the eggs through the bottom of the nest; these are from two to four in number, of a whitish colour, besprinkled with small dark brown spots, which are somewhat clustered at the thicker end.

These birds feed on insects, and watch for their prey from some elevated perch; as soon as it is perceived, the bird gives chase, catching it on the wing and nearly always returning to the same perch. This species is said to be particularly destructive to bees.

The iris is reddish orange, the legs are brownish black, and the bill is black.

[The Damara race of *Dicrurus musicus* appears to be identical with that which occurs in Natal, and slightly smaller than that which inhabits the Cape Colony, from which, however, it does not differ except in its smaller size and inferior song, which latter distinction can, of course, only be appreciated by a naturalist who, like Mr. Andersson, had listened to the notes of both races in a state of nature.

I am indebted to the kindness of Viscount Walden, who has paid much attention to this group, for the following remarks, and for permission here to quote them:—" The Damara *Dicrurus* is a puzzler: if we take the Cape *Dicrurus musicus* as our standard, we find that throughout the whole extent of Africa it preserves representative forms; besides these, and coexisting with them, are to be found totally different species, such as *D. Ludwigi, atripennis, coracinus, modestus*, &c. Now the utmost that can be said about the Damara-Land *Dicrurus* is that it is a representative form of *D. musicus*, just as *D. divaricatus*, Licht. (= *canipennis*, Swains.), is the Senegambian representative. The Damara bird is smaller than that of the Cape; the bill in the Cape bird is longer, and the tail more deeply forked, *i. e.* the outer tail-feathers are longer than in the Damara bird; but I do not think that much reliance can be placed on any of these characters; all that one may assert for certain is that the Damara bird is not so large as that of the Cape, and that it more nearly approaches in size that of the Gambia, namely *D. divaricatus*, Licht. I cannot detect the least difference between an example from Natal and those from Damara Land; the Natal specimen is smaller than the Cape *D. musicus.*" The specimens which I have compared from the Cape and from Damara Land quite confirm Lord Walden's observations, except as to the tail in the Cape race being more deeply forked than in birds from Damara Land, a distinction which is barely, if at all, perceptible in the examples which I have examined. The length of the wing from the carpal joint to the tip of the primaries in the Cape race exceeds the corresponding measurement in the Damara bird by about three tenths of an inch; and the tarsus of the Cape bird is fully one tenth of an inch longer than that of its Damara congener.

The late Mr. Strickland's collection, now preserved in the Museum of Zoology at Cambridge, contains a *Dicrurus* from Kordofan, labelled "*D. divaricatus;*" and to this species Mr. Strickland also referred the Damara race, from which, however, this Kordofan specimen differs in having the secondary wing-feathers about a quarter of an inch shorter, and the bill a little broader at the base.

For the reasons explained in Lord Walden's remarks, I have not treated the Damara race of this *Dicrurus* as specifically distinct from that found at the Cape.—ED.]

MUSCICAPIDÆ.

162. Melanopepla pammelæna (Stanley). Black Flycatcher.

Muscicapa lugubris, Müller's Ois. d'Afrique, pl. 2.
Melænornis ater, Layard's Cat. No. 305.
Melanopepla pammelæna, Gray's Hand-list of Birds, No. 4258.
Bradyornis pammelæna, Finsch & Hartlaub's Vögel Ost-Afrika's, p. 320.

[Drs. Finsch and Hartlaub refer (*loc. cit.*) to two examples of this species obtained in Damara Land, one of which, if not both, was collected by Mr. Andersson.

A short note in Mr. Andersson's MS., referring to a bird resembling *Dicrurus Ludwigi*, Smith, and found sparingly near the Okavango River, was probably intended to apply to this species.—ED.]

163. Bradornis mariquensis, Smith. Mariqua Flycatcher.

Bradornis mariquensis, Smith's Zool. of S. Africa, pl. 113.
Bradyornis mariquensis, Gurney, Birds Damar., Proc. Zool. Soc. 1864, p. 2.
    ,,   ,,   Andersson, ibid. p. 5.

Saxicola mariquensis, Layard's Cat. No. 204.
Bradornis mariquensis, Chapman's Travels in S. Afr., App. p. 396.
Bradyornis mariquensis, Sharpe's Cat. No. 262.

This species is very common throughout Damara and Great Namaqua Land, and is very partial to burnt ground; it usually watches for its prey from some elevated position, whence it pounces on any coming within reach. It appears to me a true Flycatcher in its habits, or a Butcher bird. I have met with its newly fledged young in March and April.

Measurements of a male:—

	in.	lin.
Entire length	6	7
Length of folded wing	3	5
,, tarsus	0	10
,, middle toe	0	6
,, tail	3	2
,, bill	0	9

164. Muscicapa griseola, Linn. European Spotted Flycatcher.

Muscicapa grisola, Gould's Birds of Europe, pl. 65.
,, ,, Gurney, Birds Damar., Proc. Zool. Soc. 1864, p. 3.
,, ,, Layard's Cat. No. 281.
Muscicapa griseola, Gray's Hand-list of Birds, No. 4811.
Butalis grisola, Sharpe's Cat. No. 391.

This species is common in Damara and Great Namaqua Land, and is found there throughout the year, either singly or in pairs, perching on some low branch of a tree, whence it makes short and rapid excursions in pursuit of such winged insects as may chance to pass within view, frequently returning to the same post of observation, and uttering at intervals a kind of chirping call. The iris is dark brown.

Measurements of a male :—

	in.	lin.
Entire length	5	11
Length of folded wing	3	5
,, tarsus	0	7
,, middle toe	0	6
,, tail	2	6
,, bill	0	9

165. Tchitrea viridis (Müll.). Tchitrec Flycatcher.

Le Tchitrec, Levaillant's Ois. d'Afr. pl. 142.
Tchitrea cristata, Layard's Cat. No. 273.
Muscipeta cristata (?), Chapman's Travels in S. Afr., App. p. 396.
Terpsiphone cristata, Finsch & Hartlaub's Vögel Ost-Afrika's, p. 304.
Terpsiphone viridis, Sharpe's Cat. No. 415.

I have only observed this very pretty and elegant Flycatcher in the neighbourhood of the Okavango River, where, however, it is scarce; at Lake Ngami it is less uncommon. The few specimens that I have personally secured were exceedingly wary and difficult to approach. These birds live in pairs and frequent the forests, perching only on the larger trees. The males are said to be very quarrelsome, and to fight with considerable ferocity and tenacity.

Measurements of a male and a female :—

	Male.		Female.	
	in.	lin.	in.	lin.
Entire length	13	3	7	3
Length of folded wing	3	8	3	3
,, tarsus	0	9	0	8
,, middle toe	0	5	0	4
,, tail	9	3	3	7
,, bill	0	10	0	10

[Mr. Sharpe possesses a specimen of this Flycatcher, obtained by Mr. Andersson at Elephant's Vley.—Ed.]

166. Platysteira pririt (Vieill.). Pririt Flycatcher.

Le Pririt, Levaillant's Ois. d'Afr. pl. 161.
Platystira pririt, Strickland & Sclater, Birds Damar., Contr. Orn. 1852, p. 144.
Platysteira strepitans, Layard's Cat. No. 271.
Platysteira pririt, Sharpe's Cat. No. 410.

[This species is only doubtfully alluded to in Mr. Andersson's MS. notes; but his last collection contained some specimens of it, four of which are now in the cabinet of Mr. R. B. Sharpe. Of these, three appear by the tickets to have been obtained at Ovaquenyama in May and June, and the fourth in September at Elephant's Vley.—ED.]

167. Platysteira affinis, Wahl. Wahlberg's Flycatcher.

Platystira affinis, Wahlberg, Öfvers. 1855, p. 214.
" " Wahlberg, Journal für Orn. 1857, p. 3.
" " Finsch & Hartlaub's Vögel Ost-Afrika's, p. 866 (sub *P. pririt*).
" " Finsch, in Trans. of Zool. Soc. vol. vii. p. 315.
Platysteira affinis, Sharpe's Cat. No. 408.

This species is very abundant in the Swakop valley; it is usually seen in pairs about large trees, which it explores carefully and systematically in search of insects. It has at times a peculiar far-sounding note, which a stranger might imagine to be uttered by a bird at a great distance, whilst, in reality, the bird is near at hand.

The iris is a light lemon-yellow on the inner edge, shading off into greenish grey on the outer circle.

[Mr. R. B. Sharpe possesses four examples of this Flycatcher collected by Mr. Andersson, two of which are marked as from the vicinity of the Swakop River, one from Objimbinque, and one "West of Tjobis." This species has not been figured.

Mr. Layard, at page 143 of his 'Catalogue of the Birds of South Africa,' mentions *Platysteira pristinaria* as a species which Mr. Andersson "brought from Damara Land;" but I do not find it noticed as a Damara species in Mr. Andersson's MS. notes, neither have I seen it in any Damara-Land collection.—Ed.]

168. Platysteira torquata (Waterh.). White-tailed Flycatcher.

Platystira albicauda, Strickland & Sclater, Birds Damar., Contr. Orn. 1852, p. 144.
    „   „   Chapman's Travels in S. Afr., App. p. 395.
Lanioturdus torquatus, Sharpe's Cat. No. 406.

I met with this fine Flycatcher in the south of Damara Land; and I also found it very common, and in many cases paired, about the Omaruru River, in October and November. In the latter locality it was very tame, and I could procure almost any number of specimens; but in the former it was shy and very restless, seeking the thickest part of the tree or bush on which it might chance to be perched immediately that it found itself pursued, and making its escape from the side opposite to that on which its pursuer might be watching for it, not, however, flying far, but settling on the nearest tree on which it deemed itself secure.

This Flycatcher presents a pleasing appearance on the wing, although its flight is slow and apparently laborious; it is gregarious in its habits, and seeks its food amongst dwarf vegetation and also on the ground. I love this little bird, it is so odd-looking, and often enlivens with its quaint appearance and movements an otherwise dreary and monotonous solitude. It has a very clear plaintive call-note; but generally its notes are querulous, and at times not unlike the distant call of a Corncrake.

The tail-feathers in this species are of extraordinary stiffness. The iris is greenish yellow, the bill almost black, the legs and toes very dark brown, with a whitish or lead-coloured tint about the joints.

[As the descriptions which have been published of this species are not readily accessible to many ornithologists, I transcribe that drawn up by the late Mr. Strickland in Strickland and Sclater's paper in the 'Contributions to Ornithology,' *loc. cit.*, which may be the more desirable as (so far as I am aware) no figure of this bird has as yet been published:—

"Front pure white, extending laterally as far as the eyes; crown, lores, and cheeks deep glossy black; a white spot on the nape, surrounded by black; back slaty grey; scapulars black externally, slaty grey within, and obscurely tipped with white; the middle and greater coverts next the body pure white; basal third of primaries and the extreme tips of the first three white; secondaries and tertials black, tipped with white; the four inner secondaries next the tertials white for one-fourth from the base; rump and upper tail-feathers thick and downy, cinereous like the back, each feather with an elongate subterminal tear-like spot of black, on the median pair nearly bisected longitudinally by the white shaft; chin, throat, and sides of neck pure white, below which is a black pectoral collar; sides of breast cinereous; lower wing-covers black; middle of breast, abdomen, and lower tail-covers pure white; feathers of the tibiæ white at the tips, black at the base; beak and legs black.

"Total length $5'' 3'''$; beak to front $6\frac{1}{4}'''$, to gape $9'''$, wide $2\frac{1}{2}'''$, high $2'''$; wing $3'' 5'''$, median rectrices $1'' 8'''$, external $1'' 7'''$; tarsus $1'' 2'''$.

"This is the largest species of *Platystira* I have seen; the beak is stronger and more compressed at the sides than in the other species, the tail shorter in proportion, and the first primary longer, being nearly two-thirds the length of the fourth."—ED.]

169. Campephaga nigra, Vieill. Black Caterpillar-eater.

L'Echenilleur noir, Levaillant's Ois. d'Afr. pl. 165 (male).
L'Echenilleur jaune, id. ibid. pl. 164 (female).
Campephaga nigra, Gurney, Birds Damar., Proc. Zool. Soc. 1864, p. 3.
„ „ Andersson, ibid. 1864, p. 6.

Campephaga niger, Layard's Cat. No. 296.
Campephaga nigra, Gray's Hand-list of Birds, No. 5058.
Lanicterus niger, Sharpe's Cat. No. 497.

I first observed this species, though but very sparingly, in the neighbourhood of the Okavango River; but I never saw it in Damara Land proper or in Great Namaqua Land. It is a migratory bird, and I found it exceedingly shy and difficult to approach. I generally observed it moving about in the upper parts of large trees.

[I have followed Mr. Layard and Mr. Gray, *loc. cit.*, in referring Levaillant's *Echenilleur jaune* to the female of this species; but it is possible that it ought rather to be referred to the female of *C. melanoxantha*, which closely resembles the female of *C. nigra*; this, however, is a point which must probably remain an open one. Mr. R. B. Sharpe possesses a specimen of this Caterpillar-eater, which was obtained by Mr. Andersson at Elephant's Vley.—ED.]

170. **Ceblepyris pectoralis,** Jard. & S.   Pectoral Caterpillar-eater.

Ceblepyris pectoralis, Jardine & Selby, Ill. of Orn. pl. 57 (male).
Campephaga Anderssoni, Sharpe, in Proc. of Zool. Soc. 1869, pl. 4 (female).
Ceblepyris pectoralis, Sharpe, in Ibis, 1870, p. 432.
Campephaga pectoralis, Sharpe's Cat. No. 503.

[Mr. Andersson's last collection contained one specimen of this species, a female, from Ovaquenyama, obtained June 25th, 1867, which was described and figured by Mr. Sharpe, *loc. cit.*, under the name of *C. Anderssoni*, but was subsequently identified as a female example of *C. pectoralis.*—ED.]

LANIIDÆ.

171. **Lanius minor,** Gmel.   European Lesser Grey Shrike.

Collurio minor, Gould's Birds of Europe, pl. 68.
Lanius minor, Strickland & Sclater, Birds Damar., Contr. Orn. 1852, p. 144.

Lanius excubitor, Chapman's Travels in S. Afr., App. p. 393.
Lanius minor, Finsch & Hartlaub's Vögel Ost-Afrika's, p. 330.
„   „   Dresser & Sharpe, in Proc. Zool. Soc. 1870, p. 599.
„   „   Sharpe's Cat. No. 483.

This species is very common in Damara Land during the rainy season; but on the return of the dry weather it mostly disappears, though I believe a few individuals remain throughout the year. These Shrikes usually perch on some conspicuous tree or other elevated object, whence they can obtain a good view of what passes around them; they feed chiefly on insects, which they catch both on the wing and on the ground. A great number of these birds are often found in a very limited space and not unfrequently on the same tree.

The iris is brown, the legs and toes bluish black.

Measurements of a male:—

	in.	lin.
Entire length	8	4
Length of folded wing	4	6
„ tarsus	1	0
„ middle toe	0	7
„ tail	3	8
„ bill	0	10

[Mr. Andersson procured specimens of this Shrike, which are now in the collection of Mr. R. B. Sharpe, in Ondonga, at Elephant's Vley, and at Objimbinque.—ED.]

172. Enneoctonus collurio (Linn.). Red-backed Shrike.

L'Ecorcheur, Levaillant's Ois. d'Afr. pl. 64.
Enneoctonus Anderssoni, Strickland & Sclater, Birds Damar., Contr. Orn. 1852, p. 145.
Enneoctonus collurio, Gurney, Birds Damar., Proc. Zool. Soc. 1864, p. 3.
Enneoctornis collurio, Layard's Cat. No. 310.
Lanius collurio, Chapman's Travels in S. Afr., App. p. 393.
„   „   Sharpe's Cat. No. 475.

This Shrike is pretty common in Great Namaqua and Damara Land, as also in the Okavango region, where it breeds. It is migratory, and returns to Damara Land at the approach of the rainy season. It watches from some elevated position for its prey, which usually consists of insects.

The iris is very dark brown.

Measurements of a male:—

		in.	lin.
Entire length		7	5
Length of folded wing		3	9
,, tarsus		1	0
,, middle toe		0	7
,, tail		3	2
,, bill		0	10

[The type specimen of Messrs. Strickland and Sclater's *Enneoctonus Anderssoni*, which I have examined at the Museum of Zoology at Cambridge, appears to me to be an immature male of the present species, though in an unusual and, probably, exceptional state of plumage.—ED.]

173. Fiscus collaris (Linn.). Fiskal Shrike.

Le Fiscal, Levaillant's Ois. d'Afr. pls. 61 & 62.
Lanius collaris, Strickland & Sclater, Birds Damar., Contr. Orn. 1852, p. 144.
,, ,, Layard's Cat. No. 306.
Fiscus collaris, Gray's Hand-list of Birds, No. 5942.

This Shrike is common in the southern and middle parts of Great Namaqua Land, but further north it is replaced by *Lanius subcoronatus*; indeed, where the one species ceases, the other may be said to begin, as, to the best of my belief, *L. collaris* does not exist in any numbers where *L. subcoronatus* is found. South of Namaqua Land the Fiskal Shrike is very abundant, and nowhere more so than in the neighbourhood of Cape Town, where a pair may be seen in almost every garden. It is a bold, fearless, and quarrelsome bird, never allowing any bird of prey to remain long within its particular district. This species has the habit of impaling its prey, immediately after capture, on a thorn or a naturally pointed stick or branch—a custom which is said to have earned for it, from the old colonists, the name of Fiskal, derived from the title of the Dutch colonial magistrate of former times.

The food of this Shrike consists chiefly of various insects; but it will also attack young birds, rats, and mice. It perches on some branch of a tree commanding a good view, whence it precipitates itself upon its prey when perceived, whether on the ground or on the wing. It flies low, alternately rising and dipping, but invariably in a straight line, and does not move far at a time, generally only to the first convenient tree. During the breeding-season battles take place between the males; and it is at this season also that the bird is heard to sing in a strain at once varied and continuous, accompanying his warblings with many different gesticulations.

This species builds in the forks of trees, and constructs its nest of bark, moss, and flexible roots, lined inside with feathers, wool, &c. The eggs, which both parents assist in incubating, are four or five in number, broad at one end and very short, of a dusky green, with brown spots clustered thickly round the larger end.

The iris in this Shrike is black, as also is the bill; the tarsus is greyish black. There is a slight difference in the colouring of the plumage between the specimens of this species which I have obtained to the north of the Elephant River and those which I have shot near Cape Town: in the latter all the dark portions of the plumage of the upper parts, wing, and tail are a very dark black-brown, in fact almost black; whilst in the former these parts are of a very dark ashy brown, as in *L. subcoronatus*

174. Fiscus subcoronatus (Smith). Coronetted Shrike.

Lanius subcoronatus, Smith's Zool. of S. Africa, pl. 68.
"   "   Strickland & Sclater, Birds Damar., Contr. Orn. 1852, p. 144.
"   "   Andersson, in Proc. Zool. Soc. 1864, p. 8.
"   "   Layard's Cat. No. 307.
"   "   Chapman's Travels in S. Afr., App. p. 393.
Fiscus subcoronatus, Gray's Hand-list of Birds, No. 5943.
Lanius subcoronatus, Sharpe's Cat. No. 481.

This Shrike is common in the northern parts of Great Namaqua Land, and also in Damara Land and the parts adjacent to the east and north. Its mode of flight, manners, and habits are identical with those of the preceding species. The irides are dark brown; the bill brown horn-colour, but livid at the base of the lower mandible; the legs and toes are blackish brown.

175. Urolestes melanoleucus (Jard.). South-African Long-tailed Shrike.

Lanius melanoleucus, Jardine & Selby's Ill. Orn. vol. iii. pl. 115.
Lanius cissoides, Layard's Cat. No. 309.
Lanius melanoleucus, Chapman's Travels in S. Afr., App. p. 394.
Urolestes melanoleucus, Finsch & Hartlaub's Vögel Ost-Afrika's, p. 335.
Urolestes cissoides, Sharpe's Cat. No. 486.

I did not find this species in Great Namaqua Land, but first met with it in the central parts of Damara Land, from whence to the river Okavango and to Lake Ngami it is not uncommon. It is found in flocks of a few individuals, which at times create a tremendous hubbub amongst themselves, especially on the approach of a bird of prey; and the moment this alarm is sounded, every little bird in the neighbourhood precipitately betakes itself to a place of safety.

This Shrike usually seeks its food, which consists of insects, amongst the branches of trees and bushes or on the ground; but it will also watch for its prey from some elevated perch.

The irides are of a chocolate-colour, and the bill, legs, and toes of a dark horn-colour.

Measurements of a female:—

		in.	lin.
Entire length		17	4
Length of folded wing		5	3
,,	tarsus	1	3
,,	middle toe	0	9
,,	tail	12	3
,,	bill	1	1

176. Nilaus brubru (Lath.). Brubru Shrike.

Le Brubru, Levaillant's Ois. d'Afr. pl. 71.
Nilaus brubru, Strickland & Sclater, Birds Damar., Contr. Orn. 1852, p. 145.
Nilaus capensis, Layard's Cat. No. 312.
Lanius brubru, Chapman's Travels in S. Afr., App. p. 394.
Nilaus brubru, Sharpe's Cat. No. 456.

This species is widely distributed, extending from the Okavango River on the north to Lake Ngami on the east and the Orange River on the south. It is nowhere numerous, but is scarcer near the last-named river than in the other localities above referred to. Its usual resorts are forests or large trees, amongst which it hops about incessantly in search of insects. I do not think that I have ever seen more than a pair together.

The iris is a fine rich brown, with a very slight tint of red; the bill varies somewhat with age; but in a male in perfect plumage the upper mandible was very dark horn-colour, the under livid blue; the legs are greenish blue, the toes somewhat darker.

Measurements of a female:—

		in.	lin.
Entire length		5	9
Length of folded wing		3	4
,,	tarsus	0	10½
,,	middle toe	0	7
,,	tail	2	4
,,	bill	0	9

177. Eurocephalus anguitimens, Smith. Southern White-headed Shrike.

Eurocephalus anguitimens, Cat. of South-African Mus. (1837) p. 27.
,, ,, Strickland & Sclater, Birds Damar., Contr. Orn. 1852, p. 145.
,, ,, Layard's Cat. No. 318.
,, ,, Sharpe's Cat. No. 485.

This bird is pretty common in Damara Land and in the parts adjacent, including the Lake-regions. In the neighbourhood of the Okavango I observed single individuals perched on the highest and most conspicuous boughs of trees, from whence they would dart on passing winged insects; but in the other localities where I met with this species I found it associated in considerable flocks, which, when disturbed, moved leisurely to the nearest tree, rarely all at once, but one or two individuals at a time, chattering incessantly whilst moving. These flocks frequent indiscriminately bush and forest, hopping about amongst the branches and foliage in search of insects. The flight of this species is straight, with a short, rapid, and quivering motion of the wings.

The iris is brown, the bill black horn-colour, the legs and toes livid brown.

[In the very rare Catalogue of the South-African Museum, *loc. cit.*, the following remarks occur with reference to this species:—" Owing to this bird being remarkably shy, specimens were at first procured with difficulty by the Expedition, though small flights of them were frequently observed; subsequently, however, it was remarked that where one chanced to fall wounded, its companions continued hovering about and approaching it until it was removed. This suggested the plan of fixing one with a string by way of decoy; and a concealed hunter was thus sometimes enabled to kill ten or twelve in succession without having occasion to move from his hiding-place."—ED.]

178. Prionops talacoma, Smith. Smith's Helmet-Shrike.

Prionops talacoma, Smith's Zool. of S. Africa, pl. 5.
,, ,, Gurney, Proc. Zool. Soc. 1864, p. 3.
,, ,, Andersson, Proc. Zool. Soc. 1864, p. 5.
,, ,, Layard's Cat. No. 313.
,, ,, Chapman's Travels in S. Afr., App. p. 395.
,, ,, Sharpe's Cat. No. 470.

I did not observe this interesting bird until I had passed Omanbondé, in the twentieth degree of south latitude. It is always seen in flocks of from half a dozen to a dozen individuals, which frequent secluded spots, where they restlessly hop from branch to branch on the bushes and the lower boughs of the trees, never remaining long on the same tree, but hunting most systematically for insects, which, with the occasional addition of young shoots and leaves, form their food. Whilst some individuals of the flock are examining a tree in search of insects, others keep moving slowly on, but rarely going further than the next tree. When the locality is open, those which first reach a tree fix their gaze intently on the ground, and, if any prey be in sight, pounce upon it with great celerity, their companions, whilst the successful foragers are devouring their booty, continuing to move on slowly as before.

The irides in this species are of a bright lemon-colour, the fringe round the eyes orange, the bill black, and the legs flesh-coloured.

Measurements of a male:—

	in.	lin.
Entire length	8	0
Length of folded wing	4	3
,, tarsus	0	11
,, middle toe	0	6
,, tail	3	7
,, bill	1	0

[Mr. Chapman, in his 'Travels in South Africa,' vol. ii. pp. 211, 370, calls this species the "Quagga dzerra," and speaks of its "riding on the withers" of the Quagga; he also states that it "has a soft, sweet, and pleasant note."—ED.]

179. Prionops Retzii, Wahl. Retz's Helmet-Shrike.

Prionops Retzii, Wahlberg, Öfvers. 1856, p. 174.
,, ,, Wahlberg, in Journal für Orn. 1857, p. 1.
,, ,, Gurney, Birds Damar., Proc. Zool. Soc. 1864, p. 3,

Prionops Retzii, Andersson, Birds Damar., Proc. Zool. Soc. 1864, p. 5.

„ „ Finsch & Hartlaub's Vögel Ost-Afrika's, p. 366.
„ „ Sharpe's Cat. No. 471.

When encamped in the desert, a few days' journey south of the Okavango, I for the first and only time observed this fine Shrike. The flock consisted of six individuals, an adult male and female and four young birds of both sexes, all of which I secured after much running and dodging, as they were exceedingly wary and watchful, always perching on the loftiest and most exposed trees, in which respect they differed from the preceding species, though they resembled it in the manner in which they were feeding when I first saw them.

The iris in the present species is light orange; in the adults the ring round the eyes is orange-red, and the bill is of the same colour, paling off towards the extremities; the legs pale orange; the toes yellowish. In the immature birds the colours of these parts are similar, but paler.

The following description of the plumage is taken from an adult female:—The entire head and neck, throat, breast, and belly are glossed with dark green; the auriculars are rather brown; the back, rump, and upper wing-coverts slaty brown, shading into brown-green on the upper tail-coverts; the tail dark brownish bottle-green, but with the rectrices (except the two centre ones) broadly tipped with white, which extends for about an inch on the outer vanes of the two lateral rectrices.

The under tail-coverts are white, and so long as nearly to cover the dark portion of the tail; the wings are brown, but some parts of the outer vanes are glossed with green; on the inner vane of the second primary begins a broad white band, extending obliquely across the rest of the primaries; some of the secondaries and tertials are faintly tipped with light grey; the under surface of the wings has the white bar showing through; the under wing-coverts are dusky greyish brown; the vent is flossy white, tinted or faintly pencilled with grey. In a young male most of the plumage was greyish brown, but the tail and upper and lower tail-coverts were nearly as in the adult.

[As this species has not been figured, and as the original description is not readily accessible, I have transcribed in full the descriptive particulars contained in Mr. Andersson's notes.

Mr. R. B. Sharpe possesses two specimens, which were obtained by Mr. Andersson at Elephant's Vley, and are probably two of those above referred to.—ED.]

180. Laniarius atrococcineus (Burch.). Southern Crimson-breasted Shrike.

Lanius atrococcineus, Burchell's Travels, vol. i. p. 387 (note).
Malaconotus atrococcineus, Swainson's Zool. Ill. n. s. pl. 76.
Laniarius atrococcineus, Strickland & Sclater, Birds Damar., Contr. Orn. 1852, p. 145.
 „   „   Layard's Cat. No. 321.
 „   „   Chapman's Travels in S. Afr., App. p. 394.
 „   „   Sharpe's Cat. No. 464.

This brilliantly coloured Shrike is pretty generally dispersed over Damara and Great Namaqua Land, and also over the Lake-regions. It is usually found either singly or in pairs, and is a wary bird, frequenting both open woods and dense thickets, but preferring the latter. In such localities it searches industriously for insects and their larvæ, and, whilst thus occupied, occasionally utters pleasant, clear, ringing notes, but with the prelude of a harsh guttural sound.

This species builds its nest in the fork of a tree, constructing it roughly of the inner bark of trees, with a few grasses interlaced, but without any softer lining.

I have found the nest in the more northern portion of Damara Land as early as November; but one which I met with at Omapju contained two fresh eggs on the 8th of January. The eggs are white, spotted with light brown, and sometimes tinged with green.

The iris is very dark, the bill black, the legs and toes brown.

Measurements of a male and a female :—

	Male. in. lin.	Female. in. lin.
Entire length	9 1	8 9
Length of folded wing	3 11	3 10
,, tarsus	1 4	1 3
,, middle toe	0 7	0 6
,, tail	4 1	4 2
,, bill	1 0	1 0

181. Laniarius major, Hartl. Greater White-breasted Shrike.

Laniarius major, Hartlaub's Beitr. z. Orn. West-Afr. pl. 5.
Dryoscopus major, ejusd., Orn. West-Afr. No. 338.
Laniarius major, Finsch & Hartlaub's Vögel Ost-Afrika's, p. 344.
Tchagra major, Gray's Hand-list of Birds, No. 6041.
Dryoscopus major, Sharpe's Cat. No. 450.

I have received several specimens of this Shrike from the Lake-country. Its habits have not come under my personal observation; but I am told that it frequents secluded, dense, tangled brakes, and, though not particularly shy, is somewhat difficult to obtain, from the nature of its resorts—also that it utters pleasant ringing notes.

The bill and claws are black, the legs and feet brownish.

[Mr. Andersson's last collection contained two examples of this Shrike, one from Lake Ngami, the other from the river Cunéné: in the first of these the rectrices were wholly black, as is usual in this species; in the second (a male, obtained on the 25th of June, 1867) they are all black except the exterior pair, which have a very narrow white tip to the outer web, a variation which I have not met with in any other individuals of this species. Both of these specimens are now in the collection of Mr. R. B. Sharpe.—ED.]

182. Laniarius sticturus, Finsch & Hartl. Chapman's Shrike.

Laniarius sticturus, Finsch & Hartlaub's Vögel Ost-Afrika's, p. 342, pl. 5. fig. 1.
Tchagra sticturus, Gray's Hand-list of Birds, No. 6042.

[As the type specimen of this bird is stated by Drs. Finsch and Hartlaub, *loc. cit.*, to have been brought from Lake Ngami by Mr. Chapman, I here include it; but I have not found any specimen of it in Mr. Andersson's collections, nor any reference to it in his notes.—ED.]

183. Dryoscopus cubla (Shaw). Cubla Shrike.

Le Cubla, Levaillant's Ois. d'Afr. pl. 72.
Dryoscopus cubla, Strickland & Sclater, Birds Damar., Contr. Orn. 1852, p. 145.
Laniarius cubla, Layard's Cat. No. 320.
Dryoscopus cubla, Gray's Hand-list of Birds, No. 6018.
  „   „   Sharpe's Cat. No. 454.

I have observed this Shrike very sparingly in Damara Land, where I met with it first at Okamabuté, and thence northward as far as the Okavango; but I nowhere found it common. The few that I saw were either single or in pairs, perching on the larger trees and hunting on the branches with great assiduity for larvæ and insects, especially beetles and ants.

The notes of this species are harsh and loud. It is sometimes in the habit of elevating the lax feathers on the back, which, when fully expanded, almost encircle the bird; this appears to occur under the influence of amorous emotions, and also when the bird is aroused by fear or curiosity.

Measurements of a male and a female :—

	Male. in. lin.	Female. in. lin.
Entire length	7 0	6 9
Length of folded wing	3 4	3 2
,, tarsus	0 11	0 11
,, middle toe	0 7	0 7
,, tail	2 10	2 9
,, bill	0 10	0 10

184. Telophorus gutturalis (Müll.). Bacbakiri Shrike.

Le Bacbakiri, Levaillant's Ois. d'Afr. pl 67.
Telophonus bacbakiri, Layard's Cat. No. 317.
Telophorus gutturalis, Gray's Hand-list of Birds, No. 6043.
Lanius backbakiri, Chapman's Travels in S. Afr., App. p. 394.
Laniarius gutturalis, Sharpe's Cat. No. 465.

This species is sparingly met with in Damara Land, and only in a few localities; but as the traveller proceeds southward it becomes more numerous, till on reaching Cape Town it is found in every garden. In Damara Land I have usually found it very shy and retired in its habits, quickly escaping, the moment it finds itself observed, into the thickest part of the nearest bush, and thence moving hurriedly on as near to the ground as the locality will admit, so that it is often difficult to flush it a second time. At the Cape, on the contrary, it is very familiar, and is frequently seen perched on the garden walls, whilst it utters a succession of ringing calls which the Dutch liken to the word "Bacbakiri;" but its notes and calls are in fact very varied.

This Shrike feeds chiefly on insects, but is said occasionally to kill and devour young birds; it hunts for its prey amongst low bushes and on the ground, and runs with great swiftness. It is usually found in pairs, but for some time after the breeding-season the parents are accompanied by their families, which they tend with much care.

This species builds in thick bushes; and the eggs, which are four or five in number, of a greenish-blue colour spotted with reddish brown, especially round the thicker end, are incubated by both parents.

The iris is reddish brown, the bill black, the legs and toes dark lead-colour.

185. Chlorophoneus similis (Smith). Yellow-browed Shrike.

Melaconotus* similis, Smith's Zool. of S. Africa, pl. 46.
Laniarius similis, Layard's Cat. No. 322.
Melaconotus similis, Chapman's Travels in S. Afr., App. p. 395.
Laniarius sulfureipectus, Finsch & Hartlaub's Vögel Ost-Afrika's, p. 356.
Chlorophoneus similis, Gray's Hand-list of Birds, No. 6047.
Laniarius sulfureipectus, Sharpe's Cat. No. 402.

I have obtained several specimens of this species from

---

* The correct spelling of this generic name is *Malaconotus*.—ED.

Lake Ngami, but have never observed it to the west of that locality.

[Mr. R. B. Sharpe possesses a specimen of this Shrike from Lake Ngami, and another obtained by Mr. Andersson at the river Cunéné, no doubt subsequently to the date of his note above transcribed.—ED.]

186. Pomatorhynchus erythropterus (Shaw). Rufous-winged Bush-Shrike.

Pie-grièche rousse à tête noire du Sénégal, Buffon's Planches Enl. vol. i. pl. 479. fig. 1, p. 242.
Lanius erythropterus, Shaw's Gen. Zool., Aves, vol. vii. p. 301.
Lanius cucullatus, Temminck's Manuel d'Orn. vol. iv. p. 600.
Telophonus senegalus, Strickland & Sclater, Birds Damar., Contr. Orn. 1852, p. 145.
Lanius tchagra, Bree's Birds of Europe, p. 171 (plate).
Telephonus senegalensis, Andersson, in Proc. of Zool. Soc. 1864, p. 5.
Telophonus erythropterus, Layard's Cat. No. 315.
Telephonus erythropterus, Finsch & Hartlaub's Vögel Ost-Afrika's, p. 336.
Pomatorhynchus erythropterus, Gray's Hand-list of Birds, No. 6052.
Pomatorhynchus cucullatus, id. ibid. No. 6053.
Lanius erythropterus, Blanford's Geol. & Zool. of Abyssinia, p. 343.
Telephonus erythropterus, Finsch, Birds of N.E. Abyssinia, in Trans. of Zool. Soc. vol. vii. p. 254.
„   „   (part.), Sharpe's Cat. No. 489.
„   „   Gurney, Jun., Birds of Algeria, in Ibis, 1871, p. 76.

I only found this species in the northern parts of Damara Land and after I had passed the northern boundary of that country; it was very shy and never plentiful. It hops about amongst the lower branches of thickly foliaged trees, and is especially partial to low tangled brakes, where it searches incessantly for insects, larvæ, and caterpillars.

[Mr. Layard, in the 'Ibis' for 1870, p. 460, has shown that this species is quite distinct from the longer-billed *P. tschagra*

(Vieill.), figured by Levaillant under the name of *"Le tchagra,"* in the 'Oiseaux d'Afrique,' pl. 70, of which *P. longirostris* (Swains.) is a synonym.

The question as to whether the present species and *P. trivirgatus* (Smith) are really specifically distinct is a much more doubtful one, and some remarks on that subjeet will be found in a subsequent page under the head of *P. trivirgatus.*

Assuming all the black-headed specimens from different localities, which do not differ from each other except in size, to belong to the present species, it must be admitted that the differences in size which they exhibit is remarkable, as will be perceived from the following measurements of several skins, most of which are in the collection of Mr. R. B. Sharpe, and all specimens possessing the entire black hood.

	Length of wing from carpal joint. in.	Length of tarsus. in.	Length of tail. in.
Specimen A, from Algeria	$3\frac{1}{2}$	$1\frac{1}{8}$	$4\frac{1}{2}$
" B, from Abyssinia	$3\frac{5}{8}$	$1\frac{1}{4}$	$3\frac{7}{8}$
" C, from Bijook, N.E. Abyssinia, ♂	$3\frac{1}{4}$	$1\frac{1}{4}$	$3\frac{5}{8}$
" D, from Bijook, N.E. Abyssinia, ♂	$3\frac{1}{8}$	$1\frac{1}{4}$	$4\frac{1}{8}$
" E, from River Gambia	$3\frac{1}{2}$	$1\frac{1}{4}$	$4\frac{1}{8}$
" F, from River Gambia	$3\frac{1}{2}$	$1\frac{1}{8}$	$4\frac{1}{8}$
" G, from River Volta (West Africa)	$3\frac{1}{2}$	$1\frac{1}{4}$	$3\frac{7}{8}$
" H, from West Africa	$3\frac{3}{4}$	$1\frac{1}{8}$	4
" I, from West Africa	$3\frac{1}{4}$	$1\frac{1}{8}$	4
" J, from River Cunéné	$3\frac{3}{4}$	$1\frac{1}{8}$	$4\frac{5}{8}$
" K, from Elephant's Vley, Northern Damara Land	4	$1\frac{1}{8}$	$4\frac{1}{2}$
" L, from Elephant's Vley, Northern Damara Land, ♂	$3\frac{3}{4}$	$1\frac{1}{8}$	imperfect.
" M, from Damara Land	$3\frac{7}{8}$	$1\frac{1}{8}$	$4\frac{1}{4}$
" N, from Natal, ♂	$3\frac{3}{4}$	$1\frac{1}{4}$	$4\frac{1}{4}$
" O, from Natal, ♀	$3\frac{1}{2}$	$1\frac{1}{4}$	$4\frac{3}{8}$
" P, from Eland's Post, South Africa	$3\frac{3}{4}$	$1\frac{1}{4}$	$4\frac{1}{2}$
" Q, from Eland's Post, South Africa, ♀	$3\frac{3}{4}$	$1\frac{1}{4}$	$4\frac{1}{2}$
" R, from South Africa	4	$1\frac{1}{4}$	$4\frac{1}{4}$

—Ed.]

187. Pomatorhynchus trivirgatus (Smith). Three-streaked Bush-Shrike.

Telophonus trivirgatus, Smith's Zool. of S. Africa, pl. 94.
Telephonus trivirgatus, Gurney, Birds Damar., Proc. Zool. Soc. 1864, p. 3.
„   „   Andersson, ibid. p. 5.
Telophonus trivirgatus, Layard's Cat. No. 314.
Lanius trivirgatus, Chapman's Travels in S. Afr., App. p. 394.
Telephonus trivirgatus, Finsch & Hartlaub's Vögel Ost-Afrika's, p. 338.
Pomatorhynchus trivirgatus, Gray's Hand-list of Birds, No. 6053.
Telephonus erythropterus (part.), Sharpe's Cat. No. 489.

This species is by no means uncommon in Southern Damara Land, and it also occurs in the parts adjacent; it confines itself to dense thickets or low bushes, where it hops about incessantly, flitting rapidly from twig to twig, and passing out of sight directly you have caught a glimpse of it. It is generally found either singly or in pairs.

Two nests of these birds, which I found in the month of January, were composed of stalks of plants loosely put together and lined with stalks of a softer kind. One nest was placed very low in a low bush, the other about 5 feet from the ground in a thorn bush; each nest contained three eggs.

The iris in this species is purplish, with a narrow greyish ring round the pupil; the bill is black; the legs and toes brownish lead-colour.

[In the present as in the preceding species some differences of size are observable between different individuals; but the average measurements are decidedly less than those of the preceding species, as will be seen by the following Table, for the materials of which I am again chiefly indebted to the kindness of Mr. R. B. Sharpe.

	Length of wing from carpal joint. in.	Length of tarsus. in.	Length of tail. in.
Specimen A, from River Volta, West Africa	$2\frac{7}{8}$	1	$3\frac{1}{4}$
" B, from Fantee, West Africa	$2\frac{7}{8}$	$1\frac{1}{8}$	$3\frac{5}{8}$
" C, from Elephant's Vley, Northern Damara Land, ♀	$2\frac{7}{8}$	$1\frac{1}{8}$	$3\frac{1}{2}$
" D, from Elephant's Vley, Northern Damara Land, ♀	3	$1\frac{1}{8}$	$3\frac{1}{2}$
" E, from Objimbinque, Southern Damara Land	$3\frac{1}{5}$	$1\frac{1}{8}$	$3\frac{3}{4}$
" F, from Objimbinque, Southern Damara Land, ♂	3	$1\frac{1}{8}$	$3\frac{7}{8}$
" G, from Onani's Mouth, Southern Damara Land, ♂	3	1	$3\frac{5}{8}$
" H, from Onani's Mouth, Southern Damara Land, ♀	$2\frac{7}{8}$	1	$3\frac{1}{2}$
" I, from Anna's Wood, Southern Damara Land	$3\frac{1}{8}$	1	4
" J, Sir A. Smith's type, a ♀ from Kurichane, as described by him	3	1	4

All the above specimens are brown-headed birds, except the first, from the river Volta, which appears to have been killed whilst the colour of the hood was changing from brown to black. This circumstance induces Mr. Sharpe to think that *P. trivirgatus* is in reality merely the immature stage of the smaller form of *P. erythropterus*: whether such is or is not the fact must be decided by future observations, which it is to be hoped will be made by those who may hereafter have the opportunity.—Ed.]

CONIROSTRES.

CORVIDÆ.

188. Corvultur albicollis (Lath.). Southern Corbivau.

Le Corbivau, Levaillant's Ois. d'Afr. pl. 50.
Corvus albicollis, Layard's Cat. No. 330.
Archicorax albicollis, Finsch & Hartlaub's Vögel Ost-Afrika's, p. 371.

This Raven-like-looking bird is not, to my knowledge, an inhabitant of Damara Land, but is abundant in the southern portion of Great Namaqua Land, where it remains throughout the year. It is generally found singly or in pairs, but sometimes congregates in flocks. It perches indiscriminately on rocks and trees, and may occasionally also be seen perched on the backs of cattle and of many species of wild animals, whose hides it examines in search of parasitic insects; it, however, passes most of its time upon the ground, where it stalks about with ludicrous gravity, uttering at intervals cries not unlike those of a true Raven. Carrion probably constitutes its chief nourishment; but it is said also to attack with ferocity and success the young of the smaller quadrupeds. The flight of this species is powerful; and it occasionally rises to a great height.

The iris is hazel-brown, the bill very dark brown, but with the tips of the mandibles of a white horn-colour; the legs and toes are brownish black.

[Mr. Andersson's last collection did not contain a specimen of this species; but there can be no doubt of the correctness of his identification of it.—ED.]

189. Corvus scapulatus, Daud. Scapulary Crow.

La Corneille à scapulaire blanc, Levaillant's Ois. d'Afr. pl. 53.
Corvus scapulatus, Layard's Cat. No. 331.
„   „   Chapman's Travels in S. Afr., App. p. 404.
„   „   Sharpe's Cat. No. 541.

This species is found in Damara and Great Namaqua Land, and also in the Lake-regions, but is rather a local bird. It is a regular scavenger, being always present where offal is to be had, and at times resorting in numbers to the bays and inlets of the coast in search of carrion thrown up by the waves; indeed its scent and sight for carrion rival those of the Vultures, and it is not unfrequently seen about a carcass before even a single Vulture has appeared. It is quite fearless, and will approach a person within a few steps, but, if once shot at, displays considerable cunning and caution. From its commonness and sociability its presence is often disregarded; and it avails itself of such opportunities to carry off any pieces of meat and fat which may chance to be within its reach; but such little depredations are amply compensated by its usefulness as a scavenger, and also in ridding domestic as well as wild animals from the fearfully bloodthirsty ticks which infest their hides.

Some of the notes of this Crow, more especially on a raw misty morning, are absurdly singular and ridiculous.

In the heat of the day nearly all the birds of this kind found in a limited locality will join in circling round and round for hours together, sometimes ascending to a very great height.

I observed the nests of this species in Ondonga, built on palms and other trees, and looking very like Kites' nests. The eggs are of a drabbish blue spotted with pale brown, chiefly about the thicker end; but I have seen some nearly white, and with the markings more like streaks and lines than spots; they are generally of a very elongated form.

The irides in this Crow are hazel-brown, the bill, legs, and toes are black.

190. Corvus capensis, Licht. South-African Rook.

Le Corneille du Cap, Levaillant's Ois. d'Afr. pl. 52.
Corvus capensis, Lichtenstein's Doubletten, Vögel, No. 190.
Corvus segetum, Layard's Cat. No. 332.
Corvus capensis, Chapman's Travels in S. Afr., App. p. 404.
„   „   Sharpe's Cat. No. 542.

I found this Crow very common in Ondonga, where it nests. In Damara Land it is very local and nowhere common; but it is more frequent in Great Namaqua Land, especially towards the Orange River; and in the Cape Colony it abounds.

In Damara Land, where, till lately, the natives never cultivated the ground, it is a shy bird, and I never saw it there near the native villages; but in the Cape Colony it is pretty sociable, frequenting the cultivated land, where it resorts to the vicinity of domestic buildings, and sometimes follows in the wake of the plough in search of insects and larvæ, which constitute its chief food, though it is said that it will also feast upon carrion, and will sometimes plunder the crops of maize. It is usually found in small flocks, and is a clamorous bird, uttering harsh choking sounds, just as though its crop were too full.

The nests of this species, which I observed in Ondonga, were built on palm trees; the eggs have a buffy ground-colour, and are thickly sprinkled with spots, some of which are reddish and others brown, giving the entire egg more or less of a pinkish hue.

The irides are dark brown, the bill, legs, and toes dark horn-colour.

STURNIDÆ.

191. Cinnyricinclus Verreauxi (Bocage). Verreaux's Glossy Starling.

Pholidauges leucogaster, Gurney, Birds Damar., Proc. Zool. Soc. 1864, p. 3.
,, ,, Andersson, ibid. p. 6.
Juida leucogaster (part.), Layard's Cat. No. 346.
Lamprotornis leucogaster, Chapman's Travels in S. Afr., App. p. 404.
Pholidauges leucogaster (part.), Finsch & Hartlaub's Vögel Ost-Afrika's, p. 376.
Pholidauges Verreauxi, id. ibid. p. 867.
Cinnyricinclus Bocagei, Gray's Hand-list of Birds, No. 6349.
Pholidauges Verreauxii, Sharpe's Cat. No. 515.

This species is common in Damara Land and to the northward, but only as a migratory bird, arriving at the approach of the rainy season, and gradually leaving as the country dries up, though I have observed a few individuals remaining long after the general emigration was over; and these may probably stay throughout the year. The exquisitely coloured males arrive first, and, so far as I have observed, associate but little with the sombre females, from which they differ so marvellously in appearance.

The following is the description of an adult male and a female :—

Male.—Head, neck, back, and scapulars, upper tail-coverts, and the exposed parts of the wings in general, also the chin, throat, and breast of the most beautiful deep blue and violet, with dazzling purple, puce, and bronze shades when exposed to a full light; primaries dark brown; secondaries and tertials darker brown, broadly margined and glossed with the same colours as the upper part of the plumage; on the tertials there is also a greenish reflection, which is even more conspicuous in the two middle tail-feathers; the rest of the tail-feathers are brown, with violet and purple reflections, strongest on the outer vanes, except the lateral rectrices, in which the lower portion of the outer web is white; the belly, flanks, vent, and under tail-coverts are white, the thighs white mottled with fawn.

Female.—The whole of the upper part of the plumage brown, but with a spotted appearance on the head and sides of the neck, caused by the brown feathers being edged on two sides with reddish fawn, which also surrounds nearly the whole of the feathers on the back, but is there much paler. The inner vanes of the primaries are cinnamon-coloured, which becomes paler on those of the secondaries and quite obscure on those of the tertials; the secondary and tertial wing-coverts are broadly margined and edged with fawn. The whole of the under-parts are white, faintly dashed with fawn on the chin and throat; streaked with brown on the breast and legs, closely so on the belly and flanks, the under tail-coverts being marked with a few isolated and very small spots.

The irides in this species are light chrome-yellow, the bill black, the legs brownish, and the toes the same but darker.

Measurements of a male and a female:—

	Male. in. lin.	Female. in. lin.
Entire length	7 0	6 11
Length of folded wing	4 0	3 10
,, tarsus	0 11	0 10
,, middle toe	0 7	0 7
,, tail	2 4	2 4
,, bill	0 11	0 10

[Mr. R. B. Sharpe has called my attention to the circumstance that the *Pholidauges* inhabiting Damara Land is not (as, in common with some other ornithologists, including the late Mr. Andersson, I had previously supposed) *Pholidauges leucogaster*, but the nearly related *P. Verreauxii*, chiefly distinguished from *P. leucogaster* by the white ending to the external vane of the lateral rectrices in the adult male, and first noticed as a native of Benguela by the learned Portuguese ornithologist Professor Barboza du Bocage.

As I believe that no English description of this recently observed species has yet been published, I have here given that drawn up by Mr. Andersson and contained in his MS. notes. I am not aware that *Pholidauges Verreauxi* has yet been figured.

Mr. Blanford's observations on the generic affinities of the allied species *P. leucogaster*, as given in his 'Geology and Zoology of Abyssinia,' p. 367, appear to me to be well worthy of consideration.—ED.]

192. **Juida australis** (Smith). Burchell's Glossy Starling.

Lamprotornis Burchellii, Smith's Zool. of S. Africa, pl. 47.
Juida australis, Strickland & Sclater, Birds of Damar., Contr. Orn. 1852, p. 149.

Juida australis, Layard's Cat. No. 335.
Lamprotornis Burchelli, Chapman's Travels in S. Afr., App. p. 403.
Lamprotornis australis, Sharpe's Cat. No. 535.

This very handsome species abounds in the Lake-regions; in the Damara country it first becomes abundant at Schmelen's Hope, on the upper sources of the Swakop. It is found singly or in pairs, and passes most of its time on large trees, but occasionally descends to the ground in search of insects, which, with berries, constitute its chief food. It is, however, somewhat omnivorous in its habits; and I have observed its stomach to contain much sand. It is a very shy bird, but very lively, jerking its body and tail (the latter of which it can raise quite perpendicularly) and uttering all the while harsh clamorous notes.

Measurements of a male:—

	in.	lin.
Entire length	14	3
Length of folded wing	7	7
,, tarsus	1	9
,, middle toe	0	11
,, tail	7	3
,, bill	1	2

[The only Damara-Land specimen of this bird which has come under my observation is preserved in the late Mr. Strickland's collection, at the Museum of Zoology at Cambridge. Mr. Andersson's last collection contained an example from Lake Ngami.—ED.]

193. Juida Mevesii (Wahl.). Meve's Glossy Starling.

Juida Mevesii, Wahlberg, Öfvers. 1856, p. 174.
,, ,, Wahlberg, Journal für Orn. 1857, p. 1.
Lamprotornis Mevesi, Sharpe's Cat. No. 533.

[Mr. Andersson, in his MS. notes, does not allude to this

species as having come under his personal notice; but Mr. R. B. Sharpe possesses three specimens obtained by Mr. Andersson at Ovaquenyama in the months of June and July 1867.—ED.]

194. Lamprocolius phœnicopterus (Linn.). Nabirop Glossy Starling.

Le Nabirop, Levaillant's Ois. d'Afr. pl. 89.
Spreo bispecularis, Strickland & Sclater, Birds Damar., Contr. Orn. 1852, p. 149.
Lamprocolius phœnicopterus, Andersson, in Proc. Zool. Soc. 1864, p. 7.
Juida phœnicoptera, Layard's Cat. No. 339.
Juida aurata, id. ibid. No. 337.
Lamprocolius phœnicopterus, Sharpe's Cat. No. 528.

This bird is found most abundantly throughout Damara and Great Namaqua Land, in the valleys of the Okavango and of the Teoughe, and in the Lake-regions. Like our European Starling, which it very much resembles in manners and habits, it frequently congregates in large flocks; it is comparatively tame and easy to approach, and is often met with near villages. Its food is very various, consisting of berries, seeds, and insects, and it is very destructive to fruit-gardens; its flesh is not unpalatable.

This species forms its nest in the hollows of trees, lining the cavity well with feathers. The eggs are four in number, of a long oval shape, but tapering much more at one end than at the other; they are of a pale bluish green, spotted all over with small dots of light brown.

The iris is bright reddish orange, the bill, legs, and toes more or less black.

[I have examined the type specimen to which Messrs. Strickland and Sclater (*loc. cit.*) gave the name of "*Spreo bispecularis*," and which is now preserved in the Museum of Zoology at Cambridge, and I am of opinion that it is certainly referable to the present species.

Mr. Layard informs me that he believes the bird included in his Catalogue under the name of *Juida aurata* is also identical with *L. phœnicopterus*.—ED.]

195. Spreo bicolor (Gmel.). Spreo Glossy Starling.

Le Spreo, Levaillant's Ois. d'Afr. pl. 88.
Juida bicolor, Layard's Cat. No. 342.
Spreo bicolor, Gray's Hand-list of Birds, No. 6352.

I have not observed this bird to the west of Lake Ngami and to the north of the Orange River; but it was brought from the Lake-regions by Messrs. J. & H. Chapman, and to the south of the Orange River it is common everywhere.

This species is gregarious in its habits, being often seen in immense flocks and seeking its food on the ground amongst the herds of cattle. During the vintage it resorts to the gardens and vineyards, where it commits great ravages among the grapes; its flesh is much esteemed. Its notes are similar to those of the European Starling. It breeds in a variety of situations, frequently taking forcible possession of the nests of other birds, such as the Woodpecker, Bee-eater, and Swallow, or establishing itself under the roof or in a hole in the wall of some friendly farmer's homestead; in default of any of these conveniences, it contents itself by depositing its eggs in a hole in the ground. The eggs are five or six in number, and of a greenish-blue colour spotted with brown.

The iris is hazel; the bill very dark brown, but yellow posteriorly, especially on the lower mandible; the gape is also yellow; the legs and toes are nearly black.

[As I have not happened to find this species in Mr. Andersson's collections, its identification as an inhabitant of Little Namaqua Land and of the Lake-regions rests on Mr. Andersson's authority, but is no doubt perfectly correct.—ED.]

196. Amydrus caffer (Linn.). Caffre Glossy Starling.

Le Nabouroup, Levaillant's Ois. d'Afr. pl. 91.
Spreo nabouroup, Strickland & Sclater, Birds Damar., Contr. Orn. 1852, p. 149.
Juida fulvipennis, Layard's Cat. No. 344.
Lamprotornis fulvipennis, Chapman's Travel's in S. Afr., App. p. 404.
Juida (Nabouroupus) cafra, Gray's Hand-list of Birds, No. 6356.
Amydrus caffer, Sharpe's Cat. No. 520.

This is a tolerably common bird in Damara Land and in the adjacent countries to the north and east, as well as in Great and Little Namaqua Land. It is gregarious in its habits, congregating in small flocks, and is partial to rocky localities; it sometimes flies at a considerable height, and frequents the water morning and evening. It feeds on seeds, berries, insects, &c.

The irides are bright yellow, the bill, legs, and toes black.

197. Dilophus carunculatus (Gmel.). Wattled Starling.

Le Porte-lambeau, Levaillant's Ois. d'Afr. pls. 93 & 94.
Dilophus carunculatus, Strickland & Sclater, Contr. Orn. 1852, p. 149.
    „   „   Layard's Cat. No. 353.
Gracula carunculata, Chapman's Travels in S. Afr., App. p. 404.
Dilophus carunculatus, Sharpe's Cat. No. 538.

These singular birds appear in Damara and Great Namaqua Land about the beginning of the rainy season, and mostly leave again upon the return of the dry; but I suspect that a few pairs occasionally remain and breed, as young birds are to be found throughout the year.

This species is always found in flocks, often consisting of a hundred or more individuals, which greatly remind me of a flock of European Starlings, and are rather shy and difficult to approach; they feed on worms, berries, and insects, chiefly small coleoptera.

Measurements of a male and a female:—

	Male. in. lin.	Female. in. lin.
Entire length	8 0	. 7 9
Length of folded wing	4 8	4 8
,, tarsus	1 0	1 1
,, middle toe	1 0	1 0
,, tail	2 10	2 10
,, bill	1 2	1 1

198. Buphaga africana, Linn. Greater Oxpecker.

Le Pique-bœuf, Levaillant's Ois. d'Afr. pl. 97.
Buphaga africana, Strickland & Sclater, Birds Damar., Contr. Orn. 1852, p. 149.
,, ,, Layard's Cat. No. 347.
,, ,, Sharpe's Cat. No. 539.

I have only observed this species in the middle districts of the Damara country. It is generally met with in small flocks, which visit the cattle in search of the larvæ and ticks with which their hides are often abundantly supplied; and, indeed, I never saw these birds, except when they were occupied in thus searching for insects, though

Livingstone has recorded his having met with flocks of this species and of its congener, *B. erythrorhyncha*, "roosting on reeds in spots where neither tame nor wild animals were to be found"*.

The arrival of these birds is announced by a sharp cry; and the next moment they may be seen in a little flock descending fearlessly on and amongst the cattle, which are at first much alarmed, and run about in wild confusion just as they do when troubled with gadflies; but their apprehensions are soon dispelled and exchanged for sensations of evident pleasure as the Oxpeckers run over their backs, sides, and bellies, like Woodpeckers upon trees, except when an ox, by an occasional jerk or sudden twist, appears to indicate that the claws of the bird have caused something like pain by touching some spot where the skin of the animal happens to be tender.

The iris in this species is orange, but in one specimen which I obtained it was greyish green, and the bill dark horn-colour instead of the usual coloration of yellow tipped with red.

[Mr. Andersson, in his work entitled "Lake Ngami," refers at p. 214 to the habits of this bird as above detailed, and adds that it "is also a frequent companion of the rhinoceros, to which, besides being of service in ridding him of many of the insects that infest his hide, it performs the important part of sentinel. On many occasions has this watchful bird prevented me from getting a shot at that beast: the moment it suspects danger it flies almost perpendicularly up into the air, uttering sharp shrill notes that never fail to attract the attention of the rhino-

---

* Missionary Travels, p. 546.

ceros, who, without waiting to ascertain the cause, almost instantly seeks safety in a precipitate flight."

Mr. R. B. Sharpe possesses a specimen of this bird procured by Mr. Andersson at Ovaquenyama.—ED.]

FRINGILLIDÆ*.

199. Bubalornis erythrorhynchus (Smith). Buffalo Weaver bird.

Textor erythrorhynchus, Smith's Zool. of S. Africa, pl. 64.
Textor niger, Strickland & Sclater, Birds Damar., Contr. Orn. 1852, p. 150.
Textor erythrorhynchus, Layard's Cat. No 354.
    „   „   Chapman's Travels in S. Afr., App. p. 400.
Bubalornis erythrorhynchus, Gray's Hand-list of Birds, No. 6554.
Textor erythrorhynchus, Sharpe's Cat. No. 547.

This large finch-like bird is rather common in Damara Land and also in the Lake-regions, where it is known to the natives by the name of "Tsaba Gushoa." It is a noisy species, gregarious in its habits, breeding in colonies, and constructing many nests in the same tree: it seems to prefer the giraffe-acacia for the purpose of nidification; and it is curious that when these birds have used a tree for this purpose it usually withers in a short time after the building of the nest is completed; but whether the birds instinctively select such trees as have a tendency to decay, I am unable to say. The collective nests consist externally of an immense mass of dry twigs and sticks, in which are to be found from four to six separate nests or holes of an oval form, composed of grass only, but united to each other by intricate masses of sticks, defying the ingress of any intruder except a

[* I include in this family the *Ploceidæ* of some authors.—ED.]

small snake. In each of these separate holes are laid three or four eggs, exactly resembling Sparrows' eggs, but much larger. I obtained no less than forty of these eggs (all much incubated), on January 29th, from two low trees standing close together, at Amatoni, in latitude 18° south; and on the following day the birds were busy in repairing one of the collective nests, which had been injured during the collection of the eggs which it contained. I believe these nests are annually added to; for, so far as I have been able to see, the same nest is retained for several consecutive seasons.

This species is said by Dr. Smith to be a frequent attendant on the buffalo, and to feed on the parasites which infest the hide of that quadruped: I have never observed this in Damara Land, which may be owing to the buffalo being a scarce animal in that country.

The irides in this species are brown.

[Sir A. Smith's account of this bird attending on the South-African buffalo is confirmed by Dr. Livingstone, *vide* 'Missionary Travels,' p. 545.—ED.]

200. Plocepasser mahali, Smith. White-browed Weaver bird.

Plocepasser mahali, Smith's Zool. of S. Africa, pl. 65.
," ,, Strickland & Sclater, Birds Damar., Contr. Orn. 1852, p. 150.
,, ,, Layard's Cat. No. 372.
,, ,, Chapman's Travels in S. Afr., App. p. 401.
,, ,, Sharpe's Cat. No. 578.

Damara Land proper would seem to be the stronghold of this species; but I have also found it abundant at Lake Ngami and in the neighbourhood of the Okavango, and it likewise occurs, through less frequently, in Great Namaqua Land. It is gregarious in its habits and may occasionally be seen in large flocks; it usually frequents the wildest and most desolate spots, far away from either fountain or stream. It feeds chiefly on seeds and insects, which it seeks for on the ground, and, if disturbed, usually takes refuge in the nearest tree, remaining there till the supposed danger is passed, when it resumes its previous occupation. At the beginning of the rainy season this bird occasionally, though rarely, sings so melodiously that I have seldom heard any thing more exquisite. Several pairs of these birds build on the same tree, constructing large rambling nests of coarse grass near the extremities of the boughs: each nest contains two or, rarely, three eggs; and I have observed that all the *old* nests have two entrances.

The bill, legs, and toes in this species are yellowish red.

Measurements of a female:—

		in.	lin.
Entire length		7	0
Length of folded wing		4	1
,,	tarsus	1	0
,,	middle toe	0	8
,,	tail	2	9
,,	bill	0	9

201. Philetærus socius (Lath.). Social Weaver bird.

Philetærus lepidus, Smith's Zool. of S. Africa, pl. 8.
Philetærus socius, Layard's Cat. No. 371.
Euplectes lepidus, Chapman's Travels in S. Afr., App. p. 401.

Great Namaqua Land is the headquarters of this species, and the Orange River is its southern limit; in

Damara Land proper it is of somewhat rare occurrence. It congregates in large flocks; and when breeding, many pairs incubate their eggs under the same roof, which is composed by these birds of whole cartloads of grass piled on a branch of some kamel-thorn tree in one enormous mass of an irregular umbrella-shape, looking like a miniature haystack, and almost solid, but with the under surface, which is nearly flat, honeycombed all over with little cavities, which serve not only as places for incubation but also as a refuge against rain and wind. The eggs of this species are three or four in number, of a drab colour, closely speckled with minute spots of purple grey; the eggs, however, vary in the intensity of their ground-colour, and also in the degree to which they are spotted, as well as in their shape, which is in some cases very round, and in others more elongated.

The irides are dark brown, the bill pale livid horn-colour, the legs and toes light brownish horn-colour.

Measurements of a male:—

	in.	lin.
Entire length	5	2
Length of folded wing	2	9
„ tarsus	0	10
„ middle toe	0	7
„ tail	1	7
„ bill	0	8

[Mr. Andersson's collections did not contain a specimen of this well-known Grosbeak; but there can be no doubt of the correctness of his identification, which is confirmed by Mr. Layard's statement (*loc. cit.*) as to the countries which this species inhabits.—ED.]

202. Hyphantornis spilonotus (Vig.). Spotted-backed Weaver bird.

Ploceus spilonotus, Smith's Zool. of S. Africa, pl. 66. fig. 1.
Hyphantornis spilonotus, Layard's Cat. No. 360.
„ „ Finsch & Hartlaub's Vögel Ost-Afrika's, p. 393.

I am not aware that this species is an inhabitant of either Damara or Great Namaqua Land; but I have received specimens from the Lake-country.

203. Hyphantornis velatus (Vieill.). Damara Weaver bird.

Ploceus velatus, Vieillot's Tableau Encyclopédique, p. 702.
Ploceus mariquensis, Gurney, Birds Damar., Proc. Zool. Soc. 1864, p. 3.
„ „ Andersson, ibid. p. 6.
Hyphantornis capitalis (part.), Layard's Cat. No. 359.
Hyphantornis mariquensis (part.), Layard's Cat. No. 364.
Hyphantornis Cabanisi, Sharpe's Cat. No. 553.
Hyphantornis velatus, Gurney, in Ibis, 1871, p. 254.

This species is common in Damara Land and the parts adjacent; it is partial to the neighbourhood of water, where it nests, sometimes amongst reeds, and at others on the extremities of the boughs of trees overhanging permanent or temporary waters, many nests being built on the same tree. The nest is firmly composed of strong grasses; the number of eggs laid in each nest is three; and the variation of the eggs in colour, shape, and size is astonishing.

[This Weaver bird does not appear to have been figured; the male in breeding-dress closely resembles the species which I have figured (in 'The Ibis' for 1868, pl. 10) as the male in breeding-plumage of *H. mariquensis*, Smith, but is somewhat smaller, and the yellow tints are brighter and more conspicuous.

The female resembles in coloration Sir A. Smith's plate of *H. mariquensis*.

The following measurements of *H. velatus* are taken from three specimens now in the collection of Mr. R. B. Sharpe, which were obtained by Mr. Andersson, who recorded the sexes of each:—

	Male in full breeding-dress, obtained in Ondonga, January 22nd, 1867.	Male in full breeding-dress, obtained in Oudonga, January 25th, 1867.	Female obtained at Objimbinque, Feb. 16th, 1865.
	in.	in.	in.
Entire length	$5\frac{1}{4}$	$5\frac{1}{4}$	$5\frac{1}{4}$
Bill from gape	$\frac{10}{16}$	$\frac{10}{16}$	$\frac{8}{16}$
Wing from carpal joint	3	3	$2\frac{13}{16}$
Tarsus	$\frac{13}{16}$	$\frac{13}{16}$	$\frac{13}{16}$
Tail	$2\frac{1}{16}$	$1\frac{5}{16}$	$1\frac{5}{16}$

Two others of Mr. Andersson's skins of this species, which are also now in Mr. Sharpe's collection, are evidently changing from a plumage resembling that of the female bird into the male breeding-dress. From these specimens it seems probable that the males of *H. velatus*, when not in breeding-dress, resemble the females; but, unfortunately, neither the sex of these two specimens is recorded nor the dates at which they were obtained; one of them, however, is marked as having been procured in Damara Land.—ED.]

204. Euplectes capensis (Linn.). Yellow-Fink Bishop bird.

Le Grosbec du Coromandel, Buffon's Pl. Enl. vol. iv. p. 165, pl. 101. fig. 1 (male in breeding-dress).
Le Grosbec tacheté du Cap, Buffon's Pl. Enl. vol. iv. p. 179, pl. 659. fig. 1 (male not in breeding-dress).
Euplectes capensis, Swainson's Birds of West Afr. vol. i. p. 180.
Ploceus capensis, Layard's Cat. No. 308.

This is a comparatively scarce bird in Damara and Great Namaqua Land, but is very abundant at Lake Ngami; it sometimes occurs in large flocks in the open country, and is also found in small communities in the neighbourhood of water and in humid situations, where it breeds, constructing its nest of strong grasses and suspending it between the stalks of two or three reeds.

The iris in this species is dark brown; the legs and toes straw-brown; the upper mandible of the bill dark horn-colour, the lower in some specimens of the same colour as the upper, in others of a yellowish-white horn-colour.

Measurements of a male and a female:—

	Male. in. lin.	Female. in. lin.
Entire length	6 7	5 9
Length of folded wing	3 7	3 0
,, tarsus	0 11½	0 10
,, middle toe	0 11	0 9
,, tail	2 8	2 0
,, bill	0 9	0 8

[I have not seen a Damara example of this species; but, from the measurements above cited and a description contained in Mr. Andersson's MS., I have no doubt his identification of it is correct. With reference to the variation of the colouring of the under mandible in this species remarked by Mr. Andersson, it would appear, from the account given by Mr. Layard (*loc. cit.*), that it is black during the breeding-season, and horn-coloured at other times.—ED.]

205. Euplectes taha, Smith. Taha Bishop bird.

Euplectes taha, Smith's Zool. of S. Africa, pl. 7.
Ploceus abyssinicus, Strickland & Sclater, Birds Damar., Contr. Orn. 1852, p. 150.
Ploceus taha, Layard's Cat. No. 367.
Euplectes taha, Chapman's Travels in S. Afr., App. p. 401.
,, ,, Sharpe's Cat. No. 592.

I do not recollect having met with this bird in Great

Namaqua Land or in southern Damara Land; but it breeds in great abundance in Ondonga, and I have also seen specimens from Lake Ngami. It is found in flocks amongst trees, as well as on the reedy banks of rivers and in marshes, where it suspends its nest amongst the tall stalks of reeds and coarse grasses. The nest is composed of fine grass woven somewhat closely together; the eggs are six or seven in number, white, but sprinkled all over with minute brown specks.

The iris is brown, and also the bill; the legs and toes are yellowish brown.

Measurements of a male:—

	in.	lin.
Entire length	4	6
Length of folded wing	2	5
„ tarsus	0	8
„ middle toe	0	6
„ tail	1	6
„ bill	0	6

206. Pyromelana oryx (Linn.). Red-Fink Bishop bird.

Loxia orix, Shaw's Nat. Misc. pl. 240 (male in breeding-dress).
Ploceus oryx, Layard's Cat. No. 369.
Pyromelana oryx, Finsch & Hartlaub's Vögel Ost-Afrika's, p. 410.
Euplectes oryx, Sharpe's Cat. No. 590.

This very handsome bird is abundant at Lake Ngami and in Ondonga, and though rarer in Damara and Great Namaqua Land, it is found in those countries also, congregating in small communities and frequenting moist situations, where it breeds in January and February.

In Ondonga its nest may be found in those months in almost every palm bush; the nest is very pretty, airy, and graceful, somewhat oval in form, and composed of threads torn from the edges of the branches of young palms. Some nests are thickly lined, whilst others are quite bare within: in the latter the eggs may be seen from the outside; but notwithstanding the seeming looseness with which the threads are interwoven, the apparently frail structure is in reality very strong. The eggs are of a bluish colour, and from three to four in number.

The iris in this species is dark brown, as also is the bill; the tarsus is brownish.

Measurements of a male:—

	in.	lin.
Entire length	5	4
Length of folded wing	2	10
„ tarsus	0	10
„ middle toe	0	8
„ tail	1	8
„ bill	0	7

The female is a little smaller.

207. Quelea sanguinirostris (Linn.). Latham's Finch.

> *Fringilla quelea*, Vieillot's Ois. Chant. pls. 22, 23, & 24.
> *Quelea sanguinirostris*, Bonaparte's Conspectus, vol. i. p. 445.
> *Ploceus sanguinirostris*, Strickland & Sclater, Birds Damar., Contr. Orn. 1852, p. 150.
> *Ploceus Lathamii*, Layard's Cat. No. 370 (female).
> *Ploceus sanguinirostris*, Finsch & Hartlaub's Vögel Ost-Afrika's, p. 407.
> *Quelea sanguinirostris*, Gray's Hand-list of Birds, No. 6618.
> *Ploceus sanguinirostris*, Sharpe's Cat. No. 585.

This is a very common species in Damara Land, where it congregates in immense flocks after the breeding-season; and it is also common in the Lake-regions.

The iris is brown, the ring round the eyes reddish orange; the bill scarlet, merging into vermilion at the base and at the corner of the mouth; the legs and toes are flesh-coloured.

Average measurements of four males and two females:—

	Males. in. lin.	Females. in. lin.
Entire length	4 7	4 7
Length of folded wing	2 7	2 6
,, tarsus	0 8	0 8
,, middle toe	0 7	0 7
,, tail	1 6	1 5
,, bill	0 6	0 6

208. Amadina erythrocephala (Linn.). Red-headed Finch.

Loxia erythrocephala, Smith's Zool. of S. Africa, pl. 69.
Amadina erythrocephala, Andersson, in Proc. Zool. Soc. 1864, p. 7.
,,   ,,   Layard's Cat. No. 397.
,,   ,,   Chapman's Travels in S. Afr., App. p. 402.

This pretty bird occurs in Great Namaqua Land and at Lake Ngami, also in Damara Land, where, however, it is far from common, except in places, being, in fact, quite a local species. I found it congregated in large flocks on the Omaruru River at the end of October; and it was also pretty numerous at Objimbinque, where it nested under the eaves of my house and in the adjacent trees in company with the South-African Sparrow (*Passer arcuatus*), which it exactly resembles in its manners and habits; and I have found the nests of these two species on the same tree, and even on the same bough, as well as side by side under my eaves. The present species forms its nest of small sticks, slender roots, &c., and lines it with wool, feathers, or other soft and warm material. The young are fledged in June and July. It seeks its food upon the ground, usually in small flocks, and, when disturbed, takes refuge in the nearest tree or hedge; it occasionally utters a kind of chirping twitter.

The bill is pale yellowish brown, the legs and toes bright straw-yellow.

Measurements of a male:—

	in.	lin.
Entire length	5	10
Length of folded wing	3	0
,, tarsus	0	8
,, middle toe	0	7
,, tail	2	0
,, bill	0	7

The female is somewhat smaller.

209. Alario aurantia (Gmel.). Berg Canarie Finch.

Le Bouvreuil du Cap de Bonne Espérance, Buffon's Pl. Enl. vol. v. pl. 204. fig. 2 (male).
Amadina alario, Layard's Cat. No. 396.
Alario alario, Gray's Hand-list of Birds, No. 7285.

I only met with this pretty Finch in Great Namaqua Land, where it occurred in small flocks at the water.

210. Hypochera ultramarina (Gmel.). Ultramarine Finch.

Fringilla nitens, Vieillot's Ois. Chant. pl. 21.
Amadina nitens, Layard's Cat. No. 398.
Hypochera nitens, Finsch & Hartlaub's Vögel Ost-Afrika's, p. 430.
Hypochera ultramarina, Sharpe's Cat. No. 603.

[This species is not referred to in Mr. Andersson's notes; but his last collection contained a specimen obtained in Ondonga on January 22nd, 1867.—ED.]

211. Pytelia melba (Linn.). Southern Red-faced Finch.

Fringilla elegans, Vieillot's Ois. Chant. pl. 25.
Pytelia melba, Strickland & Sclater, Birds Damar., Contr. Orn. 1852, p. 150.
Estrelda melba, Layard's Cat. No. 392.
Pytelia melba, Chapman's Travels in S. Afr., App. p. 402.
„ „ Sharpe's Cat. No. 632.

This Finch is found sparingly in Damara and Great Namaqua Land, and usually occurs in pairs; its favourite resort is low bush and old abandoned village fences, whence the Damaras call it the " Kraal bird." Its food consists of seed and insects.

The iris is orange-brown, the legs and toes purplish.

Measurements of a male and a female:—

	Male.		Female.	
	in.	lin.	in.	lin.
Entire length	4	11	4	10
Length of folded wing	2	4	2	4
„ tarsus	0	7	0	8
„ middle toe	0	6	0	5
„ tail	1	11	2	0
„ bill	0	6	0	6

212. Lagonosticta minima (Vieill.). Amadavat Finch.

Fringilla minima, Vieillot's Ois. Chant. pl. 10.
Estrelda minima, Layard's Cat. No. 387.
Pytelia minima, Finsch & Hartlaub's Vögel Ost-Afrika's, p. 444.
Lagonosticta minima, Sharpe's Cat. No. 631.

I never saw but one individual of this diminutive Finch—a male obtained at Ondonga, Nov. 25th, 1866.

Measurements of this specimen:—

	in.	lin.
Entire length	4	1
Length of folded wing	1	11
„ tarsus	0	5
„ middle toe	0	5
„ tail	1	6
„ bill	0	4

[The specimen above mentioned is now in the collection of Mr. R. B. Sharpe, and also a female subsequently obtained by Mr. Andersson at Ovaquenyama on May 25th, 1867.—ED.]

213. **Sporopipes squamifrons** (Smith). Scutellated Finch.

Amadina squamifrons, Smith's Zool. of S. Africa, pl. 95.
Philetærus squamifrons, Strickland & Sclater, Birds Damar., Contr. Orn. 1852, p. 150.
Estrelda squamifrons, Layard's Cat. No. 393.
Amadina squamifrons, Chapman's Travels in S. Afr., App. p. 402.
Sporopipes squamifrons, Sharpe's Cat. No. 584.

This species is widely and commonly diffused over the middle and southern portion of Damara Land; it is also pretty common in Great Namaqua Land, in the Lake-regions, and at the river Okavango. It is a gregarious species, and is comparatively tame, often taking up its abode close to man. It feeds on grass-seeds and insects, which it chiefly seeks on the ground amongst the grass, resorting in small flocks to open localities thinly covered with dwarf vegetation.

This Finch is a very late breeder, and builds a large grass nest, which is usually placed in "hakisdom" bushes, and has the appearance externally of a bundle of grass accidentally pitched into a bush or tree, the entrance to the nest being nearly hidden by the manner in which the grass is arranged. Internally the nest is beautifully lined or, rather, padded with the softest materials, and especially with the feathers of the Guinea-fowl, and not only serves for the purpose of incubation but also as a roosting-place in the cold season, when several individuals, probably of the same brood, may be found thus snugly housed.

The female lays several eggs, of an oval form, and of a greyish or pale bluish white, thickly blotched with brown, especially at the thicker end.

The irides are dark brown; the bill yellow horn-colour, darkest above; the legs and toes yellow-brown.

214. Estrelda astrild (Linn.). Waxbill Finch.

Fringilla astrild, Vieillot's Ois. Chant. pl. 13.
Estrilda astrild, Strickland & Sclater, Birds Damar., Contr. Orn. 1852, p. 150.
Estrelda astrild, Layard's Cat. No. 381.
 „   „   Chapman's Travels in S. Afr., App. p. 402.
Habropyga astrild, Finsch & Hartlaub's Vögel Ost-Afrika's, p. 439.
Estrelda astrild, Sharpe's Cat. No. 620.

This pretty little species is common in the southern districts of Damara Land and in some parts of Great Namaqua Land, as well as at Lake Ngami. It lives in flocks, and is generally found upon the ground, where it seeks its food, which consists of seeds, insects, &c.

The iris is brown, the bill bright red, the legs and feet liver-brown.

215. Estrelda erythronota (Vieill.). Black-cheeked Finch.

Estrelda erythronota, Gray's Genera of Birds, pl. 90. fig. 1.
Estrelda lipiniana, Strickland & Sclater. Birds Damar., Contr. Orn. 1852, p. 150.
Estrelda erythronota, Layard's Cat. No. 380.
 „   „   Sharpe's Cat. No. 613.

This species is generally distributed over Damara and Great Namaqua Land, and is also found at Lake

Ngami. It resembles its congeners in manners and habits.

The iris is bright wine-coloured red; the legs and toes black, as is also the bill, except near the base, where it is whitish blue.

Measurements of a male and a female:—

	Male. in. lin.	Female. in. lin.
Entire length	5 0	4 11
Length of folded wing	2 2	2 2
„ tarsus	0 7	0 7
„ middle toe	0 5	0 5
„ tail	2 6	2 5
„ bill	0 4	0 4

216. Mariposa cyanogastra (Daud.). Southern Benghala Finch.

Loxia cyanogastra, Daudin's Tr. d'Orn. vol. ii. p. 436.
Estrelda benghala, Gurney, Birds Damar., Proc. Zool. Soc. 1864, p. 3.
„ „ Layard's Cat. No. 391.
„ „ Chapman's Travels in S. Afr., App. p. 402.
Estrelda cyanogastra, Sharpe's Cat. No. 616.

This pretty little Finch is common at Lake Ngami and in the neighbourhood of the Okavango River; but I do not think that it is found either in Damara or Great Namaqua Land.

I have heard occasionally of immense gatherings of these birds; but usually they are found in small flocks seeking on the ground for their food, which consists almost entirely of the seeds of grasses.

On the 2nd of February, 1867, I found in Ondonga a nest of this species in a palm bush about six feet from the ground; it was constructed of grass, and had no internal lining; the eggs were five in number.

The iris is cherry-colour.

Measurements of a male and a female :—

	Male. in. lin.	Female. in. lin.
Entire length	4 10	4 7
Length of folded wing	2 2	2 2
,, tarsus	0 7	0 7
,, tail	2 0	1 11
,, bill	0 4	0 4

[*Mariposa cyanogastra* is the southern form of the equatorial *M. phœnicotis*, from which, according to the testimony of M. Jules Verreaux, it is always distinguished by the absence of the purple ear-coverts in the male bird. The southern race, I believe, has not been figured.—ED.]

217. Uræginthus granatinus (Linn.). Grenadin Finch.

Fringilla granatina, Vieillot's Ois. Chant. pls. 17 & 18.
Estrelda granatina, Strickland & Sclater, Birds Damar., Contr. Orn. 1852, p. 150.
,, ,, Hartlaub's Orn. Westafrica's, No. 438.
,, ,, Layard's Cat. No. 304.
,, ,, Chapman's Travels in S. Afr., App. p. 402.
,, ,, Sharpe's Cat. No. 618.

This species is not uncommon in Damara Land, but is more abundant further north, and also at Lake Ngami. It congregates in small flocks, and feeds on little seeds, which it seeks upon the ground.

The iris is red; and the naked ring round the eye is usually red also, but sometimes of a greyish drab; the bill purplish, but red at the extremity; the legs are purplish grey.

Measurements of a male and a female :—

	Male. in. lin.	Female. in. lin.
Entire length	5 11	5 7
Length of folded wing	2 4	2 1
,, tarsus	0 7	0 7
,, middle toe	0 5	0 5
,, tail	3 2	2 9
,, bill	0 5	0 5

218. Vidua regia (Linn.). Shaft-tailed Widow bird.

Emberiza regia, Vieillot's Ois. Chant. pls. 34 & 35.
Vidua regia, Strickland & Sclater, Birds Damar., Contr. Orn. 1852, p. 150.
    „   „   Layard's Cat. No. 374.
    „   „   Sharpe's Cat. No. 601.

This species visits Damara Land and the parts adjacent during the wet season, and is then by no means uncommon.

219. Vidua principalis (Linn.). Dominican Widow bird.

Vidua erythrorhyncha, Swainson's Birds of West Africa, vol. i. pl. 12 (male).
Vidua principalis, Layard's Cat. No. 375.
Estrelda carmelita, Hartlaub, in Ibis, 1868, p. 46 (immature).
Vidua principalis, Finsch & Hartlaub's Vögel Ost-Afrika's, p. 428.
    „   „   Finsch, in Trans. Zool. Soc. vol. vii. p. 265.
    „   „ .   Sharpe, in Proc. Zool. Soc. 1870, pp. 144, 149.
    „   „   Sharpe's Cat. No. 600.

This is rather a scarce species in Damara and Great Namaqua Land, much more so than *Vidua regia*. It only occurs during the rainy season, and is generally seen in pairs or in very small flocks. Its food consists of the seeds of grass.

The iris is dark brown, the bill red, the legs light reddish brown.

220. Vidua paradisea (Linn.). Paradise Widow bird.

La grande Veuve d'Angola, Buffon's Pl. Enl. vol. iv. p. 343, pl. 104 (male, in spring and change).
Vidua paradisea, Swainson's Birds of West Africa, vol. i. pl. 11 (male, in spring).
Steganura paradisea, Strickland & Sclater, Birds Damar., Contr. Orn. 1852, p. 157.
Vidua paradisea, Finsch & Hartlaub's Vögel Ost-Afrika's, p. 424.
    „   „   Sharpe's Cat. No. 602.

This species occurs in Damara Land in the wet season, but is even scarcer than *Vidua principalis* in that country, though not unfrequent at Lake Ngami. The bill is black, the legs and toes brown.

221. Crithagra angolensis (Gmcl.). Black-throated Bullfinch.

Linnets from Angola, Edwards's Birds, pl. 129 (male and female).
Fringilla angolensis, Gmelin's Syst. Nat. p. 918.
Poliospiza angolensis, Hartlaub's Orn. Westafrica's, No. 456.
Fringilla angolensis, Layard's Cat. No. 404.
Crithagra angolensis, Sharpe's Cat. No. 637.

I have met with this species both in Damara Land and further northward; it congregates in large flocks, and often associates with *C. chrysopyga*.

The iris is brown, the upper mandible flesh-coloured, the lower mandible and also the legs and toes of a dirty flesh-colour.

Measurements of a male and a female:—

	Male.	Female.
	in. lin.	in. lin.
Entire length	4 8	4 3
Length of folded wing	2 8	2 7½
,, tarsus	0 6	0 6½
,, middle toe	0 5	0 6
,, tail	1 11	1 9
,, bill	0 4	0 4

[The male specimen of which Mr. Andersson here records the measurements is now in the collection of Mr. R. B. Sharpe. —Ed.]

222. Crithagra chrysopyga, Swains. Golden-rumped Bullfinch.

Crithagra chrysopyga, Swainson's Birds of W. Afr. vol. i. pl. 17.
Crithagra Hartlaubii, Gurney, Birds Damar., Proc. Zool. Soc. 1864, p. 3.

Crithagra chrysopyga, Layard's Cat. No. 444.
Crithagra butyracea (part.), Finsch & Hartlaub's Vögel Ost-Afrika's, p. 455.
Crithagra chrysopyga, Sharpe's Cat. No. 638.

This species is common in the neighbourhood of the Okavango, and is also found in Damara Land, extending to the south of that country. On January 5th, 1867, I found a nest of this Finch at Omapju; it was situated in a kamel-thorn bush, about four feet above the ground, and was composed of grass, lined with softer grass internally; it contained three eggs much incubated.

223. Crithagra chloropsis, Cab. Decken's Bullfinch.

Crithagra chloropsis, Cabanis in Decken's Reisen, vol. iii. pl. 9.
 „   „   Finsch & Hartlaub's Vögel Ost-Afrika's, p. 454.
 „   „   Sharpe's Cat. No. 642.

[Mr. R. B. Sharpe possesses a specimen of this Finch, obtained by Mr. Andersson in Damara Land; but unfortunately no record has been preserved of the exact locality where it was obtained, and I do not find any reference to it in Mr. Andersson's MS. notes.—ED.]

224. Poliospiza gularis (Smith). Streaky-headed Grosbeak.

Crithagra gularis, Catalogue of South-Afr. Mus. 1837, p. 32.
Linaria gularis, Smith's Append. to Report of Exp. p. 49.
Fringilla gularis, Layard's Cat. No. 402.
Fringilla striaticeps, Layard's Cat. No. 403.
Poliospiza gularis, Sharpe's Cat. No. 645.

I have only observed this species in Great Namaqua Land, and there very sparingly.

Measurements of a male:—

	in.	lin.
Entire length	6	0½
Length of folded wing	3	0
„ tarsus	0	9
„ middle toe	0	6
„ tail	2	6
„ bill	0	7

[The specimen from which the above measurements were taken by Mr. Andersson is now in the collection of Mr. R. B. Sharpe.
This species has not been figured.—ED.]

225. Poliospiza crocopygia, Sharpe. Damara Yellow-rumped Grosbeak.

Poliospiza crocopygia, Sharpe, in Ibis, 1871, p. 101.
„ „ Sharpe's Cat. no. 646.

This species is sparingly met with in Damara and Great Namaqua Land; it is generally seen about rocks in the immediate neighbourhood of water, to which it resorts in small flocks in the morning and evening to quench its thirst. Its food consists of seeds and berries.

The iris is dark brown; the upper mandible very pale brown, and the lower still paler; the legs and toes slaty brown.

Measurements of a male and a female:—

	in.	lin.	in.	lin.
Entire length	6	0	6	0
Length of folded wing	3	4	3	2
„ tarsus	0	10	0	9
„ middle toe	0	7	0	7
„ tail	2	3	2	3
„ bill	0	7	0	7

[The female specimen from which Mr. Andersson took the measurements above recorded is now, together with two other Damara examples, in the collection of Mr. R. B. Sharpe. This species has not been figured.—ED.]

226. Petronia petronella (Licht.). Southern Yellow-throated Grosbeak.

Xanthodira flavigula, Sundevall, Öfvers. 1850, p. 98.
    "   "   Gurney, Birds Damar., Proc. Zool. Soc. 1864, p. 3.
    "   "   Layard in Ibis, 1869, p. 371.
Petronia petronella, Gray's Hand-list of Birds, No. 7244.

I have only met with this species near the river Okavango, where I obtained a few specimens.

[As this species has not been figured, I here transcribe Professor Sundevall's description of it, which is not very generally accessible, and may therefore here be useful:—" Grisescens, dorso fusco-maculato, vittâ superciliari latâ albidâ, maculâ juguli flavâ. Capillities fusca, alarum tectrices limbo pallescente, non albido: gastræum pallidum, ventre in mare pure albo. ♂ et ♀ in ceteris similes. 6-pollicaris. Ala 92 mm., tars. 21: tomia superiora medio obsoletè angulata."—ED.]

227. Passer arcuatus, Gmel. Cape Sparrow.

Le Moineau du Cap de Bonne Espérance, Buffon's Pl. Enl. vol. iv. p. 204, pl. 230. fig. 1.
Passer arcuatus, Layard's Cat. No. 408.
    "   "   Sharpe's Cat. No. 657.

This Sparrow is very abundant all over Damara and Great Namaqua Land, and extends as far south as the Cape, where, however, it seems to be a trifle larger than in Damara Land; it takes up its abode near to man, and in all its habits exactly resembles the House-Sparrow of Europe. It makes a large rambling nest of grass and sticks, lined with feathers, down, &c., and placed either on a tree or bush or under the eaves of thatched roofs; the eggs vary wonderfully in size and colour. The male bird is rather larger than the female. The bill is black in the male, and livid brown in the female; the legs and toes are olive-brown, darker in the male than in the female; the iris a very dark brown.

228. Passer motitensis, Smith. Greater South-African Sparrow.

Passer motitensis, Smith's Zool. of S. Africa, pl. 114.
"   "   Layard's Cat. No. 406.
"   "   Sharpe's Cat. No. 658.

This species is found at Lake Ngami; and I also met with it at Objimbinque, at Omapju, and near Otaiovapa. I first met with its nest at Omapju, on January 5th, 1867; it was placed on the top of a thorn bush, about seven feet from the ground; and I subsequently met with several other nests during the same month. The nest is large, and is composed of coarse grass outside, and fine grass within, and has an additional lining of feathers and other soft and warm materials; it is furnished with a long entrance, sometimes a foot or more in length, which ends in a deep and wide hollow. The eggs are from three to six; but the most frequent number is four; they are larger than the eggs of *P. arcuatus*, and are invariably covered with a glutinous matter which it is difficult to remove, even with the help of water.

The iris is dark brown, the bill liver-brown; the legs and toes resembling the bill in colour, but paler.

Measurements of a male and a female:—

	Male. in. lin.	Female. in. lin.
Entire length	6 3	6 3
Length of folded wing	3 3½	3 2
,, tarsus	0 9	0 10
,, middle toe	0 7	0 8
,, tail	2 6	2 3½
,, bill	0 7½	0 8

229. Passer diffusus, Smith. Southern Grey-headed Sparrow.

Passer diffusus, Smith's Append. to Report of Exp. p. 50.
,, ,, Gurney, Birds Damar., Proc. Zool. Soc. 1864, p. 3.
,, ,, Layard's Cat. No. 407.
Passer simplex, ibid. No. 409.
Passer Swainsoni (part.), Finsch and Hartlaub's Vögel Ost-Afrika's, p. 450.
Passer diffusus, Sharpe, in Proc. Zool. Soc. 1870, p. 143.
,, ,, Gray's Hand-list of Birds, No. 7283.
,, ,, Sharpe's Cat. No. 655.

I met with this species very sparingly in Damara Land, but found it pretty common on approaching the river Okavango; it frequents the water in the morning and evening, and feeds on seeds and berries.

In some specimens of this bird the bill is quite black, in others pale greyish brown with yellow near the base of the lower mandible; the legs and toes are pale greyish brown.

Measurements of a male and a female:—

	Male. in. lin.	Female. in. lin.
Entire length	6 6	6 3
Length of folded wing	3 4	3 1
,, tarsus	0 9	0 9
,, middle toe	0 5	0 6
,, tail	2 10	2 7
,, bill	0 6	0 6

[Mr. Layard (*loc. cit.*) mentions having seen *Passer simplex*, which is a West-African race, from Damara Land; but it seems probable, as suggested by Mr. Sharpe (*loc. cit.*), that in this instance *P. diffusus* may have been mistaken for *P. simplex*. These two Sparrows are in fact very nearly allied; and Drs. Finsch and Hartlaub (*loc. cit.*) refer them both, and also the Abyssinian *P. Swainsoni*, to one and the same species. I, however, agree with Mr. Sharpe in considering *P. Swainsoni*, *P. simplex*, and *P. diffusus* as forming three specifically distinct local races, though the differences between them are but slight.— ED.]

230. Fringillaria flaviventris (Vieill.). Southern Yellow-bellied Bunting.

Emberiza flavigaster, Rüppell's Atlas, pl. 25.
Fringillaria flaviventris, Gurney, Birds Damar., Proc. Zool. Soc. 1864, p. 3.
„   „   Layard's Cat. No. 410.
Emberiza flavigaster, Chapman's Travels in S. Afr., App. p. 403.
Emberiza flaviventris, Finsch & Hartlaub, Vögel Ost-Afrika's, p. 458.
Fringillaria flaviventris, Sharpe's Cat. No. 667.

This very handsome bird is pretty common in northern Damara Land and thence onward to the Okavango. I have usually met with it in well-wooded localities; and I obtained my specimens by quietly awaiting their arrival at some small water, which they were sure to visit in the morning and evening, especially the former; they are fearless birds, and consequently easy to secure.

The upper mandible is dark liver-brown, the under mandible salmon-yellow; the legs yellowish brown, and the toes also, but more dusky.

231. Fringillaria impetuani (Smith). Lark-like Bunting.

Fringillaria impetuani, Hartlaub's Orn. Westafrica's, No. 463.
„   „   Layard's Cat. No. 412.
„   „   · Sharpe's Cat. No. 668.

This species is common in Damara and Great Namaqua Land; but as it frequently resorts to the ground in search of seeds and insects, it often thus escapes observation; it is gregarious, and is partial to broken ground or its immediate vicinity, and also to the neighbourhood of water, which it appears to require pretty constantly.

The irides are dark brown; the legs, toes, and upper mandible of the bill a dusky flesh-colour, the lower mandible livid horn-colour.

Measurements of a male and a female:—

	Male. in. lin.	Female. in. lin.
Entire length	5 9	5 4
Length of folded wing	3 0	2 10
„ tarsus	0 7	0 8
„ middle toe	0 6	0 $6\frac{1}{2}$
„ tail	2 6	2 3
„ bill	0 5	0 $4\frac{1}{2}$

[The male specimen of which the measurements are given above by Mr. Andersson is now preserved in the collection of the Rev. H. B. Tristram.—ED.]

ALAUDIDÆ.

232. Pyrrhulauda Smithi, Bon. Smith's Finch Lark.

Pyrrhulauda leucotis, Smith's Zool. of S. Africa, pl. 26.
Pyrrhulauda Smithi, Bonaparte's Conspectus, vol. i. p. 512.
Pyrrhulauda leucotis, Layard's Cat. No. 420.
„   „   Chapman's Travels in S. Afr., App. p. 400.

Coraphites *Smithi*, Finsch & Hartlaub's Vögel Ost-Afrika's, p. 468.
Pyrrhulauda *Smithii*, Gray's Hand-list of Birds, No. 7833.
,, ,, Sharpe's Cat. No. 666.

I have observed this species, though very sparingly, in Damara Land, and also to the northward, but not in Great Namaqua Land; it is seen in small flocks, which frequent the ground in open localities covered with grass and scanty dwarf vegetation, amongst which they run with great rapidity, taking flight for a short distance only when disturbed.

The iris is dark brown, the legs and toes flesh-brown. Measurements of a male and a female:—

	Male. in. lin.	Female. in. lin.
Entire length . . .	5 0	4 9
Length of folded wing	3 5	3 0½
,, tarsus . .	0 8½	0 7½
,, middle toe	0 5½	0 5½
,, tail . . .	2 2	1 10
,, bill . . .	0 6	0 6

[I have not seen a Damara example of this species; but, from the description preserved in Mr. Andersson's notes, I have no doubt that his identification of it is correct, except that he was not aware of the distinction between the southern *P. Smithii* and the true *P. leucotis* of northern Tropical Africa.—ED.]

233. **Pyrrhulauda verticalis,** Smith.   Grey-backed Finch Lark.

Pyrrhulauda *verticalis*, Smith's Zool. of S. Africa, pl. 25.
,, ., Layard's Cat. No. 421.
,, ,, Chapman's Travels in S. Afr., App. p. 400.
,, ,, Sharpe's Cat. No. 665.

This species is pretty commonly diffused over Damara and Great Namaqua Land, especially in the latter country; and I generally observed it abundant in the rainy season at Objimbinque, during my residence there. It congregates in large flocks, and is always found on the ground, unless disturbed, when it flies but a short distance before it again alights, scattering widely amongst the grass in search of its food, which consists almost entirely of seeds. Its flight is undulating; and occasionally an individual may be seen to soar above the rest, descending again very abruptly; it may sometimes be heard to utter a shrill chirping cry. The nest is constructed on the ground, under a tuft of grass.

The iris is dark brown, the legs and toes purplish flesh-colour.

Measurements of a male:—

	in.	lin.
Entire length	4	10
Length of folded wing	3	2
,, tarsus	0	9
,, middle toe	0	5
,, tail	1	11
,, bill	0	6

234. Pyrrhulauda australis, Smith. Dark-naped Finch Lark.

Pyrrhulauda australis, Smith's Zool. of S. Africa, pl. 24.
,, ,, Layard's Cat. No. 422.
,, ,, Chapman's Travels in S. Afr., App. p. 400.

This species is not uncommon in Damara and Great Namaqua Land, where it occurs in small flocks during the rainy season. Its habits are similar to those of *P. verticalis*.

The bill is light horn-colour, darkest above; the legs and toes are light flesh-colour.

Measurements of a male:—

	in.	lin.
Entire length	4	6
Length of folded wing	3	0
„ tarsus	0	8
„ middle toe	0	4
„ tail	1	10
„ bill	0	6

[As I have not seen a specimen of this bird from Damara Land, its identification rests on Mr. Andersson's authority; but, from the description left in his MS. notes, I have no doubt that the identification is correct.—ED.]

235. Alauda conirostris, Sund. Pink-billed Lark.

Alauda conirostris, Sundevall, Öfvers. 1850, p. 99.
„ „ Sharpe's Cat. No. 689.

This Lark is not uncommon in Ondonga; before the breeding-season they are seen in small flocks, but are less observable after they are paired.

A pair which I watched occupied about eight days in constructing their nest and in the female bird laying her eggs, which I took on the 31st of March, after they had been incubated about three days. The nest is formed in a hole excavated to the depth of about two and a half inches and thickly lined with decomposed grasses.

The iris in this species is brown, the legs and bill flesh-coloured, but the lower mandible paler and more livid than the upper.

Measurements of a female:—

	in.	lin.
Entire length	4	4
Length of folded wing	2	10
„ tarsus	0	9
„ middle toe	0	6
„ tail	1	6
„ bill	0	5

[This species has not been figured; Professor Sundevall's description of it is as follows :—" Subtus fulva, superne pectoreque antico nigro-maculata; rostro crassiore, rubro, parvo, 4½ poll.; ala 74 mm.; tars. 20; d. m. 11, c. u. 15. Unguis post. subrectus, digito vix longior. Nares setulis tectæ. Ala brevis, remigibus intus fulvis. Penna spuria in nostris (ob mutationem pennarum?) deest; 2da extus albo-marginata. Lora et gula alba, striis malaribus, utrinque 2, striaque lori nigris. Cauda brevior, subæqualis, pennis utrinque 2 extus late albis. ♂ superne rufescenti-varius, alias similis feminæ. In Caffraria superiore campestri."

The female specimen of which Mr. Andersson gives the dimensions, as above quoted, is now in the collection of Mr. R. B. Sharpe, who also possesses a male obtained by Mr. Andersson at Objimbinque. This male specimen only differs from the female in having no black malar stripe, and in the bill being about one third longer than that of the female and less decidedly red, the colour of the bill in this male bird being tawny instead of a pinkish red as in the female specimen. The two specimens have been carefully compared by M. Verreaux, who assures me that he feels confident that they are male and female of the same species, of which I had previously some doubts, owing to the differences above mentioned.—ED.]

236. Alauda Grayi, Wahl. Gray's Lark.

Alauda Grayi, Wahlberg, Öfvers. 1855, p. 213.
" " Wahlberg, Journal für Orn. 1857, p. 2.
Alauda nævia, Chapman's Travels in S. Afr., App. p. 400.
Alauda Grayi, Sharpe's Cat. No. 683.

This Lark is found on the barren plains in the neighbourhood of Walwich Bay, as well as some distance inland; but I have not observed it so far east as Objimbinque. It feeds on seeds and insects, and is comparatively tame, but where grass abounds it is difficult to secure. A few individuals are usually found together.

o

Measurements of a male and a female :—

	Male. in. lin.	Female. in. lin.
Entire length	5 7	5 4
Length of folded wing	3 2	3 1
,, tarsus	0 10	0 10
,, middle toe	0 6	0 6½
,, bill	0 8	0 7¾

[As this species has not been figured, and as Professor Wahlberg's description of it will not be readily accessible to many readers, I have here transcribed it :—

"Griseo-isabellina (immaculata), subtus cum fronte lorisque albida. Latera juguli fusco-maculata. Remiges pallide fuscæ, limbis externis isabellinis (1^a et 2^a exceptis) 1^a–6^m apicem versus saturatiores; 7^a et seqq. apicibus subcordatis, albidis. Rectrices 2 intermediæ isabellinæ; laterales nigro-fuscæ, basi, margine exteriore apiceque albidæ. Rem. 1^a spuria; 2^a–5^m fere æquales. ♂. Rostr. a fr. 13·5 millim.; alt. rostri ante nares 5; ala 84; cauda 50; tars. 21; dig. med. c. ungue 15·5; pollex c. ungue 14·5. ♀. R. a fr. 12 mm.; alt. rostri ante nares 4; ala 78; cauda 50; tars. 20; dig. med. c. u. 15; poll. c. u. 12."
—Ed.]

237. Alauda erythrochlamys, Strickl. Rufous-mantled Lark.

Alauda erythrochlamys, Strickland & Sclater, Birds Damar., Contr. Orn. 1852, p. 151.
,, ,, Sharpe's Cat. No. 685.

[I am uncertain whether Mr. Andersson's note under the head of "*Alauda erythrochlamys*" really refers to that species; and as he appears himself to have been doubtful on the point, I have not transcribed it. I have only seen two examples of this Lark, one of them being Mr. Strickland's type specimen, now preserved in the Museum of Zoology at Cambridge, and the other a specimen which was also obtained by Mr. Andersson in Damara Land, and is now in the possession of Mr. R. B. Sharpe.

As this species has not been figured, and Messrs. Strickland

and Sclater's description of it (*loc. cit.*) is not very easy of access, I here transcribe it :—" Whole upper parts and wings nearly uniform ferruginous, becoming paler on the upper tail-covers and medial pair of rectrices; all the wing-feathers narrowly margined with cream-colour externally; the remiges light fuscous on their shafts and inner webs; a streak above and a spot below the eye pale cream-colour; ear-covers pale rufous; chin white; breast and lower parts pale cream-colour; beak long; gonys nearly straight, flesh-coloured; legs apparently flesh-coloured; hind claw short, straight. Total length 6″ 5‴; beak to front 7‴, to gape 8‴, high 2‴; wing 3″ 7‴, rectrices 2″ 8‴; tarsus 1″ 1‴, hind claw 2¼‴."—Ed.]

238. Calendula crassirostris (Vieill.). South-African Thick-billed Lark.

L'Alouette à gros bec, Levaillant's Ois. d'Afr. pl. 193.
Alauda crassirostris, Layard's Cat. No. 208.
Calendula crassirostris, Gray's Hand-list of Birds, No. 7773.
Alauda crassirostris, Sharpe's Cat. No. 679.

This species is pretty generally diffused over Damara and Great Namaqua Land; it frequents localities covered with grass and dwarf vegetation, and frequently those which are of a rocky character.

[A specimen of this Lark from Komaggos, in Little Namaqua Land, was contained in Mr. Andersson's last collection, and is now in the possession of Mr. R. B. Sharpe.—Ed.]

239. Megalophonus sabota (Smith). Sabota Lark.

Mirafra sabota, Smith's Zool. of S. Africa, pl. 89. fig. 2.
Megalophonus sabota, Layard's Cat. No. 429.
Alauda sabota, Sharpe's Cat. No. 687.

I met with this Lark at various places north of the Omaruru River, where, although rather local, it is abundant in some localities, and especially in Ondonga.

On the 25th of April I found its nest in the last-named locality, containing three eggs of a very elongated form and much pointed at one end. Other nests which I met with on various occasions were either empty or contained young birds.

The nest is composed of fine grasses, and is built upon the ground, into which the lower portion of it is sunk to the depth of two or three inches; it is placed between and resting against two stout plants, and is round and dome-shaped, with one large opening to the southward, the side freest from the wind and rain.

I am always delighted to hear this bird as it makes the welkin ring with its sonorous song and call, composed of a series of notes, which it utters either when perched on a tree or when moving from one spot to another. Whilst thus flying and singing it may be seen alternately to open and close its wings, sometimes almost bringing them to meet over its back, when it appears for the moment to be suspended in the air.

The iris in this species is brown; the legs and toes pale flesh-colour; the bill horn-colour, but more or less brown along the edges of the upper and at the point of the lower mandible; the gape and corner of the mouth are pale yellow.

Measurements of a male obtained in Ondonga:—

		in.	lin.
Entire length		6	6
Length of folded wing		3	5
,,	tarsus	1	1
,,	middle toe	0	6
,,	tail	2	2
,,	bill	0	9

[Mr. Andersson's last collection contained specimens of this Lark, including two from Ondonga, which are now in the possession of Mr. R. B. Sharpe, and from one of which the above measurements, given by Mr. Andersson, were taken. Both of these examples are much greyer and less rufous in their general colouring than other specimens with which they have been compared, from Damara Land proper and also from Trans Vaal; but, notwithstanding this difference, they do not appear to be specifically distinct.

Sir A. Smith figures this species and *M. cheniana* on the same plate; but *M. sabota* is incorrectly referred to figure no. 1 and *M. cheniana* to figure no. 2. It will be seen by a reference to the descriptive letterpress that this is an error, the upper and smaller figure, marked no. 1, representing *M. cheniana*, whilst the lower and larger one, marked no. 2, represents *M. sabota.*—ED.]

240. Megalophonus cinereus (Vieill.). South-African Rufous-capped Lark.

> *La petite Alouette à tête rousse*, Levaillant's Ois. d'Afr. pl. 199.
> *Alauda spleniata*, Strickland & Sclater, Birds Damar., Contr. Orn. 1852, p. 152.
> *Megalophonus cinereus*, Layard's Cat. No. 426.
> *Alauda spleniata*, Chapman's Travels in S. Afr., App. p. 400.
> *Alauda cinerea*, Sharpe's Cat. No. 682.

This Lark is very abundant in some parts of Damara and Great Namaqua Land. I have observed that it uses much gravel with its food.

The irides are dark brown, the bill yellowish brown, the legs and toes brownish or reddish brown.

Measurements of a male and a female:—

	Male.		Female.	
	in.	lin.	in.	lin.
Entire length	6	3	6	1
Length of folded wing	3	7	3	4
,, tarsus	0	10	0	10
,, middle toe	0	7	0	6
,, tail	2	6	2	4
,, bill	0	7	0	7

[I have ascertained the identity of Strickland's *Alauda spleniata* with the present species by examination of the type specimen of the former at the Museum of Zoology, Cambridge.—ED.]

241. Megalophonus Anderssoni, Tristr. Andersson's Lark.

Megalophonus Anderssoni, Tristram, in Ibis, 1869, p. 434.
Calandrella Anderssoni, Blanford's Geol. & Zool. of Abyssinia, p. 389.
Megalophonus Anderssoni, Tristram, in Ibis, 1870, p. 444.
Alauda Anderssoni, Finsch, in Trans. Zool. Soc. vol. vii. p. 237.

[This small Lark has been recently described by Mr. Tristram in the 'Ibis' for 1869, from a single female specimen contained in Mr. Andersson's last collection and labelled as having been obtained at Objimbinque on March 13th, 1865. Mr. Andersson does not allude to this species in his MS. notes; but Mr. Blanford, who met with it in Abyssinia, states (*loc. cit.*) that he found it "abundant on stony ground near Senafé, keeping much in large communities, and highly social, precisely like *Calendrella brachydactyla.*" This species, which has not been figured, is very nearly allied to *M. cinerea*, but is decidedly smaller. Mr. Tristram's notice of it in the 'Ibis' for 1869 includes a description of its nest and eggs, which accompanied the type specimen.—ED.]

242. Megalophonus africanoides (Smith). Objimbinque Lark.

Mirafra africanoides, Smith's Zool. of S. Africa, pl. 88. fig. 2.
Megalophonus africanoides, Layard's Cat. No. 430.
Alauda africanoides, Sharpe's Cat. No. 681.

This bird is very frequent in the neighbourhood of Objimbinque, and is not uncommon in various other parts of Damara and Great Namaqua Land. It is easily distinguished from most of the other Larks by its reddish appearance. It is found in pairs, and is comparatively tame, flying but a short distance when disturbed, and settling on the ground or on a bush or tree; it has a kind of chirping song.

The iris is brown; the upper mandible of the bill and the point of the lower are dark brown, the remainder of the lower mandible is horn-colour; the legs and toes are pale straw-colour, with a faint purple blush.

Measurements of a male:—

	in.	lin.
Entire length	6	5
Length of folded wing	3	8
,, tarsus	0	11
,, middle toe	0	5
,, tail	2	8
,, bill	0	8

[The specimen of which the measurements are given above by Mr. Andersson is now in the collection of Mr. R. B. Sharpe. —Ed.]

243. Megalophonus nævius (Strickl.). Dark-lined Lark.

Alauda nævia, Strickland & Sclater, Birds Damar., Contr. Orn. 1852, p. 152.
,, ,, Sharpe's Cat. No. 688.

This is a very common bird about Objimbinque; it is by no means timid, and settles on trees and bushes as well as on the ground. Its food consists of seeds and insects.

The iris is light brown; the upper mandible horn-coloured, fading into white on the edges and on the lower mandible; legs and toes pale straw-colour, with a tint of pale flesh-colour on the toes.

Measurements of a male and a female:—

	Male.		Female.	
	in.	lin.	in.	lin.
Entire length	6	6	6	4
Length of folded wing	3	6	3	5
,, tarsus	1	0	1	0
,, middle toe	0	7	0	6
,, tail	2	3	2	3

[The type specimen of Messrs. Strickland and Sclater's *Alauda nævia*, which I have examined at the Museum of Zoology at Cambridge, appears to me to be identical with specimens contained in Mr. Andersson's last collection, from two of which, now in the collection of Mr. R. B. Sharpe, Mr. Andersson's measurements, above quoted, were taken.

As this Lark has not been figured, it may be desirable to add the following copy of the description given of it by Messrs. Strickland and Sclater (*loc. cit.*) :—" Crown greyish brown; back of neck paler; back and wing-covers fulvous; the feathers of all these parts with a broad, distinct longitudinal fuscous streak on each; remiges fuscous, margined externally and internally towards the base with pale fulvous; upper tail-covers rufofulvous; rectrices deep fuscous; the middle and external pairs broadly, the rest narrowly, margined with fulvous; a whitish streak above and below the eye; cheeks and ear-covers light brown; chin whitish; breast pale cream-colour, with a small longitudinal fuscous streak on each feather; belly and vent pale cream-colour; beak corneous, paler towards the base; margin nearly straight, gonys curved upwards, and culmen equally so downwards; feet and claws flesh-coloured, tinged with light brown; hind claw short, slightly curved.

"Total length 6''; beak to front 6''', to gape 7''', high $2\frac{1}{2}'''$, broad $2\frac{1}{4}'''$; wing 3'' 4''', all the rectrices 2'' 3'''; tarsus 9''', hind claw 3'''."—ED.]

244. Megalophonus lagepa (Smith). Lagepa Lark.

Alauda lagepa, Smith's Zool. of S. Africa, pl. 87. fig. 2.
Megalophonus lagepa, Layard's Cat. No. 432.

I observed this species pretty frequently in Little Namaqua Land, but not to the north of the Orange River; it perches on bushes as well as on the ground.

[This species, during Mr. Andersson's residence at the Cape, was labelled in the Cape Museum as *Megalophonus guttatus*, on which account Mr. Andersson alludes to it in his MS. under that name; but having, through the kindness of Mr. Layard, had the opportunity of examining specimens of it from the Cape Museum, I have no doubt that the species alluded to by Mr. Andersson under the name of *M. guttatus* is in reality *M. lagepa*, although I have not seen an example actually obtained by him.—ED.]

245. Certhilauda rufula (Vieill.). Garrulous Lark.

Anthus rufulus, Vieillot's Gal. des Ois. pl. 161.
Certhilauda garrula, Smith's Zool. of S. Africa, pl. 106.
  „   „   Layard's Cat. No. 435.
  „   „   Chapman's Travels in S. Afr., App. p. 400.
Certhilauda rufula, Sharpe's Cat. No. 675.

I have not observed this species in Damara Land; but it is common in some parts of Great Namaqua Land, frequenting very open localities covered with grass and bush, where it runs about with great swiftness.

[Specimens of this Lark obtained by Mr. Andersson in Great Namaqua Land are smaller than examples procured at Colesburg, but agree in size with a specimen obtained by Mr. Ayres in the neighbourhood of Potchefstroom.—ED.]

246. Certhilauda semitorquata, Smith. Grey-collared Lark.

Certhilauda semitorquata, Smith's Zool. of S. Africa, pl. 106. fig. 2.
  „   „   Layard's Cat. No. 436.
  „   „   Chapman's Travels in S. Afr., App. p. 400.
  „   „   Sharpe's Cat. No. 677.

These Larks are not uncommon in the middle and southern parts of Great Namaqua Land; and I have also observed a considerable number of them together, between Wilson's Fountain and Tjobis, in Damara Land.

This species feeds on insects, seeds, &c. The legs are flesh-coloured grey, the toes horn-colour; the bill horn-colour, tinged with a livid hue on the lower mandible.

Measurements of a female:—

	in.	lin.
Entire length	8	1
Length of folded wing	4	3
„ tarsus	1	1
„ middle toe	0	8
„ tail	3	3
„ bill	1	0

[The specimen of which the measurements are given above by Mr. Andersson is now in the collection of Mr. R. B. Sharpe. *Certhilauda subcoronata*, of Smith's Zool. pl. 90. fig. 2, is probably not distinct from this species, with which Messrs. Verreaux, Layard, and Sharpe inform me that they identify it.—ED.]

MUSOPHAGIDÆ.

247. Colius capensis, Gmel. White-backed Coly.

Le Coliou à dos blanc, Levaillant's Ois. d'Afr. pl. 257.
Colius erythropus, Strickland & Sclater, Birds Damar., Contr. Orn. 1852, p. 151.
„ „ Layard's Cat. No. 447.
Colius capensis, Sharpe's Cat. No. 97.

This species is somewhat widely diffused over Great Namaqua Land and Southern Damara Land, through a portion of the Lake-country, and in the valley of the river Okavango. It is gregarious in its habits, being found in flocks by day and also when roosting at night. Its flight is short and feeble, seldom extending beyond the nearest bush or tree, on reaching which it usually perches on one of the lower branches and then gradually glides and creeps upwards through the foliage, using both bill and feet for that purpose. It is essentially a fruit-eating bird; but I believe, when hard pressed for its regular food, it does not despise insects and the young shoots of plants. Its flesh is palatable. On September 26th, 1866, I obtained three eggs of this Coly from a nest built in a small bush, and composed externally of grass and twigs, lined internally with softer grass; the eggs were white, but dull. On October 16th I met with several nests of this bird on the same tree; but all contained young, invariably three in number. On December 29th I obtained a nest with three eggs.

The iris in this species is intensely dark brown; the bill bluish white, except the tip of the upper mandible, which is bluish black; the lower mandible is lighter than the upper; the legs are bright red.

248. Colius erythromelon, Vieill. Quiriwa Coly.

Le Coliou Quiriwa, Levaillant's Ois. d'Afr. pl. 258.
Colius erythromelon, Vieillot's Nouv. Dict. vol. vii. p. 378.
Colius maerurus and *senegalensis*, Strickland & Sclater, Birds Damar., Contr. Orn., 1852, p. 151.
Colius capensis, Layard's Cat. No. 448.
Colius erythromelas, Finsch & Hartlaub's Vögel Ost-Afrika's, p. 469.
Colius maerurus, Chapman's Travels in S. Afr., App. p. 403.
Colius erythromelon, Sharpe's Cat. No. 100.

This species is scarce in Damara Land; but I met with it about the Swakop River, and also near Okambute, at Objimbinque, and near Ojipatera. It lives principally on the shoots and fruit of a bush of a yellowish-green colour bearing red berries, and resembles in its habits the preceding species. The males are larger and handsomer than the females. The bill is black anteriorly, and posteriorly red, which is also the colour of the lore and naked space about the eye.

Measurements of a male and a female:—

	Male.		Female.	
	in.	lin.	in.	lin.
Entire length	13	7	13	0
Length of folded wing	3	8	3	7
,, tarsus	0	9	0	9
,, middle toe	0	8	0	7
,, tail	9	0	8	7
,, bill	0	8	0	8

[I have examined, at the Museum of Zoology at Cambridge, the Coly from Damara Land cited by Messrs. Strickland and Sclater (*loc. cit.*), and find that it unquestionably belongs to this species; probably it was in consequence of this example having been misnamed that Dr. Hartlaub quotes Damara Land as a locality for *Colius senegalensis,* in his 'Ornithologie Westafrika's,' p. 155.—ED.]

249. Schizorhis concolor (Smith). Whole-coloured Plantain-eater.

Chizærhis concolor, Smith's Zool. of S. Africa, pl. 2.
Schizorhis concolor, Layard's Cat. No. 452.
Chizærhis concolor, Chapman's Travels in S. Afr., App. p. 403.
Schizorhis concolor, Finsch & Hartlaub's Vögel Ost-Afrika's, p. 479.
   ,,   ,,   Sharpe's Cat. No. 94.

This is one of the commonest birds in Damara Land; and it is also found abundantly in the Lake-regions and at the river Okavango. It is partial to localities abounding in large trees; and when prominently perched, with crest erect, it is not unlike a gigantic Coly; it also climbs and flies like the Colies, which it strongly resembles in its general habits. It is usually found in small flocks and feeds on berries and seeds, especially those of a mistletoe and of other parasitical plants, and also on fruits, young shoots, and insects. The Damaras call this species "Ongoro-oroquēna," from the extraordinary and almost human cry with which it frequently startles the traveller who is passing near its perch. It is sometimes very easy to approach these birds, whilst at other times they are so shy that they will defy the utmost exertions which may be made to obtain them.

On January 5th, 1867, I obtained three eggs of this species, of a dull bluish-white colour, at Omapju, from a boy, who told me that the nest which contained them was composed of sticks roughly put together, and situated in a tree at some distance from the ground; and on March 1st I met with a nest in Ondonga placed in a tree, but at no great elevation, which also contained three eggs, but much incubated.

The iris is brown, the legs are purplish black.

Measurements of a male:—

	in.	lin.
Entire length	19	2
Length of folded wing	8	·5
„ tarsus	1	8
„ middle toe	1	6
„ tail	9	9
„ bill	1	2

I think I have seen another species of this genus to the north of Damara Land, in Ondonga, and near the river Okavango.

BUCEROTIDÆ.

250. Bucorvus abyssinicus (Gmel.). Ground-Hornbill.

Le Calao caronculé, Levaillant's Ois. d'Afr. pls. 230, 231, 232.
Bucorvus abyssinicus, Layard's Cat. No. 458.

Tmetoceros abyssinicus, Finsch & Hartlaub's Vögel Ost-Afrika's, p. 480.
Bucorvus abyssinicus, Blanford's Geol. & Zool. of Abyssinia, p. 330.

This species is common in Ondonga, but very wild; and I have also observed it sparingly in the desert near the Okavango.

The ground seems to be the chief resort of these birds, and I have seen them running at a tremendous rate; but they also perch on trees when flushed. They utter at times an incessant kind of booming cry, which may be heard a long way off. They are generally seen in small flocks of from three to six individuals.

The Ovampos seem to have a superstition about this curious bird, which I cannot fathom. On asking Chijkongo, for instance, to get me the eggs, he said it was not to be done, as they were soft to the touch, and would fall in pieces on the least handling!

[Professor Schlegel, in his 'Muséum des Pays-Bas,' *Buceros*, p. 19, expresses his opinion that the Abyssinian, the West-African, and the South-African races of this bird are always distinguishable from each other; but Drs. Finsch and Hartlaub (*loc. cit.*) maintain the contrary view. I have not had the opportunity of comparing a sufficient number of specimens from these various quarters to enable me satisfactorily to offer an opinion on the subject. Mr. Layard (*loc. cit.*) refers to a native superstition respecting this bird, probably somewhat similar to that alluded to by Mr. Andersson.—ED.]

251. Tockus nasutus (Linn.). Tock Hornbill.

Le Calao nasique, Levaillant's Ois. d'Afr. pls. 236 & 237.
Buceros nasutus, Strickland & Sclater, Birds Damar., Contr. Orn. p. 155.
Buceros epirhinus, Sundevall, Öfvers. 1850, p. 108.

Buceros nasutus, Layard's Cat. No. 457.
Buceros hastatus, Chapman's Travels in S. Afr., App. p. 405.
Buceros nasutus, Finsch & Hartlaub's Vögel Ost-Afrika's, p. 486.
„   „   Finsch, in Trans. Zool. Soc. vol. vii. p. 277.
„   „   Sharpe's Cat. No. 73.

This species does not occur in Great Namaqua Land, but is found in Southern Damara Land, where, however, it is far from common, and excessively shy; in the northern parts of Damara Land it is less shy and very common; and it is also to be met with in the Lake-regions. It is seen in small families, rarely exceeding half a dozen individuals. It roosts on large trees if such be within reach, generally returning nightly to a fixed roosting-place; it usually perches upon trees about halfway up, and, unlike *T. melanoleucus* and *T. flavirostris*, rarely alights on the topmost boughs. In common with the rest of the genus it appears to suffer very much from the heat during the most trying season of the year, when it may be found perched at noon in the shadiest part of the forest, gasping as if for breath, and may then be approached and shot much more easily than at other times. When on the wing it occasionally utters short piercing cries.

This Hornbill is almost omnivorous; but its principal food consists of berries, young shoots, and insects.

The irides are reddish brown; the legs brownish black; the bill in the adult bird black, with a large triangular yellowish-white patch on the upper mandible, and a few narrow transverse bars of the same colour on the lower. In the young bird these transverse bars are also visible, and the bill is more or less black, with the basal part yellowish white, and the tips of both mandibles reddish.

252. Tockus melanoleucus (Licht.). Crowned Hornbill.

Le Calao couronné, Levaillant's Ois. d'Afr. pls. 234 & 235.
Buceros coronatus, Layard's Cat. No. 453.
,, ,, Chapman's Travels in S. Afr., App. p. 405.
Buceros melanoleucus, Finsch & Hartlaub's Vögel Ost-Afrika's, p. 485.
Tockus melanoleucus, Sharpe's Cat. No. 78.

The food of this Hornbill consists of beetles and lizards.

The iris is yellow, the tarsus brown, the bill reddish yellow.

Measurements of a male:—

	in.	lin.
Entire length	21	0
Length of folded wing	10	9
,, tarsus	1	6
,, middle toe	1	0
,, tail	9	3
,, bill along the curve	3	5

[Mr. Andersson has not recorded in his MS. the localities in which he met with this species; but in the appendix to Mr. Chapman's Travels (*loc. cit.*) it is spoken of as "not very abundant in Damara Land." An example from Ovampo Land was contained in Mr. Andersson's last collection.—ED.]

253. Tockus Monteiri, Hartl. Monteiro's Hornbill.

Toccus Monteiri, Hartlaub, in Proc. Zool. Soc. 1865, pl. 5.
Tockus Monteiri, Sharpe's Cat. No. 76.

This Hornbill is not very abundant in Damara Land; it is usually seen in pairs; but occasionally half a dozen individuals may be found in close proximity to one another. It is a shy and wary bird, and difficult to approach, except on hot days, when it appears to suffer a good deal from the heat. About 8 or 9 o'clock in the morning it may often be observed quietly resting on the top of a tree; and it will also perch in such situations at other times when alarmed, but takes its departure again on the least sign of danger. It seldom flies far at a time, but if closely pursued extends its flight each time it is dislodged, and thus soon distances its enemy. The flight of this and other Hornbills is not unlike that of a Woodpecker, dipping and rising alternately.

The present species feeds on flowers, young shoots, berries, birds' eggs, and insects; and, in fact, little comes amiss to it. I have found much gravel in its stomach, and have often flushed it from the ground, to which it resorts for the purpose of picking up sand as well as food.

The irides are nut-brown; the legs and toes brown horn-colour; the bill, which is much longer, broader, and stronger in the male than in the female bird, is yellowish red, darkest towards the extremities of the mandibles, which are dark purple, that tint being also sometimes apparent on other parts of the bill as well.

Average dimensions of fourteen males and of seven females:—

		Males.		Females.	
		in.	lin.	in.	lin.
Entire length		23	3	21	0
Length of folded wing		8	10	7	8
,,	tarsus	2	1	2	0
,,	middle toe	1	8	1	6
,,	tail	9	4	8	8
,,	bill	4	10	3	10

254. Tockus flavirostris, Rüpp.   Yellow-billed Hornbill.

Buceros flavirostris, Rüppell, Neue Wirbelth. pl. 2. fig. 1.
Toccus elegans, Hartlaub, in Proc. Zool. Soc. 1865, pl. 4.
Buceros flavirostris, Chapman's Travels in S. Afr., App. p. 405.
    „   „   Finsch & Hartlaub's Vögel Ost-Afrika's, p. 490.
    „   „   Finsch, in Trans. Zool. Soc. vol. vii. p. 279.
Tockus flavirostris, Sharpe's Cat. No. 77.

This species is the most common of the Hornbills in the middle and southern parts of Damara Land. It is found singly or in pairs, and, being a comparatively fearless bird, is easily killed, especially during the heat of the day, when it invariably perches on or near the top of a lofty tree (where such are to be found), and will remain for hours in this situation, keeping up, with short intermissions, a kind of subdued chattering note of Tŏc Tŏc Tŏc Tŏckĕ Tŏckĕ Tŏckĕ Tŏc, in a tone not unlike the quick yelpings of young puppies, and accompanied at intervals by a flapping and raising of its wings and an alternate lowering and erecting of its head.

There is a considerable difference in the size of adult birds of this species.

The irides are yellow; the legs and toes are very dark brown; the colour of the bill approaches orange-yellow, with the exception of the edges, upper ridge, and the tips of the mandibles, which are reddish brown; in the young bird the bill is sometimes very dark-coloured.

255. Tockus erythrorhynchus (Gmel.). Red-billed Hornbill.

Le Calao Toc, Levaillant's Ois. d'Afr. pl. 238.
Buccros rufirostris, Sundevall, Öfvers. 1850, p. 108.
Buceros erythrorhynchus, Sundevall, Öfvers. 1850, p. 130.
,, ,, Layard's Cat. No. 456.
,, ,, Chapman's Travels in S. Afr., App. p. 406.
,, ,, Gurney, in Ibis, 1869, p. 296.
,, ,, Finsch & Hartlaub's Vögel Ost-Afrika's, p. 491.
Tockus erythrorynchus, Gray's Hand-list of Birds, No. 7893.
Tockus leucomelas, Gray's Hand-list of Birds, No. 7895.
Tockus erythrorhynchus, Sharpe's Cat. No. 75.

This species is common in Ondonga, at the Okavango River, and for some distance to the south of that stream; and I have obtained specimens from Lake Ngami. I have also met with it in Damara Land proper, at Objimbinque and Schmelen's Hope; but specimens from these two last-named localities differ considerably from those found in more northern parts. Thus in the former the whole of the underparts, the forehead, a broad band above the eyes continued down the sides of the neck, the ears, cheeks, chin, and throat are of a uniform silky white; whilst in the more northern bird the colour of these parts, as well as of the breast, is mingled with blackish grey, and there is also less white about the wings and tail.

This Hornbill is frequently seen searching for food upon the ground; and the way in which it swallows some kinds of food is peculiar, raising its head and pitching the morsel into the air, receiving it again into its bill, and repeating the process several times, perhaps with the object of softening the food or reducing it to a pulp.

As regards the breeding-habits of this species, my own experience fully agrees with the very interesting account given by Dr. Livingstone in 'Missionary Travels,' p. 613.

[With regard to the difference between the white-cheeked and grey-cheeked specimens of this Hornbill, the examples contained in Mr. Andersson's collection fully bear out his remarks above recorded. Two of these examples (a white-cheeked bird from Objimbinque, and a grey-cheeked one from Ovampo Land) are preserved in the collection of Mr. R. B. Sharpe; and as both of these were marked as males by Mr. Andersson (no doubt from dissection), and as both of them, from the character of their bills, are evidently adult birds, it follows that the difference is not due either to age or to sex.

Professor Sundevall, from his remarks at p. 130 of the 'Öfversigt' for 1850, appears to consider the grey plumage of the cheeks to be especially characteristic of what he considers the distinct Caffrarian race, for which he has proposed the specific name of "*rufirostris.*" On the other hand, M. Jules P. Verreaux informs me that he considers that both birds are of one and the same species, and that the pure white on the cheeks and the parts adjacent is a nuptial dress annually assumed and lost by a double seasonal change.

Which of these opinions is the correct one can probably be only ascertained by further observations on these birds in a state of nature than have yet been made; and it is to be hoped that those who may have the opportunity of making the requisite investigations will give their attention to the elucidation of this somewhat obscure variation.—Ed.]

SCANSORES.

PSITTACIDÆ.

256. Poicephalus robustus (Gmel.). Levaillant's Parrot.

Le Perroquet à franges souçi, Levaillant's Perroquets, pl. 130.
Psittacus Levaillantii, Layard's Cat. No. 459.
Poicephalus robustus, Gray's Hand-list of Birds, No. 8282.
Psittacus robustus, Sharpe's Cat. No. 173.

I have only met with this Parrot in the country of Ovaquenyama, where it is very abundant, but very wild and difficult to approach; and in fact it is only to be obtained in the morning and evening, when it comes to the water during the dry season.

The iris is dark brown; the bill livid, but brown at the extremity; the tarsi are lead-coloured.

Measurements of a male and a female:—

	Male. in. lin.	Female. in. lin.
Entire length	12 9	12 0
Length of folded wing	9 0	8 6
„ tarsus	0 10	0 9
„ middle toe	2 10	2 10
„ tail	4 1	3 8

[Mr. R. B. Sharpe possesses two specimens of this Parrot obtained by Mr. Andersson, one in Damara Land and one in Ondonga.—ED.]

257. Poicephalus Meyeri (Rüpp.). Meyer's Parrot.

Psittacus Meyeri, Rüppell's Atlas, pl. 11.
Pœocephalus Meyeri, Strickland & Sclater, Contr. Orn. 1852, p. 156.
Psittacus Meyeri, Chapman's Travels in S. Afr., App. p. 406.
Pionias Meyeri, Finsch & Hartlaub's Vögel Ost-Afrika's, p. 500.
Poicephalus Meyeri, Gray's Hand-list of Birds, No. 8284.
Psittacus Meyeri, Sharpe's Cat. No. 170.

This Parrot is a rare bird in the middle portion of Damara Land; but further north, at Okamabuté, it is common, and also in the Lake-regions; with the occasional exception of a few stray individuals, it does not occur further south than Omabondé, which may be said to constitute its usual southern limit. Its habits and notes closely resemble those of the succeeding species, *P. Rüppelli.*

The irides are deep bright red-orange; the bare skin surrounding the eye is black; the bill darkish horn-colour, with a tinge of green; the legs and toes greenish black.

Measurements of a male and a female:—

	Male.		Female.	
	in.	lin.	in.	lin.
Entire length	9	4	9	0
Length of folded wing	6	1	6	1
,, tarsus	0	8	0	9
,, middle toe	1	7	1	9
,, tail	3	0	3	0
,, bill	0	9	0	9

[This species is subject, in both sexes, to a variation of colour in the plumage of the head, an irregular bar of pale yellow, which transversely crosses the crown, being present in some specimens and entirely absent in others.—ED.]

258. Poicephalus Rüppelli (Gray). Rüppell's Parrot.

Psittacus Rüppelli, Gray, in Proc. Zool. Soc. 1848, pl. 5.
Pœocephalus Rüppelli, Strickland & Sclater, Birds Damar., Contr. Orn. 1852, p. 156.
Psittacus Rüppelli, Chapman's Travels in S. Afr., App. p. 406.
Phæocephalus Rüppellii, Hartl. Orn. West-Africa's, No. 503.
Poicephalus Rüppellii, Gray's Hand-list of Birds, No. 8286.
Psittacus Rueppelli, Sharpe's Cat. No. 169.

This species is common in Damara Land, but is

chiefly found in the middle and southern parts of that country; it is always met with in small flocks of about half a dozen individuals, and seems to prefer the larger kinds of trees. It is rather shy, and when quietly perched amongst the branches is very difficult to perceive, until its presence is betrayed by the cries it utters as soon as it conceives itself to be in danger; these are at first shrill and isolated, but increase in strength and frequency till it leaves its perch, and are usually continued during its flight, which is generally short, but very rapid. It is rarely found far from water, which it usually frequents twice a day. It feeds on seeds and berries, sometimes also on the young shoots of trees and plants.

Messrs. Strickland and Sclater state that the blue on the upper and under tail-coverts is wanting in the female; but I have now lying before me more than one individual of that sex in which this blue plumage is present. I have, however, examined others in which it was absent, but which, in all other respects, appeared to be adult; whilst in other quite young specimens I have found the rump and the upper tail-coverts nearly as blue as in the adult; and I have also specimens in which the blue is present both above and below, but which have none of the usual orange-yellow on the elbow or on the thigh, but only on the under wing-coverts. Whether or not these variations are merely accidental, I have been unable to decide.

The iris in this species is orange, the bill, feet, and toes dark horn-colour.

Measurements of a female:—

	in.	lin.
Entire length	9	0
Length of folded wing	.5	7
,, tarsus	.0	7
,, tail	.2	10
,, bill	.0	9

259. Psittacula roseicollis (Vieill.). Rosy-necked Love-bird.

Psittacula roseicollis, Bourjot's Perr. pl. 91.
Agapornis roscicollis, Strickland & Sclater, Birds Damar., Contr. Orn. 1852, p. 156.
Psittacula roseicollis, Layard's Cat. No. 461.
Agapornis roseicollis, Chapman's Travels in S. Afr., App. p. 406.
Psittacula roseicollis, Finsch & Hartlaub's Vögel Ost-Afrika's, p. 501.
,,   ,,   Sharpe's Cat. No. 177.

This pretty little species is very generally distributed over Damara and Great Namaqua Land, and is also found on the Okavango and at Lake Ngami. It is always observed in small flocks, and seldom far from water, to which it resorts at least once in the day, and is consequently not a bad guide to a thirsty traveller; though, if he be inexperienced, it would hardly avail him much, as it frequently happens that the drinking-places resorted to by this and other water-loving birds are of but small compass and strangely situated.

This species is very swift of flight, and the little flocks in which it is observed seem to flash upon the sight as they change their feeding-grounds or pass to or from their drinking-places; their flight, however, is only for a comparatively short distance at a time. They utter rapid and shrill notes when on the wing, or when suddenly disturbed or alarmed. Their food consists of berries and large berry-like seeds.

This bird does not make any nest of its own, but takes possession of nests belonging to other birds, especially *Philetærus socius* and *Plocepasser mahali*. I cannot say whether it forcibly ejects the rightful owners of these nests, or merely occupies such as they have abandoned; but in the case of the first-named species, I have seen the Parrots and the Grosbeaks incubating in about equal numbers under the shelter of the same friendly roof. The egg is pure white, not unlike a Woodpecker's, but more elongated. The irides are of an intensely dark brown; the legs blue, with the faintest tint of green; the bill is greenish white.

Measurements of a male and a female :—

	Male. in. lin.	Female. in. lin.
Entire length	6 10	6 8
Length of folded wing	4 1	4 1
,, tarsus	0 7	0 7
,, middle toe	Not recorded.	1 3
,, tail	2 0	2 1
,, bill	0 6	0 7

CAPITONIDÆ.

260. Pogonorhynchus leucomelas (Bodd.). Black-throated Barbican.

Laimodon leucomelas, Layard's Cat. No. 463.
Laimodon unidentatus, Layard's Cat. No. 464.
Pogonias leucomelas, Chapman's Travels in S. Afr., App. p. 405.
Pogonorhynchus leucomelas, Sharpe's Cat. No. 135.
,, ,, Marshalls' Mon. of the Capitonidæ, pl. 62.

This peculiar but prettily marked bird is found from the northern border of Great Namaqua Land (south of which I did not meet with it) as far north as the river Okavango; and it also occurs at Lake Ngami. In the neighbourhood of Objimbinque it is rather abundant. It is found singly or in pairs, and is remarkable for its clear, ringing and far-sounding notes, which, though heard at all hours of the day, are most frequent in the early morning. Its food consists chiefly of fruit and seeds; but it will, to some extent, accommodate itself, as regards food, to the produce of the locality in which it happens to be located. It is rather a lively bird, and sometimes suspends itself below the fruit on which it is feeding, and makes its repast whilst hanging in that position.

The iris is umber-brown, the bill dark brownish horn-colour, the feet and toes slaty brown.

Measurements of a male and a female:—

	Male.	Female.
	in. lin.	in. lin.
Entire length	6 6	6 5
Length of folded wing	3 2	3 1
,, tarsus	0 11	0 10
,, middle toe	Not recorded.	1 3
,, tail	2 0	1 11
,, bill	0 11	0 10

PICIDÆ.

261. Thripias namaquus (Licht.). Bearded Woodpecker.

Le Pic à double moustache, Levaillant's Ois. d'Afr. pl. 251 (male) & 252 (female).
Dendrobates namaquus, Strickland & Sclater, Birds Damar., Contr. Orn. 1852, p. 155.
Dendropicus biarmicus, Malherbe's Monographie des Picidées, pl. 42. figs. 4, 5, & 6.
Dendrobates namaquus, Layard's Cat. No. 471.
Picus namaquus, Finsch & Hartlaub's Vögel Ost-Afrika's, p. 507.
Thripias namaquus, Gray's Hand-list of Birds, No. 8664.
Dendropicus namaquus, Sharpe's Cat. No. 160.

I have scarcely ever seen this Woodpecker in Great Namaqua Land, but have found it (very sparingly) throughout Damara Land, and as far north as the river Okavango, where, and at Lake Ngami, it is more numerous than in Damara Land, though no species of Woodpecker can be said to be common in any of the countries the ornithology of which is described in these pages.

The iris is of a red wine-colour, the bill greenish grey, and the legs and toes greyish olive.

Measurements of a male:—

	in.	lin.
Entire length	9	5
Length of folded wing	5	5
,, tarsus	0	11
,, middle toe	1	6
,, tail	3	1
,, bill	1	8

262. Dendropicus Hartlaubii, Malh. Hartlaub's Woodpecker.

Dendropicus Hartlaubii, Malherbe's Monographie des Picidées, pl. 44. figs. 1 & 2.
Picus Hartlaubi, Sundevall's Consp. Av. Pic. p. 43.

Picus Hartlaubi, Finsch & Hartlaub's Vögel Ost-Afrika's, p. 512.
Dendropicus Hartlaubii, Gray's Hand-list of Birds, No. 8651.

[I have not found this species amongst those collections of Mr. Andersson's which have come under my notice; nor do I find that it is mentioned in Mr. Andersson's MS. notes; but Professor Sundevall (*loc. cit.*) mentions a specimen obtained by the late Professor Wahlberg, at Walwich Bay, on the 27th of April, 1854; and Drs. Finsch and Hartlaub (*loc. cit.*) refer to a specimen procured in Damara Land by Mr. Andersson, and now preserved in the Museum at Bremen.—ED.]

263. Dendropicus cardinalis (Gmel.). Cardinal Woodpecker.

Le petit Pic à baguettes d'or, Levaillant's Ois. d'Afr. pl. 253.
Dendrobates fuscescens, Strickland & Sclater, Birds Damar., Contr. Orn. 1852, p. 155.
Dendropicus fulviscapus, Malherbe's Monographie des Picidées, pl. 43. figs. 1, 2, 3.
  „   „   Layard's Cat. No. 472.
Picus fulviscapus, Chapman's Travels in S. Afr., App. p. 405.
Dendropicus cardinalis, Gray's Hand-list of Birds, No. 8649.
  „   „   Sharpe's Cat. No. 164.

This pretty little Woodpecker, though it cannot be said to be abundant, is the commonest of all those found in Damara and Great Namaqua Land, and is also tolerably numerous at Lake Ngami. It is a comparatively tame species, and is sometimes found singly, but more often in pairs; it usually frequents trees of moderate size, situated in the more scanty woods or on the banks of the periodical watercourses.

The sexes in this species do not differ in size. The iris is red; the bill greenish slaty; the legs and toes green, tinged with slate-colour.

[A specimen of this Woodpecker, procured by Mr. Andersson at Objimbinque, is in the collection of Mr. R. B. Sharpe.—ED.]

264. Ipagrus capricorni, Strickl.   Capricorn-Woodpecker.

Campethera capricorni, Strickland & Sclater, Birds Damar., Contr. Orn. 1852, p. 155 (male).
  „   „   Newton, in Ibis, 1869, pl. 9 (male).
Dendrobates nigrogularis, Bocage, Jorn. Acad. Lisb. 1867, p. 336 (female).
Ipagrus capricorni, Gray's Hand-list of Birds, No. 8700.
Campethera capricorni, Sharpe's Cat. No. 149.

I never met with this species in Great Namaqua Land; and in Damara Land proper it is scarce. I do not remember to have seen it much south of Omanbondé; but on my journey to the Okavango I found it more frequent in the neighbourhood of that river, though even there it was of comparatively rare occurrence. It appears to be a migratory bird, as I never saw it during the dry season.

The iris is claret-coloured, the bill a brown slate-colour, the legs and toes lead-coloured.

[A specimen obtained by Mr. Andersson at the river Cunéné is in the collection of Mr. R. B. Sharpe; this example is a female, which, in this species, resembles the male, with the following exceptions, viz.:—only the occiput is scarlet, the remainder of the upper part of the head being black, very thickly interspersed with large white spots; a brownish-black stripe runs from the gape under the eye and ear, down the side of the neck, where it becomes intermingled with white, this stripe is broadest below the ear; lastly, the space between the rami of the lower mandible is also brownish black, which colour extends from the chin for about an inch down the throat.—ED.]

265. Ipagrus Brucei (Malh.).   Bruce's Woodpecker.

Campethera Abingoni, ♀, Strickland & Sclater, Birds Damar., Contr. Orn. 1852, p. 156.
Chrysopicus Brucei, Malherbe's Monographie des Picidées, pl. 93. fig. 1 (male).

Ipagrus Brucei, Gray's Hand-list of Birds, No. 8696.
Campethera Brucii, Sharpe's Cat. No. 153.

This species is not unfrequent in Damara Land. The iris is a pink wine-colour, the bill bluish brown, the legs and toes greyish green.

Measurements of a female:—

	in. lin.
Entire length . .	8 10
Length of folded wing	5 5
„ tarsus . .	0 9
„ middle toe	1 6
„ tail	3 0
„ bill	1 6

[Specimens of this Woodpecker were more numerous in Mr. Andersson's collections than those of any other species. It appears to me that the female of this species agrees more accurately than any other with that described by Messrs. Strickland and Sclater (*loc. cit.*), under the designation of "*Campethera Abingoni*, Smith, ♀," which name, however, is in reality a synonym of the succeeding species, and therefore not correctly applied to the present bird.

I endeavoured to refer to Mr. Strickland's original specimen when I visited the Museum of Zoology at Cambridge, in which his collection is now preserved, but I did not succeed in discovering it there.—Ed.]

266. Ipagrus variolosus (Licht.). Bennett's Woodpecker.

Chrysopicus Abingoni, Malherbe's Monographie des Picidées, pl. 95. figs. 1 & 2.
Picus Bennetti, Sundevall's Consp. Av. Pic. p. 63.
Campethera chrysura, Layard's Cat. No. 474.
Chrysoptilus Bennetti, Gurney, in Ibis, 1869, p. 296.
Ipagrus variolosa, Gray's Hand-list of Birds, No. 8693.
Campethera Bennetti, Sharpe's Cat. No. 154.

[Mr. Andersson does not mention this Woodpecker in his notes; but a specimen, marked as being from the "Lake-regions," was contained in his last collection, and is the most westerly example which I have seen of this species. This specimen is now in the collection of Mr. R. B. Sharpe.—ED.]

CUCULIDÆ.

267. Indicator minor, Steph. Little Honey-Guide.

Le petit Indicateur, Levaillant's Ois. d'Afr. pl. 242.
Indicator minor, Layard's Cat. No. 480.
    ,,   ,,   Finsch & Hartlaub's Vögel Ost-Afrika's, p. 515.
    ,,   ,,   Sharpe's Cat. No. 126.

This species is met with sparingly in Damara Land throughout the year; and I have also observed it in Great Namaqua Land. It occurs either singly or in pairs, and feeds on small bees, ants, &c. The stomach also invariably contains a white substance surrounding the other food; but what this substance is I cannot make out.

The iris is very dark brown; the upper mandible and the extremity of the lower are dark horn-colour, the rest of the lower mandible is livid; the legs and toes are lead-coloured, tinged with green in front.

Measurements of a male:—

		in.	lin.
Entire length		6	4
Length of folded wing		3	8
,,	tarsus	0	7
,,	middle toe	0	7
,,	tail	2	6
,,	bill	0	7

[According to Mr. Layard and Drs. Finsch and Hartlaub (*loc. cit.*), the Honey-Guides, like the Cuckoos, lay their eggs in the nests of other birds—a circumstance which confirms the correctness of the arrangement by which they are classified as an aberrant family of the *Cuculidæ*, from the other divisions of which they, however, in some respects differ materially.

The white substance found by Mr. Andersson in the stomach of the Little Honey-Guide appears, from the information given by Mr. Layard (*loc. cit.*), to have been due to the pollen collected by the bees on which the birds had fed.

Mr. Andersson alludes to the allied species, *Indicator major*, as not uncommon in the eastern portion of the Cape Colony, and states that it sometimes falls a prey to the attacks of the bees whose nests it seeks, in consequence of these insects " fastening on and about the eyes of the bird."—ED.]

268. Centropus senegalensis (Linn.). Senegal Spur-heeled Cuckoo.

Le Coucal Houhou, Levaillant's Ois. d'Afr. pl. 219.
Centropus Burchellii, Layard's Cat. No. 487.
Centropus senegalensis, Finsch & Hartlaub's Vögel Ost-Afrika's, p. 526.

This species is found abundantly at Lake Ngami, but I have observed it nowhere else. It occurs singly or in pairs, and perches on lofty trees—but also frequents reedy thickets, to the interior of which it usually retreats when alarmed or pursued. Its flight is heavy and clumsy. Its food consists of insects; and it is partial to locusts and grasshoppers.

[Mr. Andersson identified the *Centropus* to which the above remarks refer with *C. Burchellii* of Swainson, which, according to Drs. Finsch and Hartlaub (*loc. cit.*), is not specifically separable from *C. senegalensis*. I am unable to confirm this identification from personal observations, as no specimens of the bird were contained in Mr. Andersson's last collection, neither have I had the opportunity of comparing a sufficient number of specimens from other sources to be able to satisfy myself as to whether or not *C. Burchellii* is really distinct from *C. senegalensis*; and as this is the case, I have here followed the nomenclature adopted by Drs. Finsch and Hartlaub, as above referred to.—ED.]

269. Coccystes glandarius (Linn.). Great Crested Cuckoo.

Oxylophus glandarius, Gould's Birds of Europe, pl. 241.
„ „ Gurney, Birds Damar., Proc. Zool. Soc. 1864, p. 3.
„ „ Layard's Cat. No. 496.
Coccystes glandarius, Finsch & Hartlaub's Vögel Ost-Afrika's, p. 518.
„ „ Sharpe's Cat. No. 108.

This species is not uncommon during the wet season in Damara Land, and also about the river Okavango.

The iris is pale brown; the bill brown, tinged with yellow on the posterior portion of the lower mandible; the legs and toes bluish, with a shade of brown.

270. Oxylophus jacobinus (Bodd.). Black-and-White Cuckoo.

Le Coucou edolio (femelle), Levaillant's Ois. d'Afr. pl. 208.
Oxylophus melanoleucus, Layard's Cat. No. 499.
Oxylophus jacobinus, Gray's Hand-list of Birds, No. 9062.
Coccystes jacobinus, Sharpe's Cat. No. 109.

This is about the most common Cuckoo in Damara Land, and the first to arrive with the rainy season; I also received several specimens of it from Lake Ngami. I believe it breeds in Damara Land, having seen young birds barely able to fly.

This species has a true Cuckoo's note, and is very swift of flight and quick in its movements.

271. Oxylophus caffer (Licht.). Levaillant's Cuckoo.

Le Coucou edolio, var., Levaillant's Ois. d'Afr. pl. 209.
Coccyzus Vaillantii, Swainson's Zool. Ill. 2nd ser. pl. 13.
Oxylophus afer, Layard's Cat. No. 500.

Oxylophus afer, Finsch, in Trans. of Zool. Soc. vol. vii. p. 285.
Oxylophus caffer, Gray's Hand-list of Birds, No. 9066.
Coccystes caffer, Sharpe's Cat. No. 110.

Like the rest of the Cuckoos found in Damara Land, this species is only a periodical visitant during the rainy season, and takes its departure long before the return of the dry weather. I have heard its note on the river Okavango as early as September; but it is a scarce bird in that district, and very rare in Damara Land. It always appears in pairs, and is of a rather shy and retiring disposition.

The iris is dark olive-brown; the bill black; the legs and toes brown tinged with lead-colour.

272. Oxylophus serratus (Sparrm.). Edolio Cuckoo.

Le Coucou edolio (mâle), Levaillant's Ois. d'Afr. pl. 207.
Oxylophus edolius, Layard's Cat. No. 498.
Oxylophus serratus, Gray's Hand-list of Birds, No. 9064.
Coccystes serratus, Sharpe's Cat. No. 111.

[Mr. Andersson's MS. notes do not refer to this species as inhabiting Damara Land; but two examples of it, obtained by him in that country, are in the collection of Mr. R. B. Sharpe. —Ed.]

273. Cuculus clamosus, Cuv. Noisy Cuckoo.

Le Coucou criard, Levaillant's Ois. d'Afr. pls. 204 & 205.
Cuculus clamosus, Gurney, Birds Damar., Proc. Zool. Soc. 1864, p. 3.
„   „   Layard's Cat. No. 492.
„   „   Chapman's Travels in S. Afr., App. p. 408.

I first observed this species in the neighbourhood of the river Okavango, but only very sparingly; and the few individuals which there came under my notice were so excessively wild and wary that I only succeeded in bagging some after an immense deal of trouble and smart shooting. They were invariably perched on lofty trees, where they uttered loud cries, which were my only guide to their whereabouts; and before I got near they would leave their perches and dart with lightning speed through the neighbouring thickets. They were wild shots these! On a subsequent occasion I observed a flock of fully a dozen of these Cuckoos, creating a desperate hubbub, on the 21st of December, at Objimbinque. I have also obtained specimens of this Cuckoo from Lake Ngami.

274. Cuculus canorus, Linn. European Cuckoo.

Cuculus canorus, Gould's Birds of Europe, pl. 240.
,, ,, Chapman's Travels in S. Afr., App. p. 407.
,, ,, Finsch & Hartlaub's Vögel Ost-Afrika's, p. 520.
,, ,, Gurney, in Ibis, 1871, p. 103.
,, ,, Sharpe's Cat. No. 103.

This species is occasionally observed in Damara Land, but from its great general resemblance to the following species (*C. gularis*) may be easily confounded with that bird*.

The iris is yellow; the bill bluish black, except at the base, where it is yellowish brown; the legs and toes are bright yellow.

[Mr. R. B. Sharpe possesses two specimens of the true *Cuculus canorus,* which were both obtained by Mr. Andersson at Objimbinque, Damara Land. One of these is a male bird which has nearly attained the adult dress, but shows some remains of immature plumage on the wing-coverts and on the crown of head. This specimen was killed on the 6th of February, 1865; the other example is a fully adult female obtained on the 1st of April, 1864.—Ed.]

* See note under *Cuculus gularis* at page 228.

275. Cuculus gularis, Steph. Lineated Cuckoo.

Le Coucou vulgaire d'Afrique, Levaillant's Ois. d'Afr. pls. 200 & 201.
Cuculus lineatus, Swainson's Birds of West Afr., vol. ii. pl. 18.
Cuculus gularis, Layard's Cat. No. 491.
Cuculus lineatus, Chapman's Travels in S. Afr., App. p. 407.
Cuculus gularis, Sharpe's Cat. No. 104.

This Cuckoo is pretty common in the rainy season throughout Damara Land and in some parts of Great Namaqua Land. Its flight is very rapid and zig-zag; but it does not move far at a time, usually taking refuge, after being disturbed, in the nearest convenient tree.

The iris is yellowish brown; the tips of the mandibles dark brown, their base yellowish, shading into greenish yellow in the intermediate parts.

[The yellow or "yellowish" base of the upper mandible in this species, above referred to by Mr. Andersson, is perhaps the readiest criterion by which it may be distinguished from *Cuculus canorus*, in which the whole upper mandible is of a black horn-colour, except a very narrow yellow margin below each nostril.

Mr. R. B. Sharpe possesses a specimen of *Cuculus gularis*, obtained by Mr. Andersson at Objimbinque, and another procured by him in Ondonga.—ED.]

276. Chrysococcyx cupreus (Bodd.). Didric Cuckoo.

Le Coucou didric, Levaillant's Ois. d'Afr. pls. 210 & 211.
Chalcites auratus, Layard's Cat. No. 493.
Chrysococcyx cupreus, Finsch & Hartlaub's Vögel Ost-Afrika's, p. 522.
   „   „   Sharpe's Cat. No. 113.

This beautiful species is by no means uncommon in Little Namaqua Land; but to the north of the Orange River I have only met with it in the neighbourhood of the river Okavango, where it is both scarce and shy, and near Otniovapa, where I observed two of these birds pairing on the 11th of January.

This Cuckoo feeds on larvæ.

Measurements of a male and a female:—

	Male. in. lin.	Female. in. lin.
Entire length	7 11	7 11
Length of folded wing	4 8	4 10
„ tarsus	0 9	0 9
„ middle toe		1 4
„ tail	3 5	3 6
„ bill	1 0	1 3

277. Chrysococcyx Klaasi (Steph.). Klaas's Cuckoo.

Le Coucou de Klaas, Levaillant's Ois. d'Afr. pl. 212.
Chalcites Klaasii, Layard's Cat. No. 494.
Chrysococcyx Klaasi, Finsch & Hartlaub's Vögel Ost-Afrika's, p. 520.

I shot on the Swakop River (a day's journey from Objimbinque) a single individual of this species, which was the only one that I procured in any of the countries to the ornithology of which these pages refer.

[The specimen above referred to not having been comprised in Mr. Andersson's last collection, I am unable to confirm by personal observation his identification of it, which, however, from the detailed description contained in his MS. notes, I have no doubt is correct.

Mr. Andersson also records in his notes his belief that *C. splendidus* (*smaragdineus* of Layard's Catalogue) occurs in Damara Land during the rainy season, although he did not succeed in procuring a specimen of it in that country.—Ed.]

COLUMBÆ.

COLUMBIDÆ.

278. Phalacrotreron calva (Temm.). Bald-fronted Pigeon.

Le Pigeon à front nu, Temminck & Knip's Pigeons, pl. 7.
Phalacrotreron calva, Bonaparte's Icon. des Pigeons, pl. 3. fig. A (head).
Treron australis, Layard's Cat. No. 503.
Phalacrotreron calva, Gray's Hand-list of Birds, No. 9111.

Iris light blue; basal half of bill red, the remainder livid; feet and toes yellow.

Measurements of a male and a female:—

	Male.		Female.	
	in.	lin.	in.	lin.
Entire length	10	10	10	3
Length of folded wing	6	7	6	4
,, tarsus	0	11	0	11
,, middle toe	0	11	0	11
,, tail	4	0	3	8
,, bill	0	11	0	9

[The specimens of this species from which the above measurements were taken by Mr. Andersson were obtained in Ondonga, in November 1866, together with several other examples, one of which is now in the British Museum.

These Ondonga specimens only differ from West-African examples in having the bill very slightly more robust, and in the yellow tints of the plumage being throughout somewhat brighter.

Mr. Layard informs me that he considers the Treron recorded in his Catalogue (*loc. cit.*) as having been obtained in Damara Land by Mr. Kisch to be referable to this species, and not to *T. nudirostris*, as cited by Drs. Finsch and Hartlaub in their work on the birds of East Africa, p. 539.—ED.]

279. Stictœnas phæonotus, Gray. Roussard Pigeon.

Le Ramier roussard, Levaillant's Ois. d'Afr. pl. 265.
Columba guinea, Strickland & Sclater, Birds Damar., Contr. Orn. 1852, p. 156.
Columba guineæ, Layard's Cat. No. 505.
Columba guinea, Chapman's Travels in S. Afr., App. p. 411.
Columba trigonigera, Gurney, in Ibis, 1868, p. 164.
Columba guineensis (part.), Finsch & Hartlaub's Vögel Ost-Afrika's, p. 539.
Stictœnas phæonotus, Gray's Hand-list of Birds, No. 9262.

This Pigeon is common throughout Damara and Great Namaqua Land, and congregates in immense flocks about March, April, and May, after the breeding-season, and may then be obtained in any quantity at the expense of a little powder and shot, which these birds are well worth, as their flesh is well-tasted and gamy.

Bonaparte has endeavoured to separate the South-African bird from that of the west coast on the ground that the spots on the wings are smaller, and the colour of the lower part of the back resembles that of the head; but are these differences sufficient to warrant the establishment of two distinct species? I must say that for my own part I do not think they are.

In this Pigeon the inner ring of the iris is yellow, the outer yellowish red; the large naked spaces round the eyes, the lores, and corners of the mouth are deep red; the cere is whitish horn-colour; the bill is dark horn-colour, except the base of the lower mandible, which is livid.

Measurements of a male:—

	in.	lin.
Entire length	12	10
Length of folded wing	9	10
,, tarsus	1	0
,, middle toe	1	1
,, tail	4	8
,, bill	1	0

[I am disposed to differ from Mr. Andersson's opinion as to the propriety of uniting this Pigeon with the *S. guinea* of equatorial Africa; but at the same time I have not had the opportunity of examining a sufficient series of specimens to enable me to form a very decided opinion on the subject.

The examples contained in Mr. Andersson's last collection appeared decidedly to belong to the southern race; but I have not been able to trace the exact locality in which they were obtained, neither have I been able to identify with certainty, at the Cambridge Museum, the Damara specimen referred to by Strickland and Sclater (*loc. cit.*).—ED.]

280. **Turtur senegalensis** (Linn.). Senegal Dove.

La Tourterelle maillée, Levaillant's Ois. d'Afr. pl. 270.
Columba cambayensis, Temminck & Knip's Pigeons, pl. 45.
Turtur senegalensis, Layard's Cat. No. 511.
Columba cambayensis, Chapman's Travels in S. Afr., App. p. 411.
Turtur senegalensis, Finsch & Hartlaub's Vögel Ost-Afrika's, p. 551.
    ,,   ,, Gray's Hand-list of Birds, No. 9317.

This species is found abundantly from the Okavango River southwards throughout Damara Land and Great and Little Namaqua Land, as also at Lake Ngami.

In its habits, manners, and nidification it resembles the succeeding species; but its eggs are a trifle smaller.

281. Streptopelia damarensis (Finsch & Hartl.). Damara Dove.

La Tourterelle blonde à collier, Levaillant's Ois. d'Afr. pl. 268.
Turtur vinaceus, Strickland & Sclater, Birds Damar., Contr. Orn. 1852, p. 157.
Columba risorius (?), Chapman's Travels in S. Afr., App. p. 410.
Turtur semitorquatus, Layard, in Ibis, 1869, p. 374.
Turtur damarensis, Finsch & Hartlaub's Vögel Ost-Afrika's, p. 550.
Streptopelia damarensis, Gray's Hand-list of Birds, No. 9335.

Levaillant's figure of this Dove is very good, but the artist has evidently omitted the black streak which runs from the corner of the mouth to the eyes; the white on the under tail-coverts and tail is also scarcely enough distinguished. Levaillant considers this and another Ringdove found abundantly in the Cape Colony * as distinct, but I cannot agree with him; the size and distribution of colouring is just the same, only the colours in the Damara bird are a few shades lighter; but as this is always the case in all the land birds of the interior indigenous to that country, it cannot be admitted as a sufficient characteristic to separate species, unless extended to all others which stand in the same relation to each other.

This is the most abundant species of Dove in Damara Land and the parts adjacent. It cannot be strictly said to be a gregarious species; yet numbers are often found in close proximity both on trees and on the ground, and rise in one flock when flushed, producing a great noise by the rapid concussion of their wings above their backs.

They seek on the ground for their food, which consists almost exclusively of seeds. They build in small trees,

* *S. capicola* (Sundev.).

generally at the extremity of a bough, constructing a rough nest of a few twigs, with no lining of any kind. The eggs are two in number, of a pure white.

I have observed these Doves building on August 20th, and have found their eggs abundantly at the end of December; so that it is probable that they produce two broods in the year.

[Mr. Andersson's last collection contained specimens of this Dove, one of which was procured as far north as the river Cunéné. I have not had the opportunity of examining a sufficient series of specimens to enable me to form a confident opinion as to the propriety of separating the Damara race from the more southern *S. capicola*; but it would appear from Mr. Andersson's remarks that the difference of tint, though not great, is yet both constant and perceptible.—ED.]

282. Streptopelia semitorquata (Rupp.). Red-eyed Dove.

Columba semitorquata, Rüppell, Neue Wirbelth. pl. 23. fig. 2.
Turtur erythrophrys, Swainson's Birds of West Africa, vol. ii. pl. 22.
" "   Bonaparte's Consp. Av. vol. ii. p. 63.
Turtur vinaceus, Layard's Cat. No. 509.
Turtur erythrophrys (?), Chapman's Travels in S. Afr., App. p. 411.
Turtur semitorquatus, Finsch & Hartlaub's Vögel Ost-Afrika's, p. 541.
Streptopelia semitorquata, Gray's Hand-list of Birds, No. 9325.

I have obtained specimens of a large Dove from Lake Ngami, where it seems to be common, which agrees very nearly with Swainson's description of his "Red-eyed Dove."

This Dove is not found either in Damara or Great Namaqua Land; neither do I remember to have met with it on the Okavango.

[Mr. Andersson's last collection contained a specimen of *Streptopelia semitorquata*, obtained at the river Cunéné on June 25, 1867.—ED.]

283. Œna capensis (Linn.). Black-throated Dove.

La Tourterelle à cravatte noire, Levaillant's Ois. d'Afr. pls. 273, 274, 275.
Œna capensis, Gurney, Birds Damar., Proc. Zool. Soc. 1864, p. 3.
„   „   Layard's Cat. No. 508.
Columba capensis, Chapman's Travels in S. Afr., App. p. 411.

This exquisite little Dove inhabits most parts of Ovampo, Damara, and Great Namaqua Land, but is most abundant in the latter country, and is known by the name of "Namaqua Dove" to the Dutch Cape-colonists.

This species occurs in pairs, and is chiefly found frequenting the ground; and when disturbed, it seeks shelter in low trees or bushes, but rarely in the larger trees. It feeds on seeds; and its flesh is very palatable. It constructs its nest on a low bush of similar materials to those employed by its congeners, but with rather more care. Its two white eggs have a rosy tint, from the thinness and semitransparency of the shell.

The irides are dark brown, legs and toes light purple; the basal parts of the bill purple, the anterior parts yellow.

Measurements of a male and a female:—

		Male.		Female.	
		in.	lin.	in.	lin.
Entire length		9	9	9	4
Length of folded wing		4	4	4	2
„	tarsus	0	7	0	6
„	middle toe	0	6	0	6
„	tail	5	7	5	2
„	bill	0	7	0	7

284. Chalcopelia afra (Linn.). Emerald-spotted Dove.

La Tourterelle émeraudine, Levaillant's Ois. d'Afr. pl. 271.
Peristera afra, Strickland & Sclater, Birds Damar., Contr. Orn. 1852, p. 157.
" " Layard's Cat. No. 513.
Columba afra, Chapman's Travels in S. Afr., App. p. 411.
Chalcopeleia afra, Finsch & Hartlaub's Vögel Ost-Afrika's, p. 554.
Chalcopelia chalcospilos, Gray's Hand-list of Birds, No. 9409.

Some naturalists are for separating the South-African race of this bird from that of Western and Northern Africa, giving to the former the specific name of *chalcospilos* and to the latter of *afra*, and grounding the distinction chiefly on the colour, or rather shading, of the metallic spots on the wing-coverts. These, they maintain, are in the northern race more or less of a steel-blue passing into strong green on the borders, whilst in the southern race the green extends all over the spot and is of a clear and uniform tint. But I cannot help thinking that this distinction is too trivial to constitute a diagnostic character, especially as, according to my experience, there is a difference between the sexes of the South-African bird in this respect, the colour of the spots being in the male of a beautiful duck-green, with bronze reflections in a few individuals; in the female the spots are smaller, less bright, and with more of a purplish hue.

I have never observed this pretty Dove in Great Namaqua Land, nor in Southern Damara Land. I consider Omanbondé its southerly limit; and in travelling northwards I did not find it abundant till I reached Okamabuti, some miles to the north of Omanbondé; from thence to the Okavango River it was common.

This Dove constructs a nest of a few rough sticks in a bush or at the extremity of a bough of some low stunted tree. The sticks composing the nest are so loosely put together that a person looking at it from below may see the two white eggs through the nest. It is seldom that more than one egg is hatched. The young are usually fledged by the middle of January.

The iris in this species is dark brown, the bill black, the legs bluish flesh-colour.

[I incline to the opinion expressed by Mr. Andersson, that the South-African race of this Dove is not specifically distinct from that found to the north of the equator; but I have not examined a sufficient number of specimens to enable me to speak confidently on the subject.—ED.]

GALLINÆ.

MELEAGRIDÆ*.

285. Numida cornuta, Finsch & Hartl. Cape Guinea-fowl.

Numida mitrata, Layard's Cat. No. 519.
Numida cornuta, Finsch & Hartlaub's Vögel Ost-Afrika's, p. 569.
" " Gray's Hand-list of Birds, No. 9630.

This Guinea-fowl is the commonest game bird in Damara and Great Namaqua Land, being most abundant from the Orange River in the south to the Okavango in the north of those countries; and it is also very common in the Lake-regions. It is a highly gregarious bird, especially during the dry season, when it is not uncommonly found in flocks of several hundred individuals; and on one occasion I saw upwards of a thousand collected in one spot, which was one of the prettiest sights I have had the good fortune to witness. These wonderful congregations usually occur in the immediate neighbourhood of waters of small extent; and it is quite evident that were such a mass of birds to make a simultaneous rush for the precious liquid, there would be much confusion, and comparatively few would be enabled to have their fill. But, on the contrary, they go to work most economically and judiciously, and it is very interesting to watch the process. The first comers enter the well or hole, as the case may be, and, rapidly and dexter-

* [I have followed Doctors Finsch and Hartlaub (Vögel Ost-Afrika's, p. 568) in treating the Meleagridæ as a distinct family, and not merely as a subfamily of the Phasianidæ.—ED.]

ously taking their fill, they make their exit in a different direction, if possible, from that by which they entered; in the meanwhile the outsiders gradually and evenly approach, and the ring is gradually narrowed by a steady progressive movement of the whole. A batch of fresh comers never attempt to force their way amongst those which had previously arrived, but remain quietly on the outside of the ring until their turn comes.

I may add that I have observed the same habit amongst the Sand-Grouse.

The Guinea-fowl feeds on grass, seeds, and insects, but chiefly on a small bulb which is also eagerly sought for by all gallinaceous birds, as well as by man, and which grows very abundantly throughout the country.

These birds are great travellers, often going over fifteen or twenty miles in the course of the day, but always returning, if possible, to the water at night; so that by judiciously dodging their steps a thirsty traveller may find the desired pool, though implicit reliance should never be placed on this mode of obtaining water.

The Guinea-fowls usually rest during the heat of the day under some mimosa, resuming their journeyings when the greatest heat is passed.

A flock of these birds is in general easily discovered by their sharp, discordant, and metallic cries, something like a rapid succession of blows struck upon iron. They have many enemies, and seek security at night by roosting in tall mimosas.

The flesh of the young Guinea-fowl is very white, tender, and well flavoured, but that of the old birds is far from tempting.

The eggs of the wild Guinea-fowls are often hatched under domestic Fowls, and the young are not difficult to rear; but as they grow their propensity for roosting on high trees is rapidly developed, much to the distress of their foster-mother, which is usually unable to follow them to their lofty perch.

I have also known young chicks of this species successfully reared when captured in a wild state; but I have never known an instance of one of these birds, when tamed, having reared a brood of its own young.

The nests of this species consist of slight rounded depressions in the ground, and may be found from the end of December to May, containing from fifteen to twenty eggs, of a buffy-white or pale-buff colour, sometimes obscurely speckled with pale grey.

[I have not seen a specimen of the Damara-Land Guinea-fowl; and as Mr. Andersson appears to have been doubtful to what species it belonged, I forwarded a copy of a full description of it, contained in his MS. notes, to Mr. G. R. Gray, requesting his opinion as to the species to which the description applied. Mr. Gray has been so good as to write me in reply as follows:—"The description which you have sent me seems to agree with *Numida cornuta*, No. 9630 of my Hand-list." This identification is confirmed by Mr. Layard (*loc. cit.*).—ED.]

PTEROCLIDÆ.

286. Pterocles bicinctus, Temm. Double-banded Sand-Grouse.

Pterocles bicinctus, Temminck's Hist. Nat. des Pigeons et Gallinacés, vol. iii. p. 247.
  ,,   ,, Strickland & Sclater, Birds Damar., Contr. Orn. 1852, p. 157.
  ,,   ,, Layard's Cat. No. 536.
  ,,   ,, Chapman's Travels in S. Afr., App. p. 411.
  ,,   ,, Gray's Hand-list of Birds, No. 9460.

This is perhaps the most common species of Sand Grouse in Damara and Great Namaqua Land, where considerable numbers may be seen, during the dry season, at any of the few permanent waters which exist in those countries, and which these birds frequent in large flocks about dusk and during the early part of the night, as well as sometimes also at early dawn; they remain only a short time at the water, and announce their arrival and departure by incessant sharp cries. When dispersed on their feeding-grounds, they are generally found in pairs, or at most two or three together. They feed chiefly on the seeds of grass, as well as on other seeds and berries, and mingle with their food considerable quantities of coarse sand. Their flesh is very white but excessively tough; it may, however, be somewhat improved by divesting the bird of its skin before cooking it. The eggs of this species are from two to three in number, laid upon the bare sand, and of a pinkish-yellow colour, spotted with grey and reddish brown.

The iris is deep red, the skin round the eye chrome-yellow, the bill yellowish brown, the legs and toes pale dull yellow.

R

[This fine Sand-Grouse is at present unfigured. I have not personally examined a Damara example of this species; but a characteristic portrait of it by Mr. Baines was included in Mr. Andersson's collection of drawings of Damara birds.—ED.]

287. Pterocles variegatus, Burch. Variegated Sand-Grouse.

Pterocles variegatus, Burchell's Travels in S. Africa, vol. ii. p. 345.
,, ,, Smith's Zool. of S. Africa, pl. 10 (male).
,, ,, Strickland & Sclater, Birds Damar., Cont. Orn. 1852, p. 157.
,, ,, Layard's Cat. No. 538.
,, ,, Chapman's Travels in S. Afr., App. p. 412.
,, ,, Gray's Hand-list of Birds, No. 9465.

This species is not uncommon in the northern and middle parts of Damara Land, as well as in the Lake-regions; but I do not recollect having met with it further to the south. It feeds on seeds, berries, and roots, and frequents its drinking-places early in the morning. Its flesh is tough like that of its congeners.

The mode of drinking adopted by this and by other South-African Sand-Grouse resembles, as has been already mentioned, that which I have described in the case of the wild Guinea-fowl.

[I have not seen an example of this species from Damara Land; but Mr. Andersson's identification of it is confirmed by Messrs. Strickland and Sclater (*loc. cit.*).—ED.]

288. Pteroclurus namaqua (Gmel.). Namaqua Sand-Grouse.

Pterocles tachypetes, Layard's Cat. No. 535.
Pteroclurus namaqua, Gray's Hand-list of Birds, No. 9468.

This Sand-Grouse is very abundant in some parts of Damara Land, where these birds may be observed to make their appearance at the water about eight or nine o'clock in the morning in immense flocks, circling round the water at a considerable height before they descend, and adding to their numbers at almost every turn they take. Frequently they make no attempt at a descent until they are directly over the spot they intend to visit, when they suddenly descend with great velocity, at the same time describing more or less of a semicircle before they alight.

This species feeds on seeds, berries, and small bulbs, and swallows gravel freely to assist its digestion.

Its eggs are deposited on the sand, and are of a drabbish colour closely spotted with grey and brown; they are oval and less elongated than those of *P. bicinctus*.

The iris is very dark brown; the skin round the eye is a somewhat pale yellow; the bill is bluish, tinged with white on the lower mandible.

Measurements of a female:—

		in.	lin.
Entire length	. . .	11	0
Length of folded wing	.	6	5
,, tarsus . . .	.	0	11
,, middle toe . .	. .	0	9
,, tail		3	6
,, bill .	.	0	7

[Mr. Andersson does not allude to the occurrence of this species in Great and Little Namaqua Land; but Mr. Layard (*loc. cit.*) speaks of it as "very abundant on the arid karroo plains throughout the Colony and Namaqua Land."

This Sand-Grouse has not been figured.—ED.]

TETRAONIDÆ.

289. Pternises nudicollis (Gmel.). Crimson-throated Francolin.

Francolinus nudicollis, Layard's Cat. No. 522.
Pternises nudicollis, Gray's Hand-list of Birds, No. 9647.

[A Damara-Land specimen of this Francolin was contained in Mr. Andersson's last collection; but I have been unable to trace the exact locality where it was obtained, and the species is not referred to in Mr. Andersson's MS. notes. This fine Francolin has not been figured.—ED.]

290. Pternises Swainsonii (Smith). Swainson's Francolin.

Francolinus Swainsonii, Smith's Zool. of S. Africa, pl. 12.
Francolinus Swainsoni, Strickland & Sclater, Birds Damar., Contr. Orn. 1852, p. 157.
Francolinus Swainsonii, Layard's Cat. No. 524.
  „   „   Chapman's Travels in S. Afr., App. p. 412.
Pternises Swainsoni, Gray's Hand-list of Birds, No. 9650.

In travelling northwards through Damara Land I first met with this powerful and somewhat coarse-looking Francolin at the southern extremity of Omuveroom, where it occurred sparingly; but it became more common as I proceeded further north, till, on the banks of the Okavango River, I found it quite abundant. It frequents grassy localities sprinkled with brushwood—generally, but not always, selecting the neighbourhood of springs, streams, or marshes. It feeds in open spots, but retires to the jungle on the first approach of danger, chiefly trusting to its legs to effect its retreat. It always roosts on trees by night, and occasionally perches on them by day; in the early morning and at evening it utters frequent harsh cries.

This species feeds chiefly on small bulbs, but also eats seeds, berries, and insects.

The young of this Francolin are strong on the wing about the month of May.

The iris in this species is dark brown; the lore and the bare skin round the eyes and on the chin and throat is pale red, and the same colour tinges the lower mandible of the bill and the base of the upper, the remainder of the latter being a dark horn-colour, lightest at the tip.

[I have not seen a Damara-Land specimen of this Francolin, but Mr. Andersson's identification of it is confirmed by Messrs. Strickland and Sclater (*loc. cit.*).—ED.]

291. Scleroptera gariepensis (Smith). Orange-River Francolin.

Francolinus gariepensis,		Smith's Zool. of S. Africa, pls. 83 (male) and 84 (female).
,,	,,	Strickland & Sclater, Birds Damar., Contr. Orn. 1852, p. 157.
,,	,,	Layard's Cat. No. 527.
,,	,,	Chapman's Travels in S. Afr., App. p. 412.
,,	,,	Finsch & Hartlaub's Vögel Ost-Afrika's, p. 582.

Scleroptera gariepensis, Gray's Hand-list of Birds, No. 9657.

I only met with this beautiful Francolin on the high tablelands of Damara and Great Namaqua Land; but there it is frequently very abundant, in coveys usually of about six or eight individuals, though sometimes as few as three birds, and at others as many as fourteen compose the covey.

These Francolins invariably frequent grassy slopes sprinkled with dwarf bush; they lie very close, and, after having been once or twice flushed, are not easily found again, even with the assistance of dogs. They feed on bulbs, grass, berries, and seeds; and their flesh is very good for the table.

The iris is brown; the bill horn-colour, except near the base, where it is yellowish.

Dr. Smith's figure of the female of this species is altogether too dull, as, though the tints in the female are not so deep as in the male, they are still exceedingly rich and bright.

292. Scleroptera subtorquata (Smith). Coqui Francolin.

Francolinus subtorquatus, Smith's Zool. of S. Africa, pl. 15 (old female).
,, ,, Gurney, in Ibis, 1860, p. 215.
,, ,, Gurney, Birds Damar., Proc. Zool. Soc. 1864, p. 3.
,, ,, Andersson, Birds Damar., Proc. Zool. Soc. 1864, p. 6.
,, ,, Layard's Cat. No. 530.
Scleroptera subtorquata, Gray's Hand-list of Birds, No. 9658.

I only met with this species in the neighbourhood of the river Okavango, where it is found in coveys on grassy plains interspersed with large trees and brushwood. This Francolin lies very close and is exceedingly difficult to flush without the assistance of dogs. It roosts on the ground, and utters a shrill but not unpleasant call-note in the early morning, and also towards evening. It feeds on small bulbous roots, seeds, berries, and insects; its flesh is very good.

The iris is reddish brown; the bill dark horn-colour; the angle of the mouth lemon-colour, which is also the colour of the tarsus.

293. Scleroptera pileata (Smith). Pileated Francolin.

Francolinus pileatus, Smith's Zool. of S. Africa, pl. 14.
  „   „   Layard's Cat. No. 528.
Scleroptera pileata, Gray's Hand-list of Birds, No. 9667.

In travelling northward I first met with this Francolin on the stony and wooded slopes above Okamabuté in Northern Damara Land, and subsequently observed it to the north of that locality. It occurs in coveys, and feeds on small bulbs, seeds, and berries ; its flesh is very palatable.

The iris is brownish.

294. Scleroptera adspersa (Waterhouse). Red-billed Francolin.

Francolinus adspersus, Waterhouse, in Alexander's Exp. vol. ii. p. 267.
  „   „   Layard's Cat. No. 523.
Scleroptera adspersa, Gray's Hand-list of Birds, No. 9669.

This is the most common and abundant Francolin indigenous to Damara and Great Namaqua Land, where it is found in coveys which, in favourable seasons, not uncommonly consist of from ten to fourteen individuals. This species is seldom found at any great distance from the banks of the periodical streams, and on the least approach of danger seeks shelter in the trees and bushes with which these banks are generally studded. It lives much on trees, roosting amongst the branches by night, and also resting there during the heat of the day.

These Francolins run with extraordinary swiftness, and will not use their wings unless very hard pressed ; and when they do so, it is with the view of concealing themselves amongst the thickest of the branches of some convenient tree, where they remain perfectly motionless; and it requires an exceedingly good and practised eye to detect one of these birds after it has taken refuge in a full-foliaged tree; when the danger is passed they generally again seek the ground.

Their feeding-time is in the early morning and the cool of the evening; and their food consists of seeds, berries, and insects. The notes of these birds are harsh, and so loud that they may be heard at a great distance; they resemble a succession of hysterical laughs, at first slow, but increasing in rapidity and strength till they suddenly cease.

This species deposits its eggs in a hollow in the ground, without any lining.

The iris is dark brown, the bare skin round the eye pale yellow, the bill and legs in the adult bird are a rich warm red, the toes and the spurs of the male bird are purple. In very young birds the bill is dark purple, and the colour of the legs is much paler than in the adult.

The female birds are somewhat smaller and less robust than the males.

[This Francolin has not been figured; its geographical range appears to extend to Lake Ngami, as a specimen from thence was contained in Mr. Andersson's last collection.—ED.]

295. Coturnix communis, Bonn. European Quail.

Coturnix dactylisonans, Gould's Birds of Europe, pl. 263.
Coturnix dactylisonans ?, Strickland & Sclater, Birds Damar., Contr. Orn. 1852, p. 157.
Coturnix dactylisonans, Layard's Cat. No. 531.
Coturnix communis, Gray's Hand-list of Birds, No. 9705.

This species is not uncommon in Middle and Southern Damara Land.

During the year 1865 countless numbers of these Quails arrived in the neighbourhood of Cape Town; whilst much of the rest of the Colony, which had suffered severely from drought, was nearly denuded of these birds.

296. Coturnix Delegorguei, Deleg. Harlequin Quail.

Coturnix histrionica, Hartlaub's Beitr. z. Orn. Westafr. pl. 11.
,,   ,,   Layard's Cat. No. 532.
Coturnix Delegorguei, Finsch & Hartlaub's Vögel Ost-Afrika's, p. 591.

I killed this bird for the first-time at Ondonga on the 30th of March, 1867; Mr. Chapman obtained it at Lake Ngami.

The iris in this species is the colour of new leather when well browned by exposure; the bill is nearly black, but the point of the upper mandible is light horn-colour; the legs and toes are dusky.

297. Turnix lepurana (Smith). Kurichane Hemipode.

Hemipodius lepurana, Smith's Zool. of S. Africa, pl. 16.
Turnix lepurana, Strickland & Sclater, Birds Damar., Contr. Orn. 1852, p. 158.
,,   ,,   Layard's Cat. No. 534.
,,   ,,   Finsch & Hartlaub's Vögel Ost-Afrika's, p. 593.

This species is not uncommon in Great Namaqua Land during the rainy season; but I have never found many of these birds near together, and it is rarely that more than one of them is flushed at a time. Their favourite resorts are rank grassy spots in the neighbourhood of temporary rain-pools and periodical watercourses; here they run about with great celerity, and, when hard pressed, lie so close as almost to allow themselves to be trodden on before they take wing, after which it is nearly impossible to flush them a second time. They feed on insects and seeds.

The iris is lemon-coloured; the bill blue, with the tip of the upper mandible dark horn-colour; the tarsi and feet flesh-coloured white.

Measurements of a male:—

		in.	lin.
Entire length	.	5	8
Length of folded wing	. . . . .	2	10
,, tarsus	.	0	10
,, middle toe . . .		0	6
,, tail		1	5
,, bill		0	8

STRUTHIONES.

STRUTHIONIDÆ.

298. Struthio australis, Gurney. South-African Ostrich.

The Ostrich, Andersson's Lake Ngami, pl. 7.
Struthio camelus, Layard's Cat. No. 539.
Struthio australis, Gurney, in Ibis, 1868, p. 253.
Struthio camelus (part.), Finsch & Hartlaub's Vögel Ost-Afrika's, p. 597.

Many naturalists are of opinion that the North-African Ostrich is distinct from that of South Africa; but, from an early date in my wanderings, I suspected that there was more than one species of Ostrich indigenous to Southern Africa, though it was only in 1866 that I had the means and the necessary time to verify my surmises.

I am now able to state that there are two distinct kinds of Ostrich in Damara Land. The first species is that which is so well known throughout Southern Africa, in which the male bird is black with white tail and wings, whilst the female is of a greyish colour. Of the second species, which is quite new to science, I have not an entire specimen by me so as to enable me to produce an accurate description of the whole bird: still I possess sufficient data to be able to vouch for the correctness of my statement; for, as I write, three skins (or rather portions of skins) are lying before me, consisting of two adults, male and female, and a young bird apparently about half-grown. The male bird does not appear to differ from the well-known South-African species, except in size, being larger; but the greatest specific difference lies with the female and the young, and more especially with the former, which is jet-black like the cock bird. The young is of a sooty brown, the feathers, which are narrow and come to an acute point, being tipped with light brown; the tail similar, but interspersed with a few grey or greyish-white feathers; the wings like the tail, but of a softer texture; the thighs and neck white. The egg of this species is said to be larger than that of the ordinary one. This bird seems to be pretty commonly distributed over the boundless wastes and plains of both Great Namaqua and Damara Land, and herds with the common species as well as in separate flocks.

Many of the native hunters also speak of a third species of Ostrich, which they characterize by some very marked distinctions; thus, for instance, I am assured that it has a narrow but conspicuous bar across the back or rump, and a double row of quills on each wing, also that the colour is brownish grey in both sexes: nevertheless I mention it here more for the purpose of drawing the attention of naturalists to the subject than from any personal belief in its existence as a third distinct South-African species.

In my first publication, 'Lake Ngami,' I have devoted a chapter to the history of the Ostrich; but the following are devices for its capture which I have not previously described, and which the Namaquas pursue with considerable success. A troop of Ostriches being espied, a number of men unite and surround them; but the interval between the hunters is great, the cordon being drawn at such a distance beyond the birds (in fact out of their sight) as not to arouse their suspicions. When a sufficient time has been allowed to enable the hunters to complete the circle, a general advance is commenced, the men taking care not to appear to direct their attentions to the birds, but merely showing themselves in such a manner as to force them to give way.

As a matter of course the Ostriches make off from the supposed danger, but only to encounter a fresh one, and thus rapidly become exhausted by a constant flight from one human being to another, whilst the hunters, as soon as they perceive that the chances of escape for the Ostriches are at an end, take care to keep them moving at a rapid pace by shouting, yelling, and gesticulating violently. A very short time usually suffices to bring the exciting scene to a close by a general slaughter of the entrapped birds; and with experienced hunters this plan rarely fails to ensure success. The same people employ another method with even greater success: a whole tribe will join, mounted if possible on horseback; and the locality to be "drawn" having been decided on, the body of the horsemen station themselves across some pass, defile, or narrow part of a plain; whilst a few are dispatched in search of the Ostriches, and, on their way, station a comrade here and there at convenient intervals, and, having found their destined victims, the favourite haunts of which are well known to them, gradually and cautiously begin to drive the unsuspecting birds towards their passive confederates. As soon as they have fairly succeeded in directing the course of the Ostriches towards the desired quarter, they urge them on at a steady telling pace; and as the birds pass the mounted men, who had been left behind at intervals, each of these successively takes up the pursuit, enabling the previous pursuer to drop behind and to allow his horse to recover its wind, whilst the Ostriches, on the contrary, are never for a moment allowed to slacken their pace in their now headlong flight. The relieved hunters follow leisurely in an extended line or semicircle as the occasion may require, thus forming an effectual barrier to the retreat of the birds should they attempt such a course; whilst the Ostriches, by the time they reach the main body of horsemen, or even sooner, become so exhausted as to come to a dead halt or to fall helplessly to the ground, when the hunters slaughter them at leisure. It has happened that so great a number of Ostriches have thus been entrapped and completely tired out, that after as many had been killed as were required for food, the remainder, after being despoiled of their valuable wing- and tail-feathers, were suffered to depart without further injury.

[Mr. Andersson's work on Lake Ngami contains, at page 253, a spirited representation and description of the capture of some young Ostriches by Mr. Galton and himself; and the latter may here be quoted as an interesting addendum to the foregoing observations :—

"Ostriches are at all times more or less numerous on the Naarip Plain, but more particularly so at this season, on account of the Naras * * * * being now ripe† * * * *; and in a part of the plain entirely destitute of vegetation we discovered a male and female Ostrich with a brood of young ones about the size of ordinary barn-door fowls. This was a sight we had long been looking for, as Galton had been requested by Professor Owen to procure a few craniums of the young of this bird in order to settle certain anatomical questions; accordingly we forthwith dismounted from our oxen and gave chase, which proved of no ordinary interest.

"The moment the parent birds became aware of our intention they set off at full speed, the female leading the way, the young following in her wake, and the cock, though at some little distance, bringing up the rear of the family party. It was very touching to observe the anxiety the old birds evinced for the safety of their progeny. Finding that we were quickly gaining upon them, the male at once slackened his pace and diverged somewhat from his course; but seeing that we were not to be diverted from our purpose, he again increased his speed, and with wings drooping so as almost to touch the ground, he hovered round us, now in wide circles and then decreasing the circumference till he came almost within pistol-shot, when he abruptly threw himself on the ground and struggled desperately to regain his legs, like a bird that has been badly wounded: having previously fired at him, I really thought he was disabled, and made quickly towards him; but this was only a ruse on his part; for on my nearer approach he slowly rose and began to run in an opposite direction to that of the female, who, by this time, was considerably ahead with her charge.

"After about an hour's severe chase, we secured nine of the brood; and though it consisted of about double that num-

† The fruit of a creeping desert plant, described by Mr. Andersson in his work on Lake Ngami, p. 21.

ber, we found it necessary to be contented with what we had bagged.

"On returning to the bay, however, the next morning in a mule-cart, Mr. Galton again encountered the same birds with the remainder of the family, and, after a short race, captured six more of the chicks."

I am not able to offer any opinion as to how far Mr. Andersson was correct in believing that more than one race of Ostrich exists in Southern Africa; but that the ordinary South-African Ostrich, for which I proposed the specific name of *australis* in the 'Ibis' for 1868, p. 253, differs from the true *Struthio camelus* of North Africa, I cannot doubt.

Mr. A. D. Bartlett, the experienced and observant superintendent of the Gardens of the Zoological Society of London, informs me that the skin which is visible on the thighs and other bare parts of the northern Ostrich is always flesh-coloured, whilst in the southern Ostrich it is invariably bluish, excepting the angle of the gape, which is flesh-coloured, as are also the cere and the scutellations of the tarsi and feet.

Mr. Bartlett is also of opinion that the average height of the southern is somewhat greater than that of the northern Ostrich, and that in the male of the southern race the black portions of the plumage are deeper in colour than in the male of the northern bird.

The Rev. H. B. Tristram, in a paper on the ornithology of Northern Africa, published in the 'Ibis' for 1860, thus describes, at p. 74, the differences between the eggs of the northern and southern Ostrich:—" I may remark that the egg of the North-African Ostrich seems to differ decidedly from that of the Cape bird. I have seen hundreds of specimens, and always found them rather larger than the southern eggs which we generally see in England, and quite smooth, with an ivory-polished surface and free from any punctures. Until I found the eggs myself, I was under the impression that they might be polished by the Arabs; but this is a mistake."

Mr. P. L. Sclater, in his paper on the Struthious birds living, in 1860, in the menagerie of the Zoological Society, published in the Society's 'Transactions,' vol. iv., speaks, at p. 354, of the eggs of the southern Ostrich as being smoother than those of the northern bird; but I believe this to be a *lapsus calami*, as the series of eggs on which the observation was founded, and which are still in the possession of Mr. Bartlett, show that the contrary is the fact, and that Mr. Tristram's statement is correct.

Mr. Sclater's paper is accompanied by a fine plate of the male of *Struthio camelus*; and it is to be regretted that no equally satisfactory representation of *Struthio australis* has yet been published.—ED.]

GRALLÆ.

OTIDIDÆ.

299. Eupodotis kori (Burch.). Kori Bustard.

Otis kori, Burchell's Travels in Southern Afr. vol. i. p. 393 (descr.), p. 402 (head).
„ „ Rüppell's Mon. der Gatt. Otis, Mus. Senck. vol. ii. pl. 13.
Eupodotis cristata, Layard's Cat. No. 540.
Otis kori, Chapman's Travels in S. Afr., App. p. 413.
„ „ Finsch & Hartlaub's Vögel Ost-Afrika's, p. 611.

This splendid bird is found throughout the year in Damara and Great Namaqua Land, and is common as far as Ondonga, but is partially migratory.

Personally I have never seen one beyond thirty pounds weight; but I have been assured on good authority that in some parts of the Free States and the Trans-Vaal districts individuals are sometimes shot weighing from sixty to seventy pounds. The spread of the wings is eight feet four inches. The flesh is excellent eating.

This Bustard is usually found in pairs; but sometimes three or four are to be found together. Its flight is heavy, but is nevertheless very rapid; and at night, when changing its feeding-ground, it may be seen flying at a very great height. It feeds on insects, berries, &c., and is very partial to the sweet gummy exudations of the low mimosa thorn so abundant in Damara Land.

300. Eupodotis ruficrista (Smith). Rufous-crested Bustard.

Otis ruficrista, Smith's Zool. of S. Africa, pl. 4.
„   „   Strickland & Sclater, Birds Damar., Contr. Orn. 1852, p. 158.
Eupodotis ruficrista, Layard's Cat. No. 546.
Otis ruficrista, Chapman's Travels in S. Afr., App. p. 414.

I have met with this species pretty frequently in Great Namaqua Land, and also, but less frequently, in Southern Damara Land, to the north of which I believe it does not extend. It is usually found on open ground thinly covered with dwarf bush.

The irides are greyish brown with a whitish-yellow ring next the pupil; this ring is sometimes tinged with orange.

Measurements of a male :—

	in.	lin.
Entire length	17	9
Length of folded wing	10	8
„ tarsus	3	2
„ middle toe	1	6
„ tail	0	9
„ bill	1	11

301. Eupodotis Rüppellii (Wahl.). Rüppell's Bustard.

Otis Rüppellii, Wahlberg, in Journ. für Orn. 1857, p. 1.
Otis picturata, Hartlaub, in Proc. Zool. Soc. 1865, pl. 6.
Otis Rüppelli, Finsch & Hartlaub's Vögel Ost-Afrika's, p. 619.

This species is plentiful in Great Namaqua Land and is not uncommon in some of the more open parts of Southern Damara Land; it is partial to slightly undulating ground abounding in boulders and loose stones, and is usually found in pairs. When disturbed it utters a succession of quick harsh notes, and crows not unlike a Corncrake on taking wing, but in a much louder strain.

The irides are pale transparent grey, more or less variegated with dark brown; the legs and toes are dirty yellow.

302. Eupodotis afra (Gmel.). Cape Knorhaan Bustard.

White-eared Bustard, Latham's General Synopsis of Birds, pl. 69.
Eupodotis afra, Layard's Cat. No. 548.

I have frequently met with this bird to the south of the Orange River, and at no great distance from it; but I have never met with it to the north of that stream.

The eggs of this Bustard are of a dark greenish drab, more or less profusely spotted and blotched with brown.

[As Mr. Andersson's last collection did not contain specimens of this Bustard, its identification rests upon his authority.—ED.]

303. Eupodotis afroides (Smith). Black-and-white-winged Bustard*.

Otis afroides, Smith's Zool. of S. Africa, pl. 19.
Eupodotis afroides, Layard's Cat. No. 549.
Otis afroides, Chapman's Travels in S. Afr., App. p. 414.

This is perhaps the most common Bustard in both Great Namaqua and Damara Land. On taking wing or when disturbed the male becomes outrageously noisy and will rise vertically and to a great height, often descending as abruptly. This is a great nuisance to the hunter, who is frequently done out of his quarry by the Bustard's sudden and noisy apparition.

* [In addition to the Bustards enumerated in the text, Mr. Andersson's last collection contained a specimen of *Eupodotis Colei* (Smith) = *E. Ludwigii*, Rüppell; but I have not been able to ascertain whether this specimen was procured within the geographical limits to which the present volume is confined, and I have therefore not included it.—ED.]

The irides are brown; the upper mandible horn-colour, except the edges, which, with the under mandible, are bluish white, as also are the legs.

Two nests of this Bustard, found by me at Omapju in January 1867, each contained two eggs.

CHARADRIADÆ.

304. Cursorius senegalensis (Licht.). Senegal Courser.

Tachydromus senegalensis, Swainson's Birds of W. Afr. vol. ii. pl. 24.
Cursorius senegalensis, Schlegel's Mus. des Pays-Bas, *Cursores,* p. 13.
„ „ Layard's Cat. No. 553.

[This species is not referred to in Mr. Andersson's notes; but a male specimen obtained by him at Ondonga, on November 22nd, 1866, is in the collection of Mr. J. E. Harting.

This Courser is to be distinguished from its near ally *C. rufus,* Gould (*C. Burchellii,* Swains.), by its slightly smaller dimensions, by the somewhat brighter tints of the rufous portions of its plumage, by the greater size and intensity of the blackish abdominal patch, by the absence of grey from the occiput, and by the less extended white tipping of the secondary feathers of the wing.—ED.]

305. Cursorius bicinctus, Temm. South-African Double-collared Courser.

Cursorius bicinctus, Temminck's Manuel d'Orn. vol. ii. p. 515.
„ „ Jardine & Selby's Ill. Orn. pl. 48.
„ „ Andersson, in Proc. Zool. Soc. 1864, p. 8.
„ „ Hartlaub, in Proc Zool. Soc. 1866, p. 61.
„ „ Layard's Cat. No. 552.
„ „ Chapman's Travels in S. Afr., App. p. 414.

This Courser is not uncommon in various parts of Great Namaqua and Damara Land, chiefly during the wet season. I frequently found it plentiful in the neighbourhood of Objimbinque, and comparatively tame. It sometimes occurs singly, at others in pairs, and occasionally in small flocks, each flock probably consisting of an entire family.

This species runs with considerable celerity.

Measurements of a male:—

		in.	lin.
Entire length		8	7
Length of folded wing		6	1
,,	tarsus	2	4
,,	middle toe	0	10
,,	tail	2	8
,,	bill	0	10

[It appears that this species also inhabits Ovampo Land, as the male specimen of which the dimensions are given above by Mr. Andersson was ticketed as having been obtained in Ondonga on January 22nd, 1867.—ED.]

306. Cursorius cinctus, Heuglin. Heuglin's Courser.

Cursorius cinctus, Heuglin's Systematische Uebersicht, No. 555.
Hemerodromus cinctus, Heuglin, in Journ. für Orn. 1863, p. 25.
,, ,, Heuglin, in Ibis, 1863, pl. 1.

[No mention is made of this species in Mr. Andersson's notes; but his last collection contained two examples, a male and a female, obtained in Ondonga on January 22nd and 25th, 1867.

Both these specimens now form part of the collection of Mr. J. E. Harting, who has favoured me with the following remarks, relating to this Courser:—" This appears to be an extremely rare species. Von Heuglin described and figured it in 'The Ibis' (1863, p. 31, pl. 1) from a specimen which he considered unique, and which was obtained at Gondokoro on the White Nile, in the Bari Country, in 5° N. lat. I cannot see that it is generically distinct from *Cursorius*, any more than the better-known *C. chalcopterus.*"

Mr. Andersson's specimens agree exceedingly well with the figure in 'The Ibis' above referred to; and from examining them, I am led to the opinion that the natural position of this species is intermediate between *Cursorius chalcopterus* and *C. bicinctus.*

So far as I am aware, the type specimen and the pair in Mr. Harting's collection are the only examples known; there is no perceptible difference between the sexes.—ED.]

307. Cursorius chalcopterus, Temm. Violet-winged Courser.

Cursorius chalcopterus, Gray's Genera of Birds, vol. iii. p. 537, pl. 143.
,, ,, Layard's Cat. No. 554.
,, ,, Chapman's Travels in S. Afr., App. p. 414.
,, ,, Finsch & Hartlaub's Vögel Ost-Afrika's, p. 629.

The Violet-winged Courser arrives in Damara Land in small flocks at the beginning of the rainy season. On their first arrival they keep exclusively to the bush, but later disperse themselves more over the open. The cause of their first seeking cover is explained by the circumstance of their reaching Damara Land in a moulting condition, or about to change their plumage, when of course the bush affords them better chance of escape and concealment from their natural enemies. They are not particularly wild, yet often very difficult to obtain, as they stick to the cover with great tenacity, and when suddenly flushed the bird just darts behind the nearest bush or tree, when it alights and continues its flight by hard running, only using its wings in its utmost need. During the heat of the day they remain nearly stationary. This species feeds on insects of various kinds. The flesh is very dark-coloured but excellent eating.

The irides are dark brown, the ring round the eyes reddish yellow, the legs bright red, and the front of the toes bluish purple.

308. Glareola melanoptera, Nordm.   Nordmann's Pratincole.

Glareola Nordmanni, Layard's Cat. No. 555.
    „   „   Gurney, in Ibis, 1868, pl. 8.
Glareola melanoptera, Finsch & Hartlaub's Vögel Ost-Afrika's, p. 633.

[Mr. J. E. Harting possesses a pair of these birds obtained by Mr. Andersson at Objimbinque; and Drs. Finsch and Hartlaub (*loc. cit.*) refer to this species as occurring in Great and Little Namaqua Land.

As Mr. Andersson's notes do not contain any remarks on the habits of this Pratincole, I may here quote the following interesting account of the manner in which these birds attack the swarms of locusts which are so destructive in South Africa, communicated to 'The Field' newspaper of February 26th, 1870, by a gentleman who was engaged in a survey near the Fish River under the Colonial Engineering Department:—" The principal enemy of these great swarms (of locusts), and the valued friend of the Cape farmer, is the small locust-bird, *Glareola Nordmanni.* * * * * * These birds come, I may say, in millions, attendant on the flying swarms of locusts; indeed the appearance of a few of them is looked upon as a sure presage of the locust-swarms being at hand. Their mode of operation, as I saw it, was as follows :—They intercept a portion of the swarm and form themselves into a ring of considerable height, regularly widening towards the top, so as to present the appearance of a revolving balloon or huge spinning-top. They thus fly one over the other, and, hawking at the locusts, gradually contract their circle and speedily demolish the locusts within its limits. As their digestion, like that of all insectivorous birds, is very rapid, the form in which they thus enclose their prey is admirably adapted to enable the lower to escape the droppings of the upper birds. When they have consumed this portion of the swarm, they follow up the main body and commence another attack, and so on, until night sets in and the birds happen to lose the swarm or the locusts are all devoured. I should not forget to mention that the beak of these birds is exactly of such a shape and such dimensions that when they seize the locust the snap cuts off the four wings, and a passer-by sees a continual shower of locusts' wings falling on the ground. At another time, when I was stationed at Fort Peddie, and the country was suffering from the effects of a long drought and was overrun with unusual quantities of ants and grasshoppers, we were visited by thousands of these birds, which remained many days devouring these pests. Though the locust-birds are excellent eating, no one ever thinks of destroying them, and they were so fearless that though I often rode or ran amongst them to test their tameness, only a few in my immediate vicinity would rise, the rest continuing to feed; but every ten minutes or so the whole mass would rise of their own accord and fly, first a few yards to the right, and then to the left, in a slanting direction, presenting alternately a black and white wave of birds some miles in length, a sight never to be forgotten by the spectator."—ED.]

309. Glareola pratincola (Linn.). Collared Pratincole.

Glareola torquata, Gould's Birds of Europe, pl. 265.
 „   „   Gurney, in Ibis, 1863, p. 329.
Glareola pratincola, Finsch and Hartlaub's Vögel Ost-Afrika's, p. 630.

[I have not seen a Damara specimen of this Pratincole; but I find, by one of the MSS. left by Mr. Andersson, that of two Pratincoles contained in a collection of Damara birds sent by him to London for identification, one was ascertained to be an example of *Glareola Nordmanni*, and the other an adult of the present species.

This and an example received by myself from Natal are the only two well-authenticated instances of the occurrence of this Pratincole in Southern Africa which have come to my knowledge.

Mr. Andersson alludes to this species in his MS. notes, and

also in his notes contributed to the Appendix to Chapman's
'Travels,' p. 415, as being not uncommon at Lake Ngami; but I
believe he intended to apply these remarks to the preceding
species, as at the time he made them he seems not to have been
aware of the distinction between these two nearly allied Pratincoles, although he subsequently became acquainted with the
differences between them.—ED.]

310. Œdicnemus capensis, Licht. Spotted Thicknee.

Œdicnemus capensis, Lichtenstein's Doubletten, p. 69.
Œdicnemus maculosus, Temminck's Pl. Col. pl. 292.
  „   „   Layard's Cat. No. 550.
Œdicnemus capensis, Finsch & Hartlaub's Vögel Ost-Afrika's, p. 624.

This bird is partially migratory in Great Namaqua
and Damara Land—and though uncommon, is nevertheless very local, so that numbers of them may be found
in one small spot, whilst the country far and wide
around does not produce a single individual.

This species prefers broken ground sparingly covered
with dwarf bush; it is comparatively tame and easy of
approach, and seems to be chiefly on the move after dusk.

311. Œdicnemus vermiculatus, Cab. Vermiculated Thicknee.

Œdicnemus senegalensis, Gurney, in Ibis, 1865, p. 270.
Œdicnemus vermiculatus, Cabanis in Decken's Reisen, vol. iii. pl. 10.
  „   „   Finsch & Hartlaub's Vögel Ost-Afrika's,
              p. 622.
  „   „   Gray's Hand-list of Birds, No. 9942.

[I have not met with this species in Mr. Andersson's notes
or collections, but introduce it here on the authority of a specimen from the Orange River referred to by Drs. Finsch and
Hartlaub (*loc. cit.*) as forming part of the ornithological collection at the Stuttgard Museum.

The same authors also cite Namaqua Land as a locality for this
Thicknee, on the authority of M. Jules Verreaux.—ED.]

312. Lobivanellus lateralis (Smith). South-African Wattled Plover.

Vanellus lateralis, Smith's Zool. of S. Africa, pl. 23.
Chettusia lateralis, Gurney, Birds Damar., Proc. Zool. Soc. 1864, p. 3.
„ „ Layard's Cat. No. 557.
Lobivanellus lateralis, Finsch & Hartlaub's Vögel Ost-Afrika's, p. 643.

I have met with this large Plover on the rivers Okavango and Teoughe, at Lake Ngami, and in Ondonga in the rainy season, when it is occasionally to be found in small flocks, though more frequently in pairs or singly.

It frequents the banks of streams and the sides of marshy places; but though it appears to find its food, consisting of insects and small mollusca, in the immediate neighbourhood of such localities, I have never observed that it approaches the water very closely.

The iris is yellowish white, the legs and lower part of the wattles lemon-yellow, the upper part of the wattles red.

Measurements of a male and a female :—

	Male. in. lin.	Female. in. lin.
Entire length	13 0	13 4
Length of folded wing	8 9	9 4
„ tarsus	3 5	3 11
„ middle toe	1 7	1 6
„ tail	4 1	4 4
„ bill	1 4	1 4

313. Hoplopterus speciosus (Licht.). Blacksmith Plover.

Pluvianus armatus, Jardine & Selby's Ill. of Orn. vol. ii. pl. 54.
Charadrius albiceps, Temminck's Pl. Col. pl. 526.
Hoplopterus armatus, Layard's Cat. No. 558.

Hoplopterus albiceps, Layard's Cat. No. 559.
Charadrius armatus, Chapman's Travels in S. Afr., App. p. 415.
Hoplopterus speciosus, Finsch & Hartlaub's Vögel Ost-Afrika's, p. 639.

This is a common Plover in Damara and Great Namaqua Land, at Lake Ngami, and on the rivers Teoughe and Okavango. It is a somewhat local bird in Damara Land on account of its partiality for water; indeed I have never seen it far from the water, but generally in the immediate neighbourhood of springs, marshes, lakes, and streams. It is always found in small flocks, and, where not disturbed, is comparatively tame; when flushed it rises with short rapid jerks of the wings, but never flies far at a time. It is very noisy when on the wing; and hence probably is derived its Sichuana name of " Setula tsipi," *i. e.* "iron-hammer" or "blacksmith;" for its cries have a peculiarly metallic, ringing sound.

This Plover feeds on insects and worms; and its flesh is palatable. It breeds in Damara Land, as, although I have not met with its nest myself, one of my friends discovered one at Omanbondé, and presented the two eggs which it contained to the Cape museum. These eggs are buff-coloured, profusely spotted with dark brown; their length is 1″ 6‴, and their breadth 1″ 1‴.

The iris is red; the bill, legs, and toes are black.

314. Chettusia coronata (Gmel.). Wreathed Plover.

Pluvier du Cap de Bonne Espérance, Buffon's Pl. Enl. pl. 800, vol. ix. p. 23.
Hoplopterus coronatus, Strickland & Sclater, Birds Damar., Contr. Orn. 1852, p. 159.

Hoplopterus coronatus, Layard's Cat. No. 560.
Chettusia coronata, Finsch & Hartlaub's Vögel Ost-Afrika's, p. 636.

This handsome species is widely diffused throughout Damara and Great Namaqua Land, in the Lake-regions, and on the river Okavango; it was very plentiful at my late residence at Objimbinque, where a flock or two were generally to be found throughout the year; and I have no doubt that it breeds there, as I have found young birds in almost every stage of plumage. It is a gregarious species, not unusually associating in flocks of from thirty to forty individuals, sometimes frequenting the moist beds of periodical streams, but more often haunting districts away from the water, and of the driest and most arid character. I have often been charmed with the presence of these birds in such localities, where there was nothing else to break the monotony and the dreariness of the desolate scene.

The long stilt-like legs of this species enable it to run with great celerity; and if pursued, it invariably trusts to its legs for safety, and only takes to the wing as a last alternative: but this is not from any want of power of wing, for it flies well; and its flight is very similar to that of the European Lapwing.

These Plovers fly by moonlight as well as by day; and when flapping about either by day or night, they utter shrill cries which may be heard at a great distance.

The food of this species consists of insects; and its flesh is excellent eating.

The iris is yellow; the bill is bright red, except the

points of the mandibles, which are horn-coloured; the legs are bright red, but the toes are tinged with dark purple on their upper surface.

315. Squatarola varia (Linn.). Grey Plover.

Squatarola cinerea, Gould's Birds of Europe, pl. 290.
Squatarola helvetica, Strickland & Sclater, Birds Damar., Contr. Orn. 1852, p. 158.
 ,, ,, Layard's Cat. No. 562.
Tringa squatarola, Chapman's Travels in S. Afr., App. p. 416.
Charadrius varius, Finsch & Hartlaub's Vögel Ost-Afrika's, p. 644.
Squatarola helvetica, Sharpe & Dresser's Birds of Europe, pl. 46. fig. 2 & pl. 49. fig. 2.

This Plover is rarely seen inland in Damara or Great Namaqua Land, but is very common on the coast. It is found singly or in small flocks, searching for its food (which consists of small crustacea, insects, and worms) where the tide has receded, and sometimes wading in the water up to its belly. It runs with very great rapidity, and is of a wary and suspicious disposition, becoming extremely difficult to obtain if at all pursued. Its flesh is palatable.

This species takes its departure on the approach of the breeding-season; and I have never seen it in such dark plumage as it is represented as attaining in Europe at that season. The iris is dark brown, the bill, legs, and toes bluish black.

Measurements of a female:—

		in.	lin.
Entire length		10	10
Length of folded wing		7	5
,,	tarsus	1	10
,,	middle toe	1	3
,,	tail	3	0
,,	bill	1	5

[One of the specimens of this Plover obtained by Mr. Andersson in Damara Land was partly in breeding dress; but South-African examples of this species are usually in full winter plumage.—ED.]

316. Eudromias asiaticus (Pall.). Caspian Plover.

Charadrius asiaticus, Pallas's Reise, vol. ii. p. 715.
Charadrius caspius, Pallas, Zoogr. R.-A. vol. ii. pl. 58.
Charadrius damarensis, Strickland & Sclater, Birds Damar., Contr. Orn. 1852, p. 158.
Charadrius asiaticus, Layard's Cat. No. 570.
Eudromias asiaticus, Harting, in Ibis, 1870, pl. 5.

Small flocks of this Plover may at times be seen in Damara Land; but it is never common, and very shy. All my Damara specimens were procured at Objimbinque, in the moist bed of the river Swakop.

The iris is very dark brown, the ring round the eyes black, the legs yellow, and the toes dusky.

Individual specimens differ a good deal in size; but there is no marked distinction in the dimensions of the sexes.

The largest adult specimen I obtained measured $8'' \ 4\frac{1}{2}'''$ in length, the smallest $7''$.

Measurements of a male and a female:—

	Male. in. lin.	Female. in. lin.
Entire length	8 0	8 1
Length of folded wing	5 10	5 7
,, tarsus	1 10	1 7
,, middle toe	0 9	0 9
,, tail	2 3	2 3
,, bill	1 0	1 1

[Mr. Harting possesses two specimens obtained at the Knysna in the month of January, which are in breeding dress; but this species usually occurs in South Africa in non-breeding plumage.

Mr. Andersson obtained specimens in the latter plumage in Ondonga, in November 1866, no doubt subsequently to the date of his note transcribed above.

Mr. Harting (*loc. cit.*) also gives the Orange River as a locality for this species, on the authority of M. Jules Verreaux.—ED.]

317. Ægialites alexandrinus (Linn.). Kentish Plover.

Charadrius alexandrinus, Linnæus, Syst. Nat. p. 253.
Ægialites cantianus, Boie, in Isis, 1822, p. 558.
Charadrius cantianus, Gould's Birds of Europe, pl. 298.
    ,,   ,,   Layard's Cat. No. 564.
Charadrius littoralis, Finsch & Hartlaub's Vögel Ost-Afrika's, p. 652.

This is rather a scarce bird in Damara Land, and hardly to be found except on the sea-coast, in the neighbourhood of which it seeks its food on open ground slightly interspersed with grass and aquatic herbage. It feeds on worms and insects, and also on the sandhoppers which abound on the beach at Walwich Bay, and of which it seems to be particularly fond.

I have invariably found it in pairs, but have never met with its nest.

[I have not seen a Damara-Land example of this Plover; but a specimen from that country is mentioned by Drs. Finsch and Hartlaub (*loc. cit.*) as forming part of the collection in the Bremen Museum.—ED.]

318. Ægialites marginatus (Vieill.). Heywood's Plover.

Charadrius marginatus, Vieillot's Nouv. Dict. d'Hist. Nat. vol. xxvii. p. 138.
Charadrius leucopolius, Wagler's Syst. Av. sp. 28.
Charadrius heywoodi, Gray, in Exp. to Niger, vol. ii. p. 50.
Charadrius pallidus, Strickland & Sclater, Birds Damar., Contr. Orn. 1852, p. 158.
Charadrius nivifrons, id. ibid. p. 159.

Charadrius marginatus, Layard's Cat. No. 568.
Charadrius nivifrons, Chapman's Travel's in S. Afr., App. p. 415.
Charadrius marginatus, Finsch & Hartlaub's Vögel Ost-Afrika's, p. 654.

This species is very numerous at Walwich Bay and at many intervening points of the coast between that locality and the Cape, but I do not recollect having seen it inland. It can hardly be said to go in flocks, although a considerable number of individuals may be found scattered over a small extent of beach in close proximity to each other. It runs along the sand with great swiftness; and, unless very hard pressed, it prefers making its escape, when pursued, by trusting to its legs rather than by taking wing; when it has not been previously disturbed it is easy of approach. It utters a a low plaintive cry, and feeds on small sandflies and other minute insects, which it generally catches by running rapidly in pursuit of its prey. Its flesh is very palatable.

This Plover breeds sparingly at Walwich Bay, scooping a small round hole in the sand, but without any lining. The female lays two eggs in the month of April or early in May; these are of the usual Plover shape, but sometimes rather blunt at the smaller end; they are of a yellowish-brown colour, prettily variegated with numerous irregular spots and streaks of a dark liver-brown.

The iris is deep dark brown, the bill black, the legs and toes bluish grey.

[This Plover appears never to have been figured. The type specimen of Mr. Strickland's *Charadrius pallidus*, which is now preserved in the Museum of Zoology at Cambridge, seems to me to belong to this species, but to be a very small specimen and unusually pale and grey in its coloration.—ED.]

319. Ægialites pecuarius (Temm.). Kittlitz's Plover.

Charadrius pecuarius, Temminck's Pl. Col. pl. 183.
Charadrius Kittlitzi, Reichenbach's Syn. Av. pl. 105. fig. 1063.
Ægialites Kittlitzi, Editorial Note in Ibis, 1867, p. 251.
Charadrius Kittlitzi, Layard's Cat. No. 566.
Charadrius pecuarius, Finsch, in Trans. Zool. Soc. vol. vii. p. 297.

This pretty little Plover is not uncommon in Damara Land; but I do not think that it breeds there. It is to be seen in flocks, often composed of a considerable number of individuals, and feeds on the small insects which are to be found in the moist and humid localities to which it is partial. At some seasons I found it very abundant at Objimbinque, but I do not recollect having ever observed it on the sea-shore.

The iris is very dark brown, the tibia and toes bluish black, the tarsus greyish.

[Mr. Layard (*loc. cit.*) is of opinion that this Plover should bear the specific name of "*Kittlitzi*," and that the appellation of "*pecuarius*" should be restricted to the nearly allied but somewhat larger race which inhabits the island of St. Helena; and this view is also taken by Professor Newton in 'The Ibis' (*loc. cit.*). But it seems to me to be more probable that it is the South-African bird to which the name of "*pecuarius*" was originally applied, and therefore that the St.-Helena race, if specifically distinct, is at present unnamed. — ED.]

320. Ægialites tricollaris (Vieill.). Treble-collared Plover.

Charadrius tricollaris, Vieillot's Nouv. Dict. p. 147.
,,   ,,   Layard's Cat. No. 565.

Sea-cow Bird, Chapman's Travels in S. Afr., App. p. 371.
Charadrius tricollaris, Finsch & Hartlaub's Vögel Ost-Afrika's, p. 655.
Ægialitis tricollaris, Gray's Hand-list of Birds, No. 10002.

This species is pretty commonly dispersed over Damara and Great Namaqua Land, both on the coast and at inland waters.

It is generally seen in small flocks, and exhibits comparatively little fear of man. It feeds chiefly on small insects, and also, at the inland rills in Damara Land, on a species of aquatic worm. Its flesh is palatable.

In the egg of this Plover the ground-colour is almost hidden by a profusion of thickly clustering spots and fine eccentric lines of grey-brown and sepia-brown; in some instances these markings are arranged in a ring round the thicker end of the egg.

The irides are very dark brown; the legs are straw-yellow with a tinge of blue throughout, deepest at the joints; the upper mandible is bluish black, and the lower also, except near the base, where it is pale salmon-colour; the fringe round the eyes is bright brick-red.

Measurements of a female:—

		in.	lin.
Entire length		7	0
Length of folded wing		4	4
„	tarsus	1	0
„	middle toe	0	8
„	tail	2	8
„	bill	0	8

[I am indebted to the kindness of Mr. Layard for the information that this Plover is the bird alluded to by Mr. Chapman (*loc. cit.*), under the name of "Sea-cow Bird," as being found in apparent attendance on the Hippopotamus.

I believe that this species has not been figured.—ED.]

321. Ægialites hiaticula, Linn. Ringed Plover.

Charadrius hiaticula, Gould's Birds of Europe, pl. 296.
  „   „   Strickland & Sclater, Birds Damar., Contr. Orn. 1852, p. 90.
  „   „   Layard's Cat. No. 567.
  „   „   Chapman's Travels in S. Afr., App. p. 415.

This Plover occurs at certain seasons at Walwich Bay, but only very sparingly elsewhere.

The irides are intensely dark brown; the legs yellowish; the bill black, except the base, which is yellowish, especially on the under mandible.

[The collection of Mr. J. E. Harting comprises a specimen of this Plover, obtained by Mr. Andersson at Walwich Bay, October 22nd, 1863.—ED.]

322. Cinclus interpres (Linn.). Common Turnstone.

Strepsilas collaris, Gould's Birds of Europe, pl. 318.
Strepsilas interpres, Strickland & Sclater, Birds Damar., Contr. Orn. 1852, p. 159.
Cinclus interpres, Layard's Cat. No. 572.
Tringa interpres, Chapman's Travels in S. Afr., App. p. 416.
Strepsilas interpres, Finsch & Hartlaub's Vögel Ost-Afrika's, p. 662.
Cinclus interpres, Gray's Hand-list of Birds, No. 10068.

The Turnstone is pretty common all along the Southwest coast of Africa, and is found either in pairs or in small flocks.

I never met with the nest of this species in Africa, although I was acquainted with its mode of nidification from having seen its nests on the coast of Sweden.

The irides are dark brown, the bill black, the legs and toes yellow, with a purple tint about the joints.

[Mr. Andersson's last collection contained a male Turnstone

in full breeding dress, obtained at Walwich Bay on April 8th, 1863. The other Damara-Land examples of this species which I have seen have been in winter plumage.—ED.]

323. Hæmatopus Moquini, Bon. Moquin's Oystercatcher.

Hæmatopus Moquini, Layard's Cat. No. 571.
Hæmatopus niger, Chapman's Travels in S. Afr., App. p. 416.
Hæmatopus Moquini, Gray's Hand-list of Birds, No. 10064.

This species is not uncommon on the mainland of the South-west African coast, as well as on the adjacent islands, in which latter localities it breeds. It is usually observed in pairs; but I have seen it in considerable flocks at Walwich Bay late in October and early in November: the birds composing these flocks are always too shy to be approached within gunshot; they rise with a shrill cry, which is continued during their flight; and they generally soar to a great height before realighting, sometimes, indeed, taking their departure altogether, on which occasions they usually steer to the northward.

This species feeds on worms, insects, and mollusca, searching for the latter in crevices and under stones; and whilst thus engaged it sometimes swims a short distance from rock to rock.

It makes no nest, but deposits its eggs on the shingle of the beach; these are four in number, of a drab colour, with eccentric streaks and spots of very dark brown.

I have been told that the flesh of this Oystercatcher is excellent, but have not myself tasted it.

The irides are bright red, the ring round the eyes and the bill are orange-red, the legs are carmine red.

[I have not seen a Damara-Land specimen of this Oystercatcher; but Mr. Layard refers to it (*loc. cit.*) as an inhabitant of the coasts of that country, thus confirming Mr. Andersson's identification of the species. The Black Oystercatcher of South Africa has, I believe, not yet been figured.—ED.]

GRUIDÆ.

324. Bugeranus carunculatus (Gmel.). Wattled Crane.

Grus carunculata, Gray's Genera of Birds, pl. 148.
„   „   Layard's Cat. No. 573.
Wattled Crane, Chapman's Travels in S. Afr., App. p. 417.
Bugeranus carunculatus, Gray's Hand-list of Birds, No. 10087.

This Crane is found very sparingly in Damara Land during the rainy season; I have also observed it on the rivers Okavango, Teoughe, and Dzonga, as well as at Lake Ngami.

325. Tetrapteryx paradisea (Licht.). Stanley Crane.

Scops paradisea, Gray's Knowsley Menagerie, pl. 14.
Anthropoides Stanleyanus, Layard's Cat. No. 574.
„   „   Chapman's Travels in S. Afr., App. p. 417.
Grus paradisea, Finsch & Hartlaub's Vögel Ost-Afrika's, p. 671.
Tetrapteryx paradisea, Gray's Hand-list of Birds, No. 10093.

This very graceful Crane is not uncommon in Damara and Great Namaqua Land during the rainy season, but migrates on the return of the dry. It is found in the open country, as well as in the moist beds of periodical streams, and always in pairs. It is a rather timid bird, and is rarely to be obtained, except by the rifle. The stomachs of the few individuals which I have opened contained nothing but insects and a large quantity of gravel.

[I have not seen a Damara-Land example of this Crane; but a description of it which is contained in Mr. Andersson's memoranda leaves no doubt of his having identified the species correctly.—ED.]

326. Balearica regulorum (Licht.). Southern Crowned Crane.

Balearica regulorum, Gray's Knowsley Menagerie, pl. 13.
  „   „   Layard's Cat. No. 575.
Ardea pavonia, Chapman's Travels in S. Afr., App. p. 417.
Grus regulorum, Finsch & Hartlaub's Vögel Ost-Afrika's, p. 669.
Balearica regulorum, Gray's Hand-list of Birds, No. 10095.

I have never observed this handsome bird in Great Namaqua or Damara land; but I have met with it at Lake Ngami, and also on the rivers Botletlé, Teoughe, and Okavango. In Ondonga it is very common during the dry season, but leaves the open plains when the wet season returns. It is met with singly or in pairs, and sometimes in small flocks, and presents a very graceful appearance as it stalks leisurely about in search of its food, which consists of various insects, of the smaller reptiles, and, it is said, of fish also.

The yellow bristles of the crown of this Crane are used by the inhabitants of Ovampo Land as ornaments for their heads.

Measurements of a male:—

		in.	lin.
Entire length		40	6
Length of folded wing		22	6
„	tarsus	7	9
„	middle toe	4	0
„	tail	10	6
„	bill	2	11

CICONIIDÆ.

327. Ciconia alba, Linn.   White Stork.

<small>*Ciconia alba*, Gould's Birds of Europe, pl. 283.
,,   ,,   Layard's Cat. No. 595.
Ardea ciconia, Chapman's Travels in S. Afr., App. p. 418.</small>

This Stork is abundant at Lake Ngami and in many localities of the Lake-regions; it is also quite common during the wet season in Ondonga and in Northern Damara Land, sometimes occurring in large flocks; and it is seen occasionally during the same season in Southern Damara Land.

The iris is brown, the bill yellowish red, and the legs and toes light red.

328. Sphenorrhynchus Abdimii (Licht.).   Abdim's Stork.

<small>*Ciconia Abdimii*, Hemprich & Ehrenberg's Symb. Phys. pl. 2.
Sphenorrhynchus Abdimii, ibid., letterpress to plate.
Ciconia Abdimii, Rüppell's Atlas, pl. 8.
 ,,   ,,   Layard's Cat. No. 597.
 ,,   ,,   Chapman's Travels in S. Afr., App. p. 418.
 ,,   ,,   Finsch & Hartlaub's Vögel Ost-Afrika's, p. 721.</small>

This somewhat coarse but handsome Stork arrives in Damara Land during the rainy season, leaving it again on the approach of the dry. The more plentiful the rain the more abundant the birds, the cause being simply the greater abundance of food. This species feeds largely on locusts, but devours with equal gusto beetles of all kinds, many hundreds of which I have taken from the stomach of a single bird; it also devours small reptiles, water-rats, fish, and frogs, but appears to prefer locusts when these are to be had, chasing them on the ground as well as in the air. On such occasions these Storks may be seen in thousands gyrating in immense circles and, as it were, in different strata, the lower frequently flying within range of the fowling-piece whilst the eye rests painfully on the higher as mere specks in the air.

The iris is greyish brown, the brown predominating next the pupil; a small patch in front of the eyes, a streak above the eye and halfway from the front, a minute spot on the lower eyelid, the chin, and the inside of the ears are all vermilion-red; the front of the head is light flesh-colour, the remainder of the naked parts bluish purple; the legs greenish merging into vermilion at the joints, the toes dusky both above and below.

[A coloured drawing of this Stork, by Mr. Baines, in Mr. Andersson's collection, has the following note appended to it:—
" Flocks of this Stork wander about the flats of Lake Ngami, picking up snails and insects."—ED.]

329. Ephippiorhynchus senegalensis, Bon. Saddle-billed Stork.

Ciconia ephippiorhyncha, Temminck's Pl. Col. pl. 64.
    "   "   Rüppell's Atlas, pl. 3.
Ephippiorhynchus senegalensis, Ayres and Gurney, Ibis, 1862, p. 34.
Mycteria senegalensis, Layard's Cat. No. 600.
    "   "   Finsch & Hartlaub's Vögel Ost-Afrika's, p. 723.
Ephippiorhynchus senegalensis, Gray's Hand-list of Birds, No. 10193.

This curious-looking Stork occurs occasionally both in Damara Land and in Ondonga; but I have only once personally obtained a specimen, which I met with in

Damara Land amongst the long rank grasses growing about a temporary rain-pool. I have more than once received the bill of this species from my hunting friends, several of whom have told me that they occasionally meet with these birds; and the Rev. C. H. Hahn, a most excellent authority, informs me that, when stationed at New Barmen, he twice saw this species in that locality.

[I have noticed, in 'The Ibis' for 1862 (*loc. cit.*), the singular fact that in this Stork and in its eastern congener, *Xenorhynchus indicus*, the colour of the iris differs in the two sexes, just as it does in the Harriers amongst the birds of prey, being pale yellow in the male bird and dark brown in the female.—ED.]

330. Leptoptilus crumeniferus, Cuv. African Marabou Stork.

Ciconia argala, Temminck's Pl. Col. pl. 301.
Adjutant, Baines's South-west Africa, p. 337, pl.
Leptoptilus crumeniferus, Layard's Cat. No. 599.
Ciconia marabou, Chapman's Travels in S. Afr., App. p. 418.
Leptoptilos crumeniferus, Gray's Hand-list of Birds, No. 10194.

This species is met with in Damara Land during the rainy months, at which season it is also frequent in Ondonga; and it is abundant at Lake Ngami throughout the year. It is usually seen in flocks, sometimes upon the ground, and at others perched on trees. It will remain for hours in the same position, with one foot drawn up under its body; and a number of individuals seen in this attitude through the fantastic medium of a mirage present a singular and ludicrous appearance.

It is a true scavenger bird, feeding on carrion, as well as on the most offensive offal; but it also captures living prey, such as fish, tortoises, and snakes.

[I have not met with this bird in Mr. Andersson's collections; but a characteristic drawing of it from the pencil of Mr. Baines was comprised in the portfolio of South-west African ornithological sketches which Mr. Andersson possessed at the time of his decease.—ED.]

ANASTOMATIDÆ*.

331. Anastomus lamelligerus, Temm. African Snail-eater†.

Anastomus lamelligerus, Temminck's Pl. Col. pl. 236.
"   "   Livingstone's Missionary Travels, pp. 252 & 494.
"   "   Layard's Cat. No. 601.

This singular-looking bird is not uncommon in the Lake-regions, and it is also common in Ondonga, where it is found throughout the year. It roosts on trees, and is sometimes found solitary, but more often in large flocks. It feeds on fish, frogs, shells, &c.; and I have often found its crop and stomach crammed full of the bodies of shells in the driest time of the year; where does it get them? It may be seen poking about on perfectly arid spots; but neither there nor in moister places have I been able to obtain a single shell by the closest observation, or even by digging.

The iris and also the bill are dark brown.

* [Mr. G. R. Gray, in his 'Hand-list of Birds,' pt. iii. p. 36, and other high authorities regard the two species of *Anastomus* as forming a subfamily of the *Ciconiidæ*; but they seem to me to be sufficiently isolated and distinct to form a separate family; and I have therefore treated them as such.—ED.]

† [The name of "Snail-eater," or, rather, its Cingalese equivalent, is applied in Ceylon to the allied eastern species *Anastomus oscitans* (*vide* Layard on the Ornithology of Ceylon, in the 'Annals of Natural History' for 1854, vol. xiv. p. 115); and I have considered the appellation appropriate for the African species also (*vide* Livingstone, *loc. cit.*).—ED.]

ARDEIDÆ.

332. Ardea cinerea, Linn.   Common Heron.

Ardea cinerea, Gould's Birds of Europe, pl. 273.
„   „   Layard's Cat. No. 577.
„   „   Chapman's Travels in S. Afr., App. p. 417.

This species is rather common on and near the seashore, from the mouth of the Orange River to Walwich Bay, wherever suitable localities occur; but I have seldom seen it inland, except in Ondonga, where it makes its appearance in December; but the natives say that it does not breed there, and I think they are right.

These Herons usually occur singly or in pairs; but I have counted as many as twenty in one small spot (not extending more than two hundred yards in any direction). They will remain motionless sometimes for hours, but in the mornings and towards evening are usually on the move in search of their food, which consists in great part of fish, though they also feed on rats, mice, and reptiles.

The iris is bright yellow; the bill golden or orange yellow, except on the ridge and at the base, where it is somewhat dusky; a patch on each side of the bill in front of the lore is pale yellow, with a tinge of green; the lore and the space round the eye are brownish slate-colour, the legs and feet a dusky greenish yellow.

333. Ardea melanocephala, Vig. & Chil.   Black-throated Heron.

Ardea atricollis, Smith's Zool. of S. Africa, pl. 86.
„   „   Layard's Cat. No. 570.
„   „   Chapman's Travels in S. Afr., App. p. 417.

Ardea melanocephala, Gray's Hand-list of Birds, No. 10100.
„ „ Sharpe & Dresser's Birds of Europe, pl. 17.

This Heron is found sparingly about humid places in Great Namaqua and Damara Land, and in the lagoons on the south-west coast. Axel also found it common in Ondonga on his arrival there in the month of November: there was then no water in the vleys; and these birds were feeding exclusively in the fields, and chiefly on grasshoppers; but the ordinary food of this species also comprises fish, reptiles, rats, and mice.

These Herons perch upon trees when there are any within reach.

The irides are yellow, as is also the bare space round the eyes; the upper mandible of the bill is liver-brown; the lower mandible is of a dark horn-colour at the edges, the point is greenish, and the remainder of the mandible yellowish; the legs and toes are dark blue-black, the latter being greenish on the under surface.

334. Ardea goliat, Temm. Goliath Heron.

Ardea goliat, Temminck's Pl. Col. pl. 474.
Ardea goliath, Layard's Cat. No. 576.
„ „ Chapman's Travels in S. Afr., App. p. 418.
„ „ Finsch & Hartlaub's Vögel Ost-Afrika's, p. 674.
Ardea goliat, Gray's Hand-list of Birds, No. 10107.

These fine Herons are not uncommon in the Lake-regions, from whence they make temporary and solitary excursions into Damara Land during the wet season. I have met with them on the rivers Okavango and Teoughe, at Lake Ngami, and thence eastward along the river Botletlé.

They are generally seen singly or in pairs, and are shy and wary birds, usually able, from their great stature, to espy any threatened danger from a considerable distance, and thus to make their escape in safety. They utter a strong, hoarse, croaking sound, not unlike the bark of a dog. I believe these Herons feed almost entirely upon fish, which they transfix with great celerity and swallow entire; it is said that a half-pound fish is thus easily disposed of.

The irides are yellow; the bill black, shading into dusky yellow towards the base and on the lore; the legs and toes are black.

[I have not seen a Damara-Land specimen of this Heron; but it is mentioned by Mr. Layard (*loc. cit.*) as found in that country, which confirms Mr. Andersson's identification of the species.

These Herons sometimes transfix and swallow fish of considerably larger weight than those mentioned by Mr. Andersson, as one shot by Mr. Ayres near Potchefstroom contained a " catfish " of two pounds weight, and with a head " as broad as a man's hand."—ED.]

335. Ardea purpurea, Linn. Purple Heron.

Ardea purpurea, Gould's Birds of Europe, pl. 274.
    "   "   Layard's Cat. No. 579.
    "   "   Finsch & Hartlaub's Vögel Ost-Afrika's, p. 676.

I have not unfrequently shot this bird on the rivers Okavango and Teoughe, and at Lake Ngami, and I believe it also visits Damara Land during the rainy season; but the specimens which I obtained in the latter country were not preserved, and I am therefore unable positively to identify them.

These birds live in small flocks or societies, and frequent marshy ground and the sides of running streams; they make daily excursions from some fixed point, to which they return at nightfall.

The food of this species consists of fish, rats, mice, reptiles, and insects.

Its nest is formed on the ground, at the roots of reeds, in some swamp or dense reed-bed.

The irides are yellow; a naked narrow space from under the eyes to past the angle of the mouth is brown; a portion of the lore and a narrow space on the edges of the upper mandible near the gape are light yellow, the remainder of the upper mandible and of the lore are brown; the lower mandible is brownish yellow; the legs and toes brown.

Measurements of a male:—

	in.	lin.
Entire length	34	0
Length of folded wing	12	0
,, tarsus	4	7
,, middle toe	4	5
,, bill	5	8

336. Ardea rufiventris, Sund. Rufous-bellied Heron.

Ardea rufiventris, Sundevall, in Öfvers. 1850, p. 110.
,, ,, Ayres, in Ibis, 1871, pl. 9.

I have shot several individuals of this very handsome Heron both on the river Okavango and on the river Teoughe, as well as in the neighbourhood of Lake Ngami; but I do not recollect having met with it in Damara Land.

337. Bubulcus ibis (Hasselq.). Buff-backed Heron.

Bubulcus russatus, Gould's Birds of Great Britain, pt. xx. pl.
Ardea bubulcus, Layard's Cat. No. 581.
,,   ,,   Finsch & Hartlaub's Vögel Ost-Afrika's, p. 694.
Bubulcus ibis, Gray's Hand-list of Birds, No. 10132.

This species is exceedingly common in Ondonga, and quite tame; it perches on trees, and may be seen in flocks of from ten to fifty, frequently in company with cattle.

The iris is lemon-colour, the bill yellow, tarsus brownish yellow, toes brown above and yellow beneath.

[The Orange River is quoted by Drs. Finsch and Hartlaub (*loc. cit.*), on the authority of M. Jules Verreaux, as a locality for this species.—ED.]

338. Ardeola comata (Pall.). Squacco Heron.

Ardea comata, Gould's Birds of Europe, pl. 275.
,,   ,,   Layard's Cat. No. 585.
Ardeola leucoptera, Chapman's Travels in S. Afr., App. p. 418.
Ardea comata, Finsch & Hartlaub's Vögel Ost-Afrika's, p. 697.
Ardeola comata, Gray's Hand-list of Birds, No. 10134.

This species of Heron is found in Damara and Great Namaqua Land throughout the year, but is more numerous in those countries during the rainy season than at other times. It is very abundant in the Lake-regions, and on the rivers Botletlé, Teoughe, and Okavango. It feeds on insects, frogs, &c.

The iris is pale yellow; the bill is of a transparent whitish horn-colour at the tip; the base of the upper mandible is bright greenish yellow, and this colour extends to the base of the nostrils; the under mandible is also greenish yellow; the tibia, tarsus, and toes are dusky green, lightest on the tibia and the upper part of the tarsus; the under surface of the toes is yellowish.

339. Herodias alba (Linn.). European Greater Egret.

Ardea alba, Gould's Birds of Europe, pl. 276.
Herodias melanorhyncha, Hartlaub's Orn. Westafr. No. 630.
Ardea alba, Finsch & Hartlaub's Vögel Ost-Afrika's, p. 683.
Herodias alba, Gray's Hand-list of Birds, No. 10108.

Iris reddish yellow, bill black, naked part round the eyes light green, tarsus black.

Measurements of a male:—

	in.	lin.
Entire length	36	0
Length of folded wing	15	6
,, tarsus	6	6
,, middle toe	4	4
,, tail	6	10
,, bill	6	0

[Mr. Andersson's last collection contained a single specimen of this species, which, I infer from his notes, was killed either in Damara or in Ovampo Land, probably the latter; but I have been unable to ascertain the exact locality in which it was obtained. The bill in this example was black, as described in the note above quoted, indicating that the bird was killed in the breeding-season, the bill being black in this species at that period, and yellow in winter.—ED.]

340. Herodias intermedia (Wagl.). Short-billed Egret.

Ardea intermedia, Wagler, in Isis, 1829, p. 659.
Herodias plumiferus, Gould's Birds of Australia, vol. vi. pl. 57.
Herodias brachyrhynchos, A. Brehm, in Journal für Orn. 1858, p. 471.
Egretta flavirostris, Bonaparte's Consp. Av. vol. ii. p. 116.
Ardea egretta, Layard's Cat. No. 583.
Ardea intermedia, Finsch & Hartlaub's Vögel Ost-Afrika's, p. 686.
Herodias intermedia, Gray's Hand-list of Birds, No. 10110.

I obtained these Herons both at Lake Ngami and at

Objimbinque in Damara Land; on one occasion (on February 2nd, 1865) I killed three out of a flock of four. Their flight is heavy, and at a distance they look larger than they really are.

The iris is pale transparent yellow; the bill yellow, with tints of green in some places; lores and bare spaces about the eyes greenish; tibia yellowish flesh-colour; tarsus and toes bluish black.

Average measurements of two females:—

	in.	lin.
Entire length	27	3
Length of folded wing	12	7
,, tarsus	4	6
,, middle toe	3	5
,, tail	4	10
,, bill	3	9

341. Herodias garzetta (Linn.). European Lesser Egret.

Herodias garzetta, Boie, in Isis, 1822, p. 559.
Ardea garzetta, Gould's Birds of Europe, pl. 277.
,, ,, Layard's Cat. No. 584.
,, ,, Chapman's Travels in S. Afr., App. p. 417.

This is a scarce bird in Damara and Great Namaqua Land, and very local; but it is pretty common on the rivers flowing into and out of Lake Ngami, and it also occurs on the Orange River.

It associates in small flocks, and feeds on fish, lizards, frogs, crustacea, and aquatic insects.

The iris is yellow; the upper mandible almost black; the lower mandible black near the extremity, the remainder being livid, with a slight greenish tinge; the bare space round the eye is lead-coloured, the cere greenish yellow; tibia and tarsus black; toes yellowish green above, and yellow below.

Measurements of a female :—

		in.	lin.
Entire length		27	0
Length of folded wing		12	4
,,	tarsus	3	10
,,	middle toe	2	4
,,	tail	4	1
,,	bill	4	0

342. Ardeiralla Sturmii (Wagl.). Sturm's Dwarf Heron.

Ardea gutturalis, Smith's Zool. of S. Africa, pl. 91.
Ardetta Sturmii, Gray's Genera of Birds, pl. 150.
Ardea gutturalis, Layard's Cat. No. 590.
Ardeiralla Sturmii, Gray's Hand-list of Birds, No. 10145.

I observed several of these birds in Damara Land, and found them very common in Ondonga. Axel also found them exceedingly abundant between Ondonga and Ovaquenyama, as well as at Ovagandyaro.

The favourite haunts of this species are vleys surrounded with trees and bushes, on which it perches; and it is rarely met with at a distance from trees. It feeds on insects and the smaller crustacea, but seems to be somewhat omnivorous in its diet. I am inclined to think that it feeds by night.

It breeds in Ondonga, usually placing its nest in the lower branches of palm bushes which are partly immersed in water, a few feet above which the nest is situated; it is composed of stalks of coarse grass or of small twigs laid across each other without much care or strength, and with hardly any depression for the reception of the eggs, which are four in number.

The male bird only differs from the female in its darker tints.

The iris is red or reddish brown in adult, and yellowish in immature birds; the bill in adult birds is almost black, shading at the base into bluish green, which is the colour of the skin surrounding the eyes and at the corners of the mouth; in younger birds the bill is greenish yellow.

The tibia, the front of the tarsus, and the upper surface of the toes are yellowish brown, with a greenish tinge on the tibia; the back of the tarsus and the soles of the toes are bright yellow.

Measurements of a male and a female:—

	Male.		Female.	
	in.	lin.	in.	lin.
Entire length	13	9	13	0
Length of folded wing	6	1	6	0
,, tarsus	1	9	1	9
,, middle toe	1	8	1	10
,, tail	2	1	2	2
,, bill	2	2	2	2

343. Butorides atricapilla (Afzel.). Black-headed Dwarf Heron.

Ardea atricapilla, Layard's Cat. No. 587.
Butorides atricapilla, Gurney, in Ibis, 1870, p. 151.
  ,,   ,,   Gray's Hand-list of Birds, No. 10158.

This Heron is not uncommon at Lake Ngami and its watersheds, and also on the Okavango.

[This species has not been figured.—ED.]

344. Ardetta minuta (Linn.). European Little Bittern.

Ardea minuta, Gould's Birds of Europe, pl. 282.
  ,,   ,,   Layard's Cat. No. 588.
  ,,   ,,   Chapman's Travels in S. Afr., App. p. 418.
Ardetta minuta, Gray's Hand-list of Birds, No. 10148.

I never met with this species in Damara or Great Namaqua Land; but it is not uncommon on the rivers Okavango and Teoughe, and also at Lake Ngami. It inhabits marshy districts, where it hides closely, coming out on the approach of night to feed on small fish and reptiles, and also on insects and mollusca. It is found singly or in pairs.

The iris is yellow*, as is also the lore; the bill is yellow, but with a tinge of brownish on the upper mandible; the legs and toes are greenish yellow.

[I have not seen any examples of this bird from South-western Africa; but Mr. Andersson's memoranda contain a carefully worded description of this species, which appears to confirm the correctness of his identification, if it was taken from a specimen of his own procuring.

It is, however, very easy to mistake for this species, on a superficial examination, its nearly allied South-African congener *Ardetta podiceps* (Bon.).—ED.]

345. Nycticorax ægyptius (Hasselq.). European Night Heron.

Ardea nycticorax, Gould's Birds of Europe, pl. 279.
Nycticorax griseus, Layard's Cat. No. 592.

This species is pretty frequent in the Lake-country; it occurs in Ondonga in the wet season, and is recorded as having been obtained on the Orange River. In Damara Land I have only observed it very rarely, and always in immature plumage. It feeds on fish, reptiles, aquatic insects, slugs, &c.

In the adult bird the iris is crimson-red; the ex-

* [In another of Mr. Andersson's MS. notes he speaks of the iris of this species as being cherry-coloured; but there is no doubt that its usual colour is yellow.—ED.]

tremity of the bill is black, as is also the upper portion of the upper mandible, the remainder of the bill is yellowish green; the lore and skin round the eyes are greenish; the legs and toes a light lemon-colour; in immature birds the iris is a deep reddish orange.

Measurements of a male:—

	in.	lin.
Entire length . . .	23	0
Length of folded wing .	11	2
,, tarsus . .	3	0
,, middle toe . .	2	11
,, tail .	4	6
,, bill .	4	0

SCOPIDÆ *.

346. Scopus umbretta, Gmel. Tufted Umbrette.

L'Ombrette du Sénégal, Buffon's Planches Enl. pl. 796, vol. viii. p. 266.
Scopus umbretta, Layard's Cat. No. 593.
,, ,, Chapman's Travels in S. Afr., App. p. 418.
,, ,, Finsch & Hartlaub's Vögel Ost-Afrika's, p. 727.

This queer and sombre-looking bird is pretty generally diffused over Damara and Great Namaqua Land during the rainy season, but is nowhere numerous, and moves to permanent waters as the rainpools dry up. It feeds much on frogs and also upon fish. It is generally observed singly or in pairs, and is of a fearless disposition, allowing a person to approach within range without difficulty.

* [Mr. G. R. Gray, in his 'Hand-list of Birds,' vol. iii. p. 34, treats the Umbrette as forming a subfamily of the *Ardeidæ*; but it appears to me to be so very distinct a form, that I have here treated it as the type of a separate group, of which, however, it is the only member.—ED.]

The iris is brown; the bill, legs, and toes black, with a slight tinge of grey.

PLATALEIDÆ.

347. Platalea tenuirostris (Temm.). Slender-billed Spoonbill.

Leucorodia tenuirostris, Reichenbach's Prakt. Nat. Vög. pls. 435 & 437.
Platalea tenuirostris, Layard's Cat. No. 594.
Platalea leucorodia, Chapman's Travels in S. Afr., App. p. 419.
Platalea tenuirostris, Finsch & Hartlaub's Vögel Ost-Afrika's, p. 718.

This species occasionally visits Damara Land, chiefly during the rainy season, when I have reason to believe that it also occurs, though less frequently, in Great Namaqua Land. At Lake Ngami and its watersheds it is by no means an uncommon bird.

It is generally observed in small flocks; and where not much disturbed, it is not particularly shy. It feeds on fish, shrimps, small mollusca, &c.; and the stomach of one which I dissected contained a large number of beetles, chiefly aquatic.

In the adult of this species the forehead and chin are naked and smooth; the former is bright red, merging on the sides of the gape into yellow, which is also the colour of the chin; the upper mandible of the bill is livid, except the base and a narrow space along the edges, which are reddish; the lower mandible is a slaty black, with small yellow spots and yellow edges; the legs and toes are bright scarlet.

In the immature bird the iris is greyish; the bill bluish

red, except towards the base, where it is yellowish red on the upper and slaty on the lower mandible; the lore and bare space about the eyes and forehead bright bluish red.

Measurements of a male:—

	in.	lin.
Entire length	33	6
Length of folded wing	15	6
,, tarsus	6	3
,, middle toe	4	3
,, tail	5	6
,, bill	8	9

TANTALIDÆ.

348. Tantalus ibis, Linn. African Roseate Ibis.

L'ibis blanc d'Egypte, Buffon's Planches Enl. pl. 389, vol. viii. p. 370.
Tantalus ibis, Layard's Cat. No. 602.
,, ,, Gray's Hand-list of Birds, No. 10208.

I have once or twice observed this singular-looking Ibis in Damara Land, but I do not recollect to have met with it in Great Namaqua Land. In the Lake-regions it is not uncommon at all seasons.

It is generally seen in small flocks, either wading about in shallow water or stalking leisurely on the adjacent mud or sand-banks, in search of insects, larvæ, &c. When not molested it is comparatively tame.

Measurements of a female:—

	in.	lin.
Entire length	35	0
Length of folded wing	17	9
,, tarsus	8	4
,, middle toe	4	6
,, tail	7	3
,, bill	8	8

349. Geronticus calvus (Bodd.). Bald Ibis.

Courly à tête nue, Buffon's Planches Enl. pl. 867, vol. viii. p. 381.
Geronticus calvus, Layard's Cat. No. 606.
„   „   Gray's Hand-list of Birds, No. 10218.

[Mr. Andersson refers to this species in his notes as found on the Orange River; but he does not give any details respecting it, and I did not meet with it in any of his collections.—ED.]

350. Ibis æthiopica (Lath.). Sacred Ibis.

Ibis religiosa, Savigny's Hist. Nat. de l'Ibis, pl. 4.
„   „   Hemprich & Ehrenberg, Sym. Phys. pl. 17.
„   „   Bree's Birds of Europe, vol. iv. p. 45, plate (bird), p. 49, plate, fig. 1 (egg).
Geronticus æthiopicus, Layard's Cat. No. 320.
Ibis æthiopica, Gurney, in Ibis, 1868, p. 259.
„   „   Finsch & Hartlaub's Vögel Ost-Afrika's, p. 733.
„   „   Sclater, in Proc. Zool. Soc. 1870, p. 382, fig 2 (head).

I have never observed this species in Damara or Great Namaqua Land; but it is not uncommon in the Lake-regions, and is extremely abundant in Ondonga, especially during the rainy season, when it is comparatively tame, though wild at other times. It is sometimes met with in flocks of from fifty to a hundred individuals; it is a heavy bird, and its flesh is good eating.

The iris is dark brown, but with an outer ring which is reddish in adult and whitish in young birds; the bill is very dark brown; the bare skin of the head and neck is black, but with a pink spot under the eye in adult birds, which is almost livid in younger specimens.

A naked space under the wings is bright brick-red in adult birds, but paler in those which are immature; the legs are black.

Measurements of a male:—

	in.	lin.
Entire length . . .	26	9
Length of folded wing	. 14	0
,, tarsus	4	0
,, middle toe	3	6
,, tail	6	0
,, bill	6	9

351. **Hagedashia caffrensis** (Licht.). Hagedash Ibis.

Ibis chalcoptera, Vieillot's Gal. des Ois. pl. 246.
Geronticus hagedash, Layard's Cat. No. 605.
   ,,   ,,   Chapman's Travels in S. Afr., App. p. 419.
Hagedashia hagedash, Gray's Hand-list of Birds, No. 10228.

This Ibis is found abundantly in the Lake-regions and on the rivers Teoughe and Botletlé. It feeds on insects, but does not despise other kinds of food; it is always observed in flocks, which vary from a few individuals to a dozen or two in number.

These birds roost at night on trees in the immediate neighbourhood of water, which they leave at daybreak for their favourite feeding-grounds; these are sometimes situate in dense forest bush, sometimes in reedy thickets, and sometimes amongst rocks. They always return to the same tree at night, and thus often fall an easy prey to the marksman, who conceals himself in ambush within a convenient distance of their favourite perch, birds of this species being much sought after for their flesh, which is very palatable.

When suddenly disturbed, or when straggling back to their nightly quarters, these birds scream most vociferously; and during my arduous and tortuous ascent and descent of the Teoughe, I was not unfrequently quite startled by their dinning noise as we surprised them in their reedy resorts, or as we passed, at a sudden turn of the river, under one of their roosting-places, which had been previously hidden from our view.

This Ibis builds on trees overhanging the water; the nest is constructed of rough sticks, superficially lined with fibrous roots, tendrils, and grasses, and is so slightly depressed above that the hollow is barely sufficient to admit the eggs and to prevent them from falling out.

It is said that these birds, when not disturbed, will nest in the same tree for several successive seasons.

The iris is dark brown; the bill black, except on the ridge and along the lateral depression, which are bright carmine; the legs and toes are dull red.

SCOLOPACIDÆ.

352. Numenius arquatus (Linn.). Common Curlew.

Numenius arquatus, Gould's Birds of Europe, pl. 302.
  ,, ,, Layard's Cat. No. 607.
Numenius arquata, Chapman's Travels in S. Afr., App. p. 419.

The Common Curlew is sparingly met with in the interior of Damara and Great Namaqua Land, but is more frequent along the coast and on the islands. It is most commonly seen in pairs, but at times in small flocks. It is an exceedingly wary bird, and, from the open character of the localities it frequents, often defies the efforts of the sportsman.

It loves flat marshy lands and the open sea-beach,

where it searches for its food, which consists of small marine insects, crustacea, worms, &c.

At some seasons these birds grow very fat, and they are palatable as food, especially when they have not frequented the sea-shore for too long a period.

The Curlew swims with considerable ease, but appears not to take to the water by choice.

The iris is dark brown; the bill brown, with a tinge of ochry yellow on the basal half of the lower mandible.

353. Numenius phæopus (Linn.). British Whimbrel.

Numenius phæopus, Gould's Birds of Europe, pl. 303.
„ „ Layard's Cat. No. 608.
„ „ Chapman's Travels in S. Afr., App. p. 419.
„ „ Finsch & Hartlaub's Vögel Ost-Afrika's, p. 739.

So far as my observation goes, the Whimbrel is less common in Damara and Great Namaqua Land than the Curlew, and, in fact, is but rarely met with. Its food consists of snails, shells, crabs, insects, &c.

The iris is dark brown; the upper mandible of the bill and the extremity of the lower are brown; the remainder of the lower mandible is livid; the legs and toes are greenish blue.

354. Totanus calidris (Linn.). Common Redshank.

Totanus calidris, Gould's Birds of Europe, pl. 310.
„ „ Layard's Cat. No. 611.

[This species did not occur in Mr. Andersson's last collection; and his notes do not refer to it; but it is included by Mr. Layard (*loc. cit.*) amongst the species obtained at Lake Ngami, and Mr. J. E. Harting informs me that he has seen a specimen from Walwich Bay.—ED.]

355. Totanus glottis (Linn.). Greenshank.

Totanus glottis, Gould's Birds of Europe, pl. 312.
Glottis canescens, Strickland & Sclater, Birds Damar., Contr. Orn. 1852, p. 159.
Totanus glottis, Layard's Cat. No. 613.
 „   „   Chapman's Travels in S. Afr., App. p. 419.
Totanus canescens, Finsch and Hartlaub's Vögel Ost-Afrika's, p. 745.
 „   „   Sharpe & Dresser's Birds of Europe, pl. 42.

The Greenshank is pretty common in Damara and Great Namaqua Land, in all suitable localities; and I have reason to believe that it is abundant in the Lake-regions and on the river Okavango. It frequents, when inland, springs and small running streams, but is more common along the coast. It does not appear to breed in Damara Land, at least not in its middle and southern portion; from the sea-coast it generally disappears about December, reappearing in March and April. It is most commonly found in small flocks, but sometimes singly or in pairs. It is a shy and wary bird, and frequently most difficult to obtain. It runs with great celerity, and is very powerful on the wing, frequently flying at a great height; when, rising it utters shrill cries, which may be heard at a very great distance.

This species feeds on the fry of fish, worms, insects, crustaceans, and molluscous animals, in search of which it may frequently be seen wading up to its belly in the water.

The flesh of the Greenshank is very palatable.

The iris is a very dark claret-colour; the bill purplish, merging into bluish black towards the point, and bluish grey on the upper ridge; the legs are yellowish blue, but bluer in some specimens than in others.

[The Damara-Land specimens of the Greenshank which have come under my notice have been in winter plumage. Mr. J. E. Harting's collection contains a male from Walwich Bay, obtained in October, a male from Objimbinque, shot the 30th November, and a male and female, also from Damara Land, obtained the 11th January.—ED.]

356. Totanus stagnatilis, Bechst. Marsh Sandpiper.

Totanus stagnatilis, Gould's Birds of Europe, pl. 314.
    „   „   Layard's Cat. No. 610.
    „   „   Sharpe & Dresser's Birds of Europe, pl. 6.

This species is nowhere common in Damara Land; but I have occasionally shot it in the valley of the Swakop, and pretty frequently at Objimbinque. It frequents small streamlets and freshwater springs, and is found singly or in pairs.

The irides are intensely dark brown; the bill black, except at the base of both mandibles, where it is dusky green; the legs and toes are dusky yellowish green.

[The specimens of this Sandpiper contained in Mr. Andersson's last collection were in winter plumage; one of these was procured at Omanbondé, another at Hykomkap, on 3rd December, and a third at Objimbinque, on 2nd January.—ED.]

357. Totanus glareola (Linn.). Wood Sandpiper.

Totanus glareola, Gould's Birds of Europe, pl. 315. fig. 1.
  „ „ Strickland & Sclater, Birds Damar., Contr. Orn. 1852, p. 159.
  „ „ Layard's Cat. No. 610.
  „ „ Chapman's Travels in S. Afr., App. p. 419.
  „ „ Finsch & Hartlaub's Vögel Ost-Afrika's, p. 750.

This is not a common bird in Damara and Great

Namaqua Land; but now and then small flocks are to be met with at inland springs, streams, and marshes; in some seasons it was frequently obtained at Objimbinque, and I also found it not uncommon in Ondonga.

It occurs singly and in pairs as well as in small flocks, and, unless much disturbed, it is quite tame. Its flesh is very palatable.

The food of this species consists of worms and insects. I have never found the nest of the Wood Sandpiper in Damara Land, but have reason to think that it breeds there occasionally.

The iris is dark brown; the bill greenish black, darkest towards the point; the legs and toes are dark green.

358. Actitis hypoleucus (Linn.). Common Sandpiper.

Totanus hypoleucus, Gould's Birds of Europe, pl. 316.
Tringoides hypoleucus, Layard's Cat. No. 327.
Actitis hypoleucus, Finsch & Hartlaub's Vögel Ost-Afrika's, p. 752.

I obtained several specimens of this bird in Damara Land. It feeds on small snails, shells, &c.

The iris is dark brown; the bill brownish, tinged with yellow on the lower mandible; the legs dusky green, tinged with flesh-colour.

Measurements of a female:—

	in.	lin.
Entire length	7	7
Length of folded wing	4	3
,, tarsus	1	1
,, middle toe	0	11
,, tail	2	2
,, bill	1	1

359. Terekia cinerea, Güldenst.   Terek Sandpiper.

Limosa terek, Gould's Birds of Europe, pl. 307 (winter dress).
Terekia cinerea, Gurney, Birds Damar., Proc. Zool. Soc. 1864, p. 3.
Limosa cinerea, Layard's Cat. No. 609.
  „   „   Sharpe & Dresser's Birds of Europe, pl. 32 (breeding dress and young).

I have obtained very few specimens of this bird in Damara Land, the only places where I remember to have met with this species being Omanbondé, Objimbinque, and Hykomkap on the river Swakop; those which I have observed were always solitary and were feeding on the sedgy borders of marshy places or sluggish streamlets.

Their food consists of small insects.

The iris is dark brown, the legs and toes yellow.

[Mr. J. E. Harting possesses a female of this species obtained by Mr. Andersson at Walwich Bay, in October 1863.—ED.]

360. Philomachus pugnax (Linn.).   Ruff.

Machetes pugnax, Gould's Birds of Europe, pl. 325.
  „   „   Strickland & Sclater, Birds Damar., Contr. Orn. 1852, p. 159.
Philomachus pugnax, Layard's Cat. No. 619.

This bird generally appears in Damara Land with the return of the rainy season, when it is not uncommon, and leaves again before the ruff of the male bird is put forth; but I have reason to believe that it is to be met with in the Lake-regions during the intervening period. It is chiefly found inland and but rarely on the coast. It feeds on insects and worms, for which it seeks in moist and humid situations; but during the rainy season, when food is abundant, it may be found almost everywhere. It is a comparatively tame bird, and is generally to be observed in small flocks of from three to a dozen individuals—such flocks generally consisting of females with perhaps now and then a male, which is easily distinguished by its greater size.

This species when on the wing resembles *Tringa subarquata*, but is larger and swifter.

The iris is dark brown, the bill dark with a dull olive tint near the base; the legs are slate-coloured in some specimens, and yellowish green in others, the green tinge being especially perceptible on the bare part of the tibia.

Average dimensions of two males:—

		in.	lin.
Entire length		11	1
Length of folded wing		7	0
,,	tarsus	2	0
,,	middle toe	1	5
,,	tail	2	8
,,	bill	1	6

Average dimensions of twenty-four females:—

		in.	lin.
Entire length		9	3
Length of folded wing		5	9
,,	tarsus	1	7
,,	middle toe	1	2
,,	tail	2	3
,,	bill	1	3

[Mr. Andersson's last collection contained several examples of this species, but none in nuptial dress, though some remains of it were visible on three females killed at Objimbinque on the 12th, 15th, and 25th of August.—ED.]

361. Tringa canutus (Linn.). Knot.

Calidris canutus, Gould's Birds of Europe, pl. 324.
Tringa canutus, Layard's Cat. No. 620.

This species is of rather rare occurrence on the coast of Damara Land; the few that I observed there were generally associating with flocks of Sanderlings, Curlew Sandpipers, and Little Stints along the shallows in Walwich Bay. The Knot feeds on aquatic insects, in search of which it will wade knee-deep in the water. Its flesh is good eating. One of my specimens has the breast and part of the belly rufous.

The iris is dark, the bill very dark olive, the legs of a somewhat similar hue but more or less tinged with yellow.

[I examined two specimens of the Knot obtained by Mr. Andersson in Walwich Bay: one of them, killed on October 20th, retained some remains of the breeding dress; the other, which was obtained on November 4th, showed scarcely any.—ED.]

362. Tringa subarquata, Güld. Curlew Sandpiper.

Tringa subarquata, Gould's Birds of Europe, pl. 328.
Pelidna subarquata, Strickland & Sclater, Birds Damar., Contr. Orn. 1852, p. 159.
Tringa subarquata, Layard's Cat. No. 621.
Numenius pygmæus, Chapman's Travels in S. Afr., App. p. 420.
Tringa subarcuata, Finsch & Hartlaub's Vögel Ost-Afrika's, p. 761.

The Curlew Sandpiper is the commonest *Tringa* at Walwich Bay and all along the lagoons and shallows of the south-west coast, where it ranges southward to Table Bay. It congregates in flocks, often of many hundreds, and not unfrequently in company with the Little Stint and the Sanderling.

At some hours of the day, probably when changing their feeding-ground, and chiefly in the early morning, these birds are more on the move than at other times; and the air over the lagoon seems then literally to teem with their myriads, presenting a most animated picture as the white portions of their plumage flash with almost dazzling effect in the early tropical sunlight, especially when the brightness of the scene is enhanced by the presence in the flock of a large reinforcement of Sanderlings.

In the afternoon, if, as is the case five days out of seven at Walwich Bay, the wind blows strongly from the south-east, these birds generally retire to some little distance from the water and seek a large open flat in the immediate neighbourhood. Whilst there they are excessively shy and difficult to approach; and I may add that I have observed that this temporary wildness is common to most water-birds on the Damara coast whenever a high wind arises. On ordinary occasions the Curlew Sandpiper is comparatively tame, and numbers may be bagged without difficulty. Considerable variations of plumage are to be met with, as I have shot at the same time specimens in the grey dress and others in which the plumage has been almost of a rusty red.

These Sandpipers grow enormously fat, but are not desirable birds for the table, as their flavour is excessively fishy.

The iris is very dark brown; the bill, legs, and toes shining black. I have, however, seen some adult specimens in which the bill was of a dark flesh-grey, and others in which it was deep greenish black.

[Mr. J. E. Harting possesses a specimen of the Curlew Sandpiper obtained by Mr. Andersson in Walwich Bay on April 15th, and two others from the same locality, obtained on October 15th and 24th.—ED.]

363. Tringa Bairdii (Coues). Baird's Sandpiper.

Actodromas Bairdii, Coues, in Proc. Acad. Nat. Sci. Philad. 1861, p. 194.
Tringa Bairdi, Harting, in Ibis, 1870, p. 151.

[Mr. J. E. Harting possesses a male specimen of this American Sandpiper, obtained by Mr. Andersson in Walwich Bay on October 24th, 1863, being the only instance with which I am acquainted of the occurrence of this species in the Old World. Mr. Andersson probably did not particularly observe this specimen, as I do not find any reference to it in his notes. Mr. Harting has favoured me with the following remarks on this interesting species, which, by his kind permission, I here insert:—

"I do not think that *Tringa Bairdii* has ever been figured.

"This species is not recognized by Professor Baird in his 'Birds of North America,' although doubtless he has met with it since the publication of that work, unless we are to suppose that he has included it under the name of *Tringa Bonapartei*, Schlegel (*i. e. T. Schinzii*, Bonap. nec Brehm). This can scarcely be the case, since he describes correctly the latter species as having the upper tail-coverts white, which in *T. Bairdii* are black or nearly so, and gives other characters which are applicable to the former but not to the latter bird. Prof. Schlegel, strange to say, gives *T. Bairdii* of Coues as a synonym of *T. maculata*, Vieillot (*i. e. T. pectoralis*, Say and Bonap.), of which species he considers it a small variety (Mus. P.-B. *Scolopaces*, p. 39) *. Bonaparte,

[* There is a remarkable similarity in the general appearance of *T. Bairdii* and *T. maculata*; but the former may be readily distinguished by its much shorter toes. The middle toe, without the claw, measures 0·45 of an inch in *T. Bairdii*, and 0·85 in *T. maculata.*—ED.]

while correctly adopting *T. pectoralis*, Say, as a synonym of *T. maculata*, Vieillot, adds also as a synonym *T. Bonapartei*, Schlegel, which of course is quite distinct (*cf.* Compt. Rend. 1856, p. 596, and Rev. Zool. 1857, p. 120). Other authors, *e.g.* Cassin, have designated *T. Bairdii* as *T. Bonapartei*.

"Messrs. Sclater and Salvin observe (P. Z. S. 1868, p. 144) that this species appears to be the *Chorlito lomo negro* of Azara; and if so, Mr. Coues's name will have to give way to *Tringa melanota*, Vieillot (*T. dorsalis*, Meyen and Licht.).

"Notwithstanding the confusion which appears to exist with regard to this species, it may be readily recognized by any one who has read the remarks of Mr. Elliott Coues in his Monograph of the Tringæ of North America (Proc. Acad. Nat. Sci. Philad. 1861); Mr. Coues has therein clearly pointed out the distinctive characters of this species.

"Adopting the genus *Actodromas*, which was proposed by Kaup in 1829, for what he considers 'a well marked and very natural group of Sandpipers,' he includes therein the five:—*A. maculata*, Vieillot; *A. Bairdii*, Coues; *A. minutilla*, Vieillot; *A. Bonapartei*, Schlegel; and *A. Cooperi*, Baird.

"After pointing out that *A. Bairdii* is intermediate in size between *maculata* and *minutilla*, he gives the characters by which it may be distinguished from the species to which it is most nearly allied in form and colour and with which it has been frequently confounded. These characters may be conveniently stated as follows:—

"*A. Bairdii*, Coues.	*A. Bonapartei*, Schleg.
Length about 7·25 inches.	Length about 7·50 inches.
Bill slender, entirely black.	Bill stout, flesh-colour at base below.
Feathers extending on the lower mandible much beyond those on the upper.	Feathers extending but little if any beyond those on the upper.
Upper tail-coverts much lengthened, *black*; central tail-feathers projecting but little; the emargination of the tail slight.	Upper tail-coverts moderate, *white*; the central tail-feathers projecting considerably, and tail deeply emarginate.

"To the general and particular description of *T. Bairdii* as given by Mr. Coues, there can be but little to add; but as Prof. Baird has not noticed the species, and as Mr. Coues considers its habitat 'coextensive with that of *T. Bonapartei*, and probably restricted to North America east of the Rocky Mountains,' I may add some remarks on its geographical range which may be of use.

"The localities noticed by Mr. Coues are Nebraska (Lieut. Warren), Fort Kearney (Dr. Cooper), Zuni river (Dr. Woodhouse), Great Slave Lake (Messrs. Kennicott and Ross).

"The additional localities which have since been recorded are California, Mexico (Boucard), Panama, New Granada (Salvin), Tambo valley, Peru (Whitely), Conchitas, Argentine Republic (W. H. Hudson), and Santiago, Chili (Leybold).

"To these localities we have now to add Walwich Bay (Andersson).

"The egg of *Tringa Bairdii* has been described by Professor Newton in the 'Proceedings of the Zoological Society' 1871, p. 57."—ED.]

364. Tringa minuta, Leisl. Little Stint.

Tringa minuta, Gould's Birds of Europe, pl. 332.
Pelidna minuta, Strickland & Sclater, Birds Damar., Contr. Orn. 1852, p. 159.
Tringa minuta, Layard's Cat. No. 622.
,, ,, Chapman's Travels in S. Afr., App. p. 420.
,, ,, Finsch & Hartlaub's Vögel Ost-Afrika's, p. 764.
,, ,, Sharpe & Dresser's Birds of Europe, pl. 55. fig. 1 (winter dress) & pl. 58. fig. 2 (breeding dress).

This species is common throughout Damara and Great Namaqua Land, and also occurs in the Lake-regions; it is found in small flocks and frequents alike the sea-coast and freshwater pools, springs, and streamlets, where it eagerly seeks for the minute insects, crustacea, and aquatic worms which are more or less abundant in such localities. The flesh of this Sandpiper is excellent; and as it is a very tame bird, numbers may be bagged without much exertion.

The iris is dark brown, the bill, legs, and toes are black.

[Mr. Andersson's memoranda contain descriptions of this Sandpiper in winter dress, killed in December and January, and in nuptial plumage obtained in April. A specimen in his last collection, obtained on October 7th, retained a considerable portion of its breeding-plumage. Another specimen, obtained at Walwich Bay, on November 26th 1863, in full winter plumage, is figured by Messrs. Sharpe and Dresser (*loc. cit.*).—ED.]

365. Calidris arenaria (Linn.). Sanderling.

Arenaria calidris, Gould's Birds of Europe, pl. 335.
Calidris arenaria, Layard's Cat. No. 623.
Tringa calidris, Chapman's Travels in S. Afr., App. p. 416.
Calidris arenaria, Finsch & Hartlaub's Vögel Ost-Afrika's, p. 767.

The Sanderling is very common on the coast of Damara Land, but is only sparingly met with inland. It is found in great flocks, and associates with the Curlew Sandpiper; but whilst the latter, as it searches for its food, hunts and ranges not only along the beach, but at some distance from it, the Sanderling, on the contrary, scarcely if ever leaves the immediate edge of the water, where it is amusing enough to observe it feeding along a beach on which the surf is breaking, now running away from the threatening waters, then turning as if by instinct the moment they have spent their fury, closely following the receding waves and rapidly seizing, amongst their foam and spray, the minute marine animals upon which this bird subsists. The Sanderlings when thus engaged appear as if they must be overwhelmed by the seething billows; but in some marvellous manner they always escape, and it is very rarely that they are even obliged to have recourse to their wings to expedite their retreat. The flocks of Sanderlings afford a pretty sight on a sunny morning, when in their evolutions on the wing they eccentrically wheel and twist in the bright light, looking not unlike silver clouds against the clear blue sky.

The cry of this species is a kind of chirping call, low and short, but shrill. Its flesh is very palatable; and being plump little birds, they are worth the trouble of shooting and cooking.

The iris is dark brown, the bill, legs, and toes shining black.

[Specimens of the Sanderling contained in Mr. Andersson's last collection were in winter dress; but one, a male obtained at Walwich Bay on May 18th, had begun to assume the nuptial plumage, though only to a slight extent.—ED.]

366. Gallinago major (Gmel.). Solitary Snipe.

Scolopax major, Gould's Birds of Europe, pl. 320.
Gallinago major, Gurney, in Ibis, 1868, p. 261.

[Mr. Andersson's last collection contained a single specimen of this Snipe (a male), obtained in Ondonga on the 6th of February, 1867.

This species is a regular migrant to Natal, and also occurs, but less numerously, in the Republic of Trans Vaal; it arrives in Natal in September or October, and leaves in January or February.

Mr. Andersson's MS. contains the following note, which may perhaps relate to this species, but which, I think, more probably refers to *Gallinago macrodactyla*, Bon. (= *æquatorialis*, Rüpp., = *nigripennis*, Bon.):—"Once, whilst encamped at Omanbondé, awaiting the falling of the rains to enable me to penetrate to the distant interior, I frequently obtained a meal by invading the marsh in pursuit of Ducks and Snipes, which were tolerably abundant, especially the latter; but, singularly enough, I never saw these birds again in Damara Land, and, unfortunately, at Omanbondé I neglected to preserve specimens. At this distant day it is impossible for me to identify the species with any certainty. I simply took these birds to be the common Snipe (*Gallinago scolopacina*); but, on reflection, I think it equally possible that they were examples of the South-African *Gallinago nigripennis*."—ED.]

367. Rhynchæa capensis (Linn.). African Painted Snipe.

Rhynchæa capensis, Layard's Cat. No. 625.
" " Finsch & Hartlaub's Vögel Ost-Afrika's, p. 774.
" " Shelley's Birds of Egypt, pl. 11 (male & female).

The Painted Snipe is sparingly found in Great Namaqua Land; but in Damara Land it is very common, a pair or two being almost always to be found wherever the ground is swampy. It is also common on all the watersheds north and east of Damara Land; and it is pretty common in Ondonga, where it breeds, making no nest, but usually laying its eggs near the water; these are from three to four in number, of a very dark colour, freely blotched with black.

Though partial to marshy ground, this species is also found on the sides of little rills and running springs. It lies close, like a Snipe, but is very different on the wing, its flight being heavy and comparatively slow; moreover it flies but a short distance before it alights. It lives singly or in pairs; but a dozen birds may sometimes be found scattered over a small marsh within a short distance of each other. It seems to feed chiefly on insects; and its flesh is very palatable.

The iris is brown.

Measurements of a male:—

		in.	lin.
Entire length		9	2
Length of folded wing		4	9
,,	tarsus	1	9
,,	middle toe	1	5
,,	tail	1	9
,,	bill	1	10

RECURVIROSTRIDÆ*.

368. Recurvirostra avocetta, Linn. European Avocet.

Recurvirostra avocetta, Gould's Birds of Europe, pl. 308.
,,   ,, Layard's Cat. No. 617.
,,   ,, Chapman's Travels in S. Afr., App. p. 420.
,,   ,, Finsch & Hartlaub's Vögel Ost-Afrika's, p. 755.

This handsome and peculiar bird is occasionally found on the south-west coast of Africa, and also occurs, though less frequently, inland. In the Cape Colony, however, I have found the case, as regards its distribution, slightly reversed.

I may mention as inland localities for this species Objimbinque, where I have seen it once or twice, and Ondonga, where it was shot by Axel.

* [Mr. G. R. Gray, in his 'Hand-list of Birds,' which I cite as the most modern work on ornithological systematic arrangement, includes this group as a subfamily of the *Scolopacidæ*; but it seems to me that it may more naturally be regarded as forming a distinct and independent family, and I have therefore so treated it.—ED.]

At certain seasons the Avocet is not uncommon on the coast, at Walwich Bay, Sandwich Harbour, Angra Pequeña, &c.; but it usually disappears from Damara Land during the breeding-season, though I have little doubt that a few pairs remain to nest there, as I have occasionally met with very young birds during the dry time of the year.

The Avocet is generally observed in small flocks, and is on the whole a shy and wary bird. It is an interesting object to the ornithologist, to whom its graceful figure, as it quietly skirts the glassy pool or wades amongst the shallows on the sea-shore, never fails to be a source of pleasure. It feeds on insects, worms, thin-skinned crustacea, &c., which it seeks when they are left exposed on the mud or sand by the receding tide, and also by wading knee-deep in shallow water.

The flesh of the Avocet is not unpalatable.

The iris is light cherry-colour; the bill is black.

Measurements of a female:—

	in.	lin.
Entire length	15	9
Length of folded wing	8	0
,, tarsus	3	3
,, middle toe	1	6
,, tail	3	2
,, bill	3	4

369. Himantopus autumnalis. European Stilt.

Himantopus melanopterus, Gould's Birds of Europe, pl. 289.
,, ,, Strickland & Sclater, Birds Damar., Contr. Orn. 1852, p. 159.
Himantopus candidus, Layard's Cat. No. 618.
Charadrius himantopus, Chapman's Travels in S. Afr., App. p. 420.

This species is sparingly met with in the middle and northern parts of Damara Land, but more frequently in the Lake-regions and on the river Okavango.

I have always found it singly or in pairs. It feeds on insects, snails, shells, &c., and is a conspicuous and interesting object, being lively and graceful in its actions, both when running (which it does with considerable celerity) about the sides of marshes and streams, and when wading quietly in shallow water.

The iris is red.

Measurements of a male:—

		in.	lin.
Entire length	.	13	3
Length of folded wing		9	2
,,	tarsus . .	4	10
,,	middle toe . .	1	6
,,	tail	3	3
,,	bill	2	10

[Mr J. E. Harting's collection contains a male of this species, obtained by Mr. Andersson in Ondonga, 6th November, 1866.—Ed.]

RALLIDÆ.

370. Rallus cærulescens, Gmel. Caffer Rail.

Rallus aquaticus, Gurney, in Proc. Zool. Soc. 1864, p. 3.
Rallus cærulescens, Layard's Cat. No. 629.
,, ,, Finsch & Hartlaub's Vögel Ost-Afrika's, p. 777.

I found this Rail plentiful at Omanbondé; and it is not uncommon in marshy localities in Damara Land and the parts adjacent, more especially in the central and northern portions of the country; it frequents reedy thickets bordered by other rank aquatic herbage, amongst which it searches for the insects, worms, and seeds of water-plants which constitute its food. It runs with great swiftness, but does not refuse to take wing when pursued. Its flesh is good.

Measurements of a female:—

	in.	lin.
Entire length	9	10
Length of folded wing	4	6
,, tarsus	1	9
,, middle toe	1	11
,, tail	1	8
,, bill	2	1

[This species has not been figured.—ED.]

371. Ortygometra pygmæa, Naum. Baillon's Crake.

Zaporna baillonii, Gould's Birds of Europe, pl. 344.
Ortygometra bailloni, Gurney, in Proc. Zool. Soc. 1864, p. 3.
       ,,   ,,   Andersson, ibid. p. 7.
Ortygometra minuta, Layard's Cat. No. 633.
Crex baillonii, Chapman's Travels in S. Afr., App. p. 421.
Zapornia pygmæa, Gray's Hand-list of Birds, No. 10461.

This pretty species is an inhabitant of the few marshes existing in Damara Land. At Omanbondé, where it breeds, I found it plentiful; it is also common in the marshy districts about Lake Ngami, and on the rivers Teoughe and Okavango; and I likewise obtained a specimen in Ondonga. It frequents alike the rank vegetation of the stagnant pools and the more scantily sheltered rills, searching industriously for insects, worms, slugs, snails, &c. When surprised it takes wing more readily than most of its congeners, but flies only for a very short distance, and drops amongst the aquatic herbage at the first convenient spot, from whence, if needful, it prolongs its retreat by running.

This Crake constructs its nest of pieces of stalks of reeds, rushes, and other vegetable substances. The eggs are six or seven in number, of a brownish-buff or olive-brown colour, closely spotted with obscure markings of a darker hue, and are rather larger than the eggs of the Starling.

The flesh of this species is very tender and delicate.

The irides are of a dull yellowish brown, but in one of my specimens were of a yellowish red; the bill is green, darkest on the ridge of the upper mandible; the legs and feet are dusky green.

Measurements of a male:—

	in.	lin.
Entire length	6	10
Length of folded wing	3	1
„ tarsus	1	8
„ middle toe	1	4
„ tail	1	9
„ bill	0	10

[Mr. Layard has been so good as to inform me that this is the same species as was included by him in his 'Catologue of the Birds of South Africa' (*loc. cit.*), under the erroneous name of *Ortygometra minuta.*—ED.]

372. Ortygometra marginalis (Hartl.). Olive-margined Crake.

Porzana marginalis, Hartlaub's Orn. West-Afrika's, No. 685.
„ „ Taczanowski, in Journ. für Orn. 1870, p. 54.

Ondonga, February 6, 1867.—A single specimen was brought to me by an Ovampo yesterday; I cannot remember to have seen it before.

Ondonga, February 23.—Another specimen; also eggs said to belong to this bird, of a yellowish ground-colour almost hidden near the thicker end by a broad zone of light brownish red.

Ondonga, March 1.—I have ascertained that the above is correct; I myself found to-day a nest containing four eggs, situated just on the edge of a marsh, in a dryish tuft of grass. I got Axel to watch the bird, which he saw several times, though unable to secure it.

Ondonga, March 2.—Another nest was brought by a native, with the bird (which he captured upon it) and four eggs.

Ondonga, March 26.—An abandoned nest with five eggs, far from the water, and had the surrounding grass tied above it, as in the nest of *Gallinula angulata*.

The iris is brown, tinged with reddish yellow; the eyelid yellow; the basal part of the bill green, merging into bluish at the extremity; the ridge of the upper mandible dark brown; legs and feet dusky green, with a slight bluish tint on the upper portion of the uncovered part of the tibia.

This Gallinule is new to me.

Measurements of a male:—

		in.	lin.
Entire length		7	0
Length of folded wing		3	6
„	tarsus	1	7
„	middle toe	1	8
„	tail	1	6
„	bill	0	9

Measurements of three females:—

	in. lin.	in. lin.	in. lin.
Entire length	8 3	8 0	7 10
Length of folded wing	4 1	4 0	3 11
„ tarsus	1 6	1 5	1 6
„ middle toe	1 8	1 6	1 8
„ tail	2 3	2 0	2 1
„ bill	0 10	0 9	0 11

[I have copied from Mr. Andersson's note-book the above memoranda, which I have reason to believe refer to this species, two specimens of which (apparently an adult and an immature bird) were contained in his last collection. Both of these specimens were purchased by Mr. Frank, the well-known dealer in objects of natural history, by whom the younger bird was resold to the Leyden Museum, where it was kindly shown to me by Professor Schlegel, who agreed with me in referring it to the present species. This Crake has not been figured.—ED.]

373. **Alecthelia dimidiata** (Temm.). Temminck's Crake.

Gallinula dimidiata, Smith's Zool. of S. Africa, pl. 20.
Corethrura dimidiata, Gurney, Birds Damar., in Proc. of Zool. Soc. 1864, p. 3.
  „   „   Andersson, ibid. p. 8.
  „   „   Layard's Cat. No. 635.
Gallinula dimidiata, Chapman's Travels in S. Afr., App. p. 421.

I have only found this species at Omanbondé, where it is not uncommon and breeds. It frequents stagnant waters thickly fringed and studded with aquatic herbage, amongst the ever progressive decay of which it loves to disport itself and to search for food. It is very shy and retired in its habits, seldom going far from effective cover, and gliding through the mazes of the rank vegetation with astonishing ease and swiftness.

The bill is reddish brown, tinged with yellowish on

the lower mandible; the legs and toes are yellowish brown.

374. Limnocorax niger (Gmel.). Black Crake.

Gallinula flavirostra, Swainson's Birds of West Africa, vol. ii. pl. 28.
Gallinula niger, Layard's Cat. No. 642.
Rallus niger, Chapman's Travels in S. Afr., App. p. 421.
Ortygometra nigra, Finsch & Hartlaub's Vögel Ost-Afrika's, p. 779.
Limnocorax niger, Gray's Hand-list of Birds, No. 10458.

I have not unfrequently met with this species in Damara Land in suitable localities, such as Objimbinque, Schmelin's Hope, Omanbondé, &c.; and it is by no means uncommon on the rivers Okavango and Teoughe and in the Lake-regions, though, from its excessively shy habits and its partiality for dense reedy thickets, it is difficult to obtain. The surest way of procuring specimens is to lie in ambush near one of their favourite haunts; but even thus success is not always certain.

The iris is red; the bill in the adult bird is yellow, with a greenish tinge; in the immature bird it is green; the legs are bright red in the adult, but paler in younger specimens; the red on the legs fades to almost a yellow after death.

GALLINULIDÆ.

375. Gallinula angulata, Sund. South-African Lesser Waterhen.

Gallinula pumila, Sclater, in Ibis, 1859, pl. 7.
" " Gurney, Birds Damar., Proc. Zool. Soc. 1864, p. 3.
Gallinula angulata, Layard's Cat. No. 641.

Ondonga, February 6th, 1867.—I find the Lesser

Waterhen literally swarming in all the vleys of this country, where it breeds most abundantly. Its flesh is much esteemed by the natives, who make up great hunting parties to chase these birds out of the water on to the dry land, where, as they unwillingly take wing and try to conceal themselves in the bushes and grass, they are easily secured, being sometimes shot with arrows and sometimes taken alive.

I have examined a number of specimens; and, as far as I can see, the dark slaty blue ones are the adult males. In the females the plumage is more or less light grey beneath, and nearly white on the chin and throat.

In the adult males the frontal shield is bright red and the bill bright yellow; the legs and toes in some specimens are grass-green, in others drab or flesh-coloured, tinged with light green; in the adult females the bill resembles that of the males, but the frontal shield is hardly so bright, and is tinged with orange next the feathers; in younger birds the shield is orange-red and the plumage still lighter than in the old females. I infer, from the development of the sexual organs, that the young birds breed before they have arrived at maturity.

The eggs of this Waterhen are from five to six in number, of a yellowish white, freely covered with small spots of light brown, with here and there a blotch of the same colour. The nest is a mass of grass, with its foundation laid on the water, and composed of standing stalks bent downwards, with some loose ones added; the hollow in which the eggs are laid is three or four inches deep and has somewhat the appearance of a shallow inverted sugar-loaf: after the nest has been completed, the bird binds the tops of the surrounding grasses and ties them together so as to form a partial shelter against the sun, as well as to afford concealment.

Ondonga, February 24th.—No abatement in the quantity coming in of the eggs of the Lesser Waterhen.

Ondonga, March 4th.—The Lesser Waterhen continues to lay, probably for the second time this season; but it is rarely that four eggs can now be found in any nest; they have probably not time to lay the full number before they are robbed.

Measurements of a male and a female:—

	Male. in. lin.	Female. in. lin.
Entire length	11 1	10 3
Length of folded wing	5 9	5 3
,, tarsus	1 9	1 7
,, middle toe	2 1	1 11
,, tail	2 9	2 5
,, bill	0 11	0 10

376. Gallinula chloropus (Linn.). British Waterhen.

Gallinula chloropus, Gould's Birds of Europe, pl. 342.
,, ,, Gurney, Birds Damar., Proc. Zool. Soc. 1864, p. 3.
,, ,, Andersson, ibid. p. 7.
,, ,, Layard's Cat. No. 640.
,, ,, Chapman's Travels in S. Afr., App. p. 421.
,, ,, Finsch & Hartlaub's Vögel Ost-Afrika's, p. 787.

This species is common in all suitable localities throughout Damara Land and the adjacent countries. It breeds in February and March, usually forming its nest amongst the rank vegetation bordering on its favourite resorts, which are stagnant pools and other still waters overgrown with weeds and aquatic plants.

A nest taken by me on 18th February contained two eggs.

Another, taken 24th February, contained four eggs.
,, ,, 2nd March, contained two eggs.
,, ,, 4th ,, ,, three ,,
,, ,, 5th ,, ,, four ,,
,, ,, 18th ,, ,, two ,,
,, ,, 19th ,, ,, three ,,

This species swims and dives with great expertness, and may be observed nodding its head first on one side and then on the other as it swims in the more open parts of the water, picking up vegetable substances, insects, and other food as it passes onwards. In the early morning and in the evening it may be seen away from the water searching amongst the grass for worms, slugs, and larvæ, in addition to which it also feeds on grass and seeds. When on land it frequently twitches and jerks the tail, exhibiting the white under-coverts; and if suddenly disturbed will occasionally take wing for a short distance, flying with its long legs hanging downward, but more frequently prefers to seek its safety by running to, and concealing itself in, the nearest suitable reedy or marshy thicket.

In the adult bird the shield on the forehead is sealing-wax red; the bill a darker red, except the tips of the mandibles, which are greenish yellow.

The feet are green, as are also the legs, with the exception of a red garter, shading into yellow at the lower edge, which surrounds the bare part of the tibia.

In immature birds the frontal shield is but slightly developed, the bill much tinged with greenish yellow, the legs dull green, and the garter greenish yellow.

Measurements of a female:—

		in.	lin.
Entire length		12	4
Length of folded wing		6	0
,,	tarsus	1	9
,,	middle toe	2	1
,,	tail	3	0
,,	bill	1	1

[It would appear, from Mr. Andersson's memoranda above transcribed, that the number of eggs laid by this species in South-west Africa is considerably less than the usual complement which it produces in Great Britain.—ED.]

377. Porphyrio smaragnotus, Temm. Green-backed Porphyrio.

Porphyrio erythropus, Layard's Cat. No. 639.
Fulica porphyrio, Chapman's Travels in S. Afr., App. p. 421.
Porphyrio smaragnotus, Finsch & Hartlaub's Vögel Ost-Afrika's, p. 783.

This splendid bird is rather scarce in Damara and Great Namaqua Land, but is pretty abundant in the Lake-regions and on the rivers Teoughe and Okavango; it is also not uncommon, during the rainy season, in Ondonga, where the inhabitants call it "King of the Waterhens," and declare that the moment it utters its deep guttural notes every Waterhen within hearing immediately responds by its own peculiar cry. The only spot in Damara Land proper where I found this species at all common was the great reedy marsh of Omanbondé; but there it was very timid, and consequently most difficult to approach. It seldom ventured into the open, but would warily skirt the dense reedy recesses which formed its favourite haunts, and into which it would precipitately retreat on the slightest sign of danger.

At Lake Ngami and on the river Botletlé I found it less difficult to obtain, probably on account of its greater abundance.

It lies close during the day, and is usually only to be seen in the early morning and in the cool of the afternoon.

Its food is very various, and consists of seeds of aquatic plants, mollusks, fish, eggs, and, I have no doubt, even young birds. In a domesticated state it will eat meat readily. If captured young it becomes very tame, and may be trusted at large, when it will freely associate with common poultry.

This species has a heavy unwieldy flight, and has recourse to its wings only as a last chance of making its escape. It dives when in water deep enough to allow of its doing so, and it runs with great rapidity amongst the tangled reedy brakes of its native haunts.

The iris is red; the bill red, but darker at the extremity; the tarsus red, tinged with blue.

[Drs. Finsch and Hartlaub (*loc. cit.*) quote the Orange River, on the authority of M. Verreaux, as a locality for this species. This Porphyrio has, I believe, not been figured. The nearly allied Madagascar race, figured by Buffon in the 'Planches Enluminées,' pl. 810, and considered by many naturalists to be identical with the present bird, appears to me to be specifically distinct (*vide* ' Ibis,' 1868, p. 470).—Ed.]

378. Porphyrio Alleni, Thoms. Allen's Porphyrio.

Poryhyrio Alleni, Thomson, in Ann. & Mag. Nat. Hist. 1842, vol. x. p. 204.
" " Gray's Genera of Birds, pl. 162.
" " Finsch & Hartlaub's Vögel Ost-Afrika's, p. 785.

Ondonga, February 5th, 1867.—Axel brought home a pair of this lovely Waterhen, the first we have seen; it is evidently scarce in these parts.

Bill, legs, and toes bright red.

Measurements of a female:—

		in.	lin.
Entire length		10	6
Length of folded wing		5	8
"	tarsus	2	0
"	middle toe	2	0
"	tail	2	9
"	bill	1	0

[It would appear by the note here transcribed that Mr. Andersson had not, at the time of writing it, become acquainted with the specific name of this Porphyrio; but two specimens of it, which I believe to be the identical two above referred to, were contained in his last collection, and I thus had the opportunity of inspecting them before they were disposed of. One of these specimens was adult, and the other immature; the latter was subsequently added to the collection of the Leyden Museum.—ED.]

379. Fulica cristata, Gmel. Rufous-knobbed Coot.

Fulica cristata, Bree's Birds of Europe, vol. iv. p. 83, pl.
" " Layard's Cat. No. 643.
" " Chapman's Travels in S. Afr., App. p. 421.

This species is common in suitable localities in Damara and Great Namaqua Land, but is more abundant in the Lake-regions.

These Coots may often be observed congregated in

large numbers on open sheets of water, where they might easily be mistaken for a flock of Ducks, except that they do not "pack" like wild fowl.

If disturbed, they will sometimes, if near a reedy brake, seek safety by hiding there; but more frequently they have recourse to their wings, when they exhibit great powers of flight.

They build their nests of, and amongst, reeds, rushes, and grasses, usually selecting the most retired spots, though I have also found their nests in most exposed situations. A few old reed-stalks serve as a footing for the nest, which is roughly but firmly constructed, and is raised, though sometimes only a few inches, above the surface of the water.

The eggs are from seven to ten in number, of a buff colour, freely speckled with minute spots of pale brown, with larger spots at wider intervals. The young birds follow their parents very soon after being hatched.

PARRIDÆ.

380. Parra africana, Gmel. Greater African Jacana.

Parra africana, Swainson's Zool. Ill. (2nd series), pl. 6.
,, ,, Gurney, Birds Damar., Proc. Zool. Soc. 1864, p. 3.
,, ,, Layard's Cat. No. 626.
,, ,, Finsch & Hartlaub's Vögel Ost-Afrika's, p. 781.

I have never seen this curious species in Great Namaqua Land; and it is a comparatively scarce bird in Damara Land, but pretty common on the rivers

Okavango, Teoughe, and Botletlé, and also at Lake Ngami.

It is found in pairs or in small flocks, frequenting stagnant pools or still waters on the sides of lakes and rivers, where it runs about on the decayed semifloating herbage, and also on the large-leaved lotus plants, which generally abound in such situations—a feat which it accomplishes without sinking, by means of its long and wide-spreading toes.

These birds are of a lively disposition, and frequently chase one another about. When they have not been previously disturbed they are generally easy to approach; and their vivacious habits, elegant forms, and handsome colouring add much to the interest of the scene.

Measurements of two males:—

	in. lin.	in. lin.
Entire length	9 2	10 6
Length of folded wing	5 6	5 11
,, tarsus	2 4	2 6
,, middle toe	2 6	2 1
,, tail	1 10½	2 1
,, bill	1 3	1 4

Measurements of two females:—

	in. lin.	in. lin.
Entire length	10 10	11 7
Length of folded wing	6 3	6 6
,, tarsus	2 8½	2 9
,, middle toe	2 6½	2 6
,, tail	2 0¼	2 3
,, bill	1 6	1 3

381. Parra capensis, Smith. Lesser African Jacana.

Parra capensis, Smith's Zool. of S. Africa, pl. 32.
„ „ Andersson, in Proc. Zool. Soc. 1864, p. 7.
„ „ Layard's Cat. No. 627.

This Jacana is common in the Lake-regions and on the Okavango, where it breeds; but it is very rare in Damara Land, and, I believe, is never seen in Great Namaqua Land.

[' The Ibis ' for 1864 contains at p. 360 the following note on this species by Mr. Thomas Ayres, who met with it in Natal:—
" Iris light hazel; bill bright brown; tarsi and feet light greenish brown. The male is precisely similar in size and plumage to the female. I found numbers of these beautiful Jacanas on Sea-cow Lake. In habits they much resemble the larger kind, running with ease on the weeds which appear on the surface; they are rather shy. If, in searching for food, they happen to approach a large Jacana (*Parra africana*), they are immediately chased away; and as both kinds are plentiful in that locality, and feed all day long, there is constant squabbling amongst them. There is one habit they have which I have not noticed in the other Jacanas, viz. the dipping the head up and down, like some of the smaller Plovers."—ED.]

ANSERES.

PHŒNICOPTERIDÆ.

382. Phœnicopterus erythræus, Verr.   Greater South-African Flamingo.

Phœnicopterus erythræus, J. & E. Verreaux, in Rev. et Mag. de Zool. 1855, p. 221.
  „   „ Andersson, in Ibis, 1865, p. 64.
  „   „ Layard's Cat. No. 644.
  „   „ Chapman's Travel's in S. Afr., App. p. 420.
Flamingo, Baines's Explorations in South-west Africa, frontispiece.
Phœnicopterus erythræus, Gray, in Ibis, 1869, pl. 14. fig. 5 (head).
  „   „ Finsch & Hartlaub's Vögel Ost-Afrika's, p. 795.

This Flamingo is very abundant at Walwich Bay, Sandwich Harbour, Angra Pequena, and the mouth of the Orange River; it is also met with in a few inland localities, such as Lake Ngami, Lake Onondava, &c., all of which are more or less impregnated with saline substances, to which this species seems to be attached.

With rare exceptions (and these not well authenticated, but merely surmised from young birds being sometimes found barely able to fly) the Flamingoes do not breed in any of the parts of the coast above particularized; and, indeed, the only locality where I know for a certainty that they nest is the inland one of Lake Ngami.

On the approach of the breeding-season they leave the coast of Damara Land, and wing their way to the northward; they take their departure about the month of February and return about the latter end of October and during November, the old birds being the first to arrive.

The Flamingo feeds both during the day and the night: but I suspect that the latter is its principal feeding time; for about sunset flocks varying from a few individuals to many hundreds may be seen pursuing their flight in various directions, and their loud croaking voices may be heard throughout the night.

The favourite resorts of these birds are shallows partially left dry by the ebbing tide; here they industriously search for the small crustacea, marine animalcula, and sea-grasses which constitute their food.

The Flamingo is strictly a wading bird, but on rare occasions will make use of its webbed feet by resorting to deep water, evidently for the sake of the fun of the thing.

Once, whilst stationed at Walwich Bay, I observed these birds for several consecutive days thus amusing themselves. About 10 or 11 A.M. they began to congregate at a particular spot, a short distance outside the lagoon, settling in about five or six feet of water; and I believe that every Flamingo for miles round must have joined in this singular assembly, for the number collected amounted to several thousands; and the scene, as may be supposed, was both pretty and striking.

This species is invariably in good condition, and is often enormously fat. The young birds are not bad eating, but, being rather fishy, require to be well cooked and spiced after all the fatty matter has been carefully removed.

In newly fledged specimens the bill and legs are of a very dark purple; but as the bird approaches its adult stage the bill becomes more elongated, with the extremities black and the remaining portions first bluish and subsequently pink, when the bird has become fully adult, at which period the legs and feet are of a scarlet-pink and the irides are orange.

383. Phœnicopterus minor, Geoff. St.-Hil. Lesser South-African Flamingo.

Phœnicopterus parvus, Temmink's Pl. Col. pl. 419.
Phœnicopterus minor, Strickland & Sclater, Birds Damar., Contr. Orn. 1852, p. 159.
,, ,, Andersson in Ibis, 1865, p. 65.
,, ,, Layard's Cat. No. 645.
,, ,, Chapman's Travels in S. Afr., App. p. 420.
,, ,, Gray, in Ibis, 1869, pl. 15. fig. 8 (head).
,, ,, Finsch & Hartlaub's Vögel Ost-Afrika's, p. 798.

This species is comparatively rare at Walwich Bay and elsewhere on the south-west coast of Africa, but at Lake Ngami it is more common.

It is a perfect gem amongst the feathered tribes; and a flock of these birds quietly feeding in some secluded nook, with their bright-coloured legs half immersed in the water and with the sunlight playing on their beautiful plumage, forms a scene which is almost fairy-like.

In the adult bird the form of the bill is precisely the same as in the newly fledged young of *P. erythrœus*; the basal part of the bill is of a dull brownish purple, the part adjoining being vermilion which deepens into crimson bordered by black, the latter being shaded off into a light horn-colour towards the extremities of the mandibles; the bare skin between the bill and the eye is of the same colour as the base of the bill.

The irides are orange, the legs and toes crimson-scarlet.

ANATIDÆ.

384. Plectropterus gambensis (Linn.). Western Spur-wing Goose.

Plectropterus gambensis, Sclater, in Proc. Zool. Soc. 1859, pl. 153. fig. 2.
      „   „   Layard's Cat. No. 646.
Anser gambensis, Chapman's Travels in S. Afr., App. p. 422.

This noble bird is not uncommon on the river Okavango and at Lake Ngami; it is also found on the river Teoughe; but I am not aware that it has been met with in either Damara or Great Namaqua Land.

It is generally found in small flocks, and presents a conspicuous object when standing erect in the marshes to which it resorts.

It is said to perch and roost occasionally on trees.

A full-grown well-conditioned male will sometimes weigh as much as fifteen pounds; but twelve pounds is a more usual weight.

The flesh of this Goose is palatable.

Measurements of a specimen believed to be a male, and of a female:—

	Male.		Female.	
	in.	lin.	in.	lin.
Entire length	44	8	37	9
Length of folded wing	21	3½	18	3
„ tarsus	5	0	3	9
„ middle toe	4	9	4	2
„ tail	10	0	7	0
„ bill	3	9	3	2

385. Sarkidiornis melanotus (Penn.). Knob-billed Goose.

Oye de la côte de Coromandel, Buffon's Pl. Enl. vol. ix. p. 396, pl. 937.
Sarkidiornis africana, Layard's Cat. No. 647.
„   „   Chapman's Travels in S. Afr., App. p. 422.
Sarcidiornis melanotus, Finsch & Hartlaub's Vögel Ost-Afrika's, p. 799.

This very handsome species is common in Damara and Great Namaqua Land during the rainy season, and is found at all seasons at Lake Ngami and on the river Okavango.

It is usually found in flocks, and may not unfrequently be seen perching on dry trees near the water; its flesh is very good.

The comb on the bill is found in the male bird only; this is black, and the bill is nearly so; the tarsi and toes are lead-coloured; the irides dark brown.

Measurements of a male and a female:—

	Male. in. lin.	Female. in. lin.
Entire length	26 9	22 0
Length of folded wing	14 3	11 4
„ tarsus	2 7	2 0
„ middle toe	3 0	2 7
„ tail	5 9	5 0
„ bill	2 7	2 3

[I believe that all the figures of this species which have been published have been taken from Indian specimens; but I agree with Drs. Finsch and Hartlaub (*loc. cit.*) in considering these to be specifically identical with African examples.—ED.]

386. Chenalopex ægyptiacus (Linn.). Egyptian Goose.

Chenalopex ægyptiacus, Gould's Birds of Europe, pl. 353.
„   „   Layard's Cat. No. 648.
Anas egyptica, Chapman's Travels in S. Afr., App. p. 421.

This is the most common species of Goose on the waters of Damara and Great Namaqua Land, where it remains throughout the year, and is invariably found either in pairs or in small flocks.

During the daytime it is not unfrequently to be seen at some distance from the water. When on the wing it utters a kind of barking quack.

Its flesh is very dark-coloured, coarse-tasted, and at times uneatable.

The irides are brownish orange, the legs bright flesh-colour.

Average dimensions of two males:—

	in.	lin.
Entire length	26	5
Length of folded wing	15	0
,, tarsus	3	5
,, middle toe	3	2
,, tail	6	0
,, bill	2	2

[As Mr. Andersson does not record the measurements of a female bird, I may add that the females of this species are considerably smaller than the males.—ED.]

387. Nettapus auritus (Bodd.). African Dwarf Goose.

Sarcelle de Madagascar, Buffon's Pl. Enl. vol. x. p. 121, pl. 770.
Nettapus madagascariensis, Layard's Cat. No. 649.
,, ,, Chapman's Travels in S. Afr., App. p. 422.
Nettapus auritus, Gray's Hand-list of Birds, No. 10593.

I have only observed this handsome little Goose on Lake Ngami and its watersheds, where it is not uncommon.

It is met with in small flocks, and is not very shy. It

is exceedingly fat at certain seasons, and is pretty good eating.

In the male the iris is bluish; the bill rich orange, inclining to livid on the edges of the lower mandible; the nail of the upper mandible horn-colour.

The legs and toes are a shining black, tinged with dusky yellowish on the outer toes and the outer side of the legs.

In the female the bill differs from that of the male in the upper mandible being dusky olive, with a greenish orange patch on the lower part at the base, and a small livid spot on each side of the nail.

Measurements of a male and a female:—

	Male.		Female.	
	in.	lin.	in.	lin.
Entire length	12	6	12	3
Length of folded wing	5	9	5	9
,, tarsus	1	1	1	1
,, middle toe	1	0	1	1½
,, tail	3	0	2	0
,, bill	1	2	1	2

[Dr. Hartlaub, in his excellent work on the birds of West Africa, alludes at p. 247 to Mr. Andersson's having obtained specimens of the eastern "*Nettapus coromandelianus*" on Lake Ngami; but as, so far as I can ascertain, the present is the only species of the genus *Nettapus* which is found in that locality, I venture to think that Dr. Hartlaub may have been accidentally misinformed in this particular.

Mr. Andersson himself seems to have suspected that some error of identification may have occurred, as I find the following memorandum among his notes:—"*Nettapus coromandelianus.* Did I really obtain this Goose in the Lake-country?"—ED.]

388. Dendrocygna viduata (Linn.). Widow Tree-Duck.

Canard du Maragnan, Buffon's Pl. Enl. pl. 808, vol. x. p. 106.
Dendrocygna viduata, Layard's Cat. No. 650.
,, ,, Chapman's Travels in S. Afr., App. p. 423.
,, ,, Finsch & Hartlaub's Vögel Ost-Afrika's, p. 806.

This Duck, which congregates in immense flocks, is exceedingly common in the Lake-regions and on the river Okavango; in the latter locality it is seen most abundantly during the annual inundation, when much of the usually dry land which abuts upon the river is converted into marshes and swamps. As these temporary resorts dry up, the Ducks of this species move eastward until they arrive at that extensive flooded country which stretches far and wide on each side of the Teoughe below Libebé.

I have never seen this Duck in Damara or Great Namaqua Land.

The iris is dark brown; the upper mandible of the bill black, darkest on the nail, which is surrounded by a livid blue patch, a similarly coloured spot being situate in front of each nostril; the lower mandible black, mingled with brown and livid blue; the legs and toes bluish lead-colour, tinged with brown on the upper surface; the webs a shade darker than the toes.

Measurements of a male:—

	in.	lin.
Entire length	17	9
Length of folded wing	9	0
,, tarsus	2	3
,, middle toe	2	7
,, tail	3	0
,, bill	2	3

[This species occurs in South America as well as in Africa and Madagascar; and Buffon's plate, the only one of this Duck which I have been able to examine, represents an American specimen.—ED.]

389. Mareca capensis (Gmel.). Cape-Wigeon.

Querquedula capensis, Eyton's Anatidæ, p. 128.
Mareca capensis, Layard's Cat. No. 652.
Anas capensis, Chapman's Travels in S. Afr., App. p. 422.
„ „ Gray's Hand-list of Birds, No. 10640.

This is rather a scarce Duck in Damara and Great Namaqua Land; but I have found it more abundant in the immediate neighbourhood of Walwich Bay than elsewhere in Damara Land.

The iris is greenish yellow; the upper mandible purplish grey, except a small yellowish-pink patch below the nostrils, which merges gradually into purplish grey; the under mandible pinkish; the legs and toes grey, mixed with brown.

Average measurement of three males:—

		in.	lin.
Entire length		15	9
Length of folded wing		7	9
„	tarsus	1	7
„	middle toe	1	11
„	tail	2	8
„	bill	2	0

[This species has not been figured.—ED.]

390. Pœcilonetta erythrorhyncha (Smith). Red-billed Teal.

Pœcilonitta erythrorhyncha, Smith's Zool. of S. Africa, pl. 104.
Anas erythrorhyncha, Layard's Cat. No. 653.
„ „ Chapman's Travels in S. Afr., App. p. 422.
„ „ Finsch & Hartlaub's Vögel Ost-Afrika's, p. 808.
Nettion erythrorhyncha, Gray's Hand-list of Birds, No. 10668.

This is the commonest Duck in Damara and Great Namaqua Land, where it is found throughout the year; and it also abounds on most of the waters to the northward. I met with several of its nests in Ondonga in the months of February and March; the largest number of eggs contained in any of these nests was ten. This species is usually observed in flocks, and, where not previously disturbed, is not very shy or difficult to obtain. Its flesh is excellent.

The iris is brown; the legs, toes, and webs blackish purple; the upper mandible is dark purple, faintly tinged with green upon the ridge, the remainder of the mandible above the edge being pinkish; the lower mandible is bluish pink.

391. Nettion hottentota (Smith). Hottentot Teal.

Querquedula hottentota, Smith's Zool. of S. Africa, pl. 105.
Querquedula hottentotta, Layard's Cat. No. 657.
      ,,   ,, Chapman's Travels in S. Afr., App. p. 422.
Nettion hottentota, Gray's Hand-list of Birds, No. 10662.

This is a rare Duck in Damara and Great Namaqua Land. I procured more individuals at Omanbondé than at any other locality; several specimens have also been brought from the Lake-country.

The iris is reddish yellow; the bill bluish red, except at the base, where it is black; the legs are purplish yellow.

Measurements of a male and a female :—

	Male.		Female.	
	in.	lin.	in.	lin.
Entire length	13	4	13	7
Length of folded wing	5	9	6	0
„ tarsus	1	2	1	2
„ middle toe	1	6	1	6
„ tail	2	11	3	0
„ bill	1	7	1	7

392. Spatula capensis (Smith). South-African Shoveller.

Rhynchaspis capensis, Smith's Zool. of S. Africa, pl. 98.
„   „   Layard's Cat. No. 658.
Spatula capensis, Gray's Hand-list of Birds, No. 10679.

This Duck is comparatively scarce in Great Namaqua and Damara Land; but I have traced it as far north as the river Okavango.

The iris is lemon-yellow; the bill deep reddish brown, almost black; the legs, toes, and webs are ochry yellow, but the latter are dusky towards the extremities.

393. Anas sparsa, Smith. White-spotted Duck.

Anas sparsa, Smith's Zool. of S. Africa, pl. 97.
Anas leucostigma, Rüppell's Syst. Uebers. pl. 48.
Anas sparsa, Layard's Cat. No. 654.
„   „   Chapman's Travels in S. Afr., App. p. 422.

I have never seen this Duck in Damara Land, and only on one or two occasions in Great Namaqua Land; but I have reason to think that it is less unfrequent during the rainy season in some parts of the latter country, chiefly along the southern course of the Great Fish River and its tributaries.

[I have not met with this species in Mr. Andersson's collections;

but his MS. notes contain a detailed description of its plumage, which leaves no doubt of the correctness of the identification. —Ed.]

394. Anas xanthorhyncha, Forst. Yellow-billed Duck.

Anas flavirostris, Smith's Zool. of S. Africa, pl. 96.
" " Layard's Cat. No. 655.
" " Chapman's Travels in S. Afr., App. p. 422.
Anas xanthorhyncha, Gray's Hand-list of Birds, No. 10639.

This fine Duck is somewhat common on Lake Ngami and the Botletlé River, but I do not remember to have met with it in Damara or Great Namaqua Land, though to the south of the Orange River it is a widely diffused and common species.

The iris is brown; the bill gamboge-yellow, with an oblong brown patch on the ridge of the upper mandible, the nail being similar.

[Mr. Andersson's MS. notes contain a description in detail of this species, from which it is clear that his identification of it is correct, though his last collection did not contain a specimen.—Ed.]

395. Aythia capensis (Cuv.). South-African Pochard.

Nyroca brunnea, Eyton's Anatidæ, pl. 23.
" " Strickland & Sclater, Birds Damar., Contr. Orn. 1852, p. 160.
" " Layard's Cat. No. 659.
Fuligula nyroca, Chapman's Travels in S. Afr., App. p. 423.
Aythia capensis, Gray's Hand-list of Birds, No. 10692.

This species is only a visitor to Damara Land, and, I suspect, a rare one; for I only remember meeting with it there on one occasion, when I obtained several individuals from a flock which had settled on a vley a day's journey from Barmen. It is, however, very common in the Ondonga country during the wet season; and whilst I was there eggs were brought to me which were said to belong to this Duck; and, though I did not succeed in identifying them with certainty, I think it probable that such was the fact, as they exceeded in size the eggs of *Anas erythrorhyncha*, from which they also differed in form and colour.

The iris is cherry-coloured; the bill lead-coloured, except the nail, which is black, as are also the legs, toes, and webs.

Measurements of a male and a female :—

	Male. in. lin.	Female. in. lin.
Entire length	19 6	18 0
Length of folded wing	9 0	8 6
,, tarsus	1 3	1 3
,, middle toe	2 2	2 2½
,, tail	3 5	3 3
,, bill	2 0	2 1

396. Thalassornis leuconota, Smith. Yellow-throated Diving Duck.

Thalassornis leuconota, Smith's Zool. of S. Africa, pl. 107.
  ,,   ,,   Layard's Cat. No. 660.
Clangula leuconota, Chapman's Travels in S. Afr., App. p. 423.

This is a comparatively scarce species in both Great Namaqua and Damara Land; but during one season I found it tolerably abundant at the large marshy vley of Omanbondé in the latter country.

It is generally found singly or in pairs. When disturbed it takes wing unwillingly and merely skims the surface of the water, settling again as soon as possible; it appears to prefer endeavouring to escape from danger by diving, in which it is very expert, being able to continue long under water.

The upper mandible of the bill is shining blackish brown, marbled with yellow on the sides, and dull on the nail; the lower mandible dusky yellow with a brown patch at the base; the legs and toes dark livid brown, the webs inky-coloured.

Measurements of a male and a female:—

	Male.		Female.	
	in.	lin.	in.	lin.
Entire length	15	1	14	9
Length of folded wing	6	6	5	4
,, tarsus	1	8	1	7
,, middle toe	3	0	2	9
,, tail	2	6	2	9
,, bill	2	0	1	10

[This species appears also to inhabit Ondonga, as a specimen obtained there on November 6th 1867, was comprised in Mr. Andersson's last collection.—ED.]

397. **Erismatura maccoa** (Smith). Maccoa Diving Duck.

Oxyura maccoa, Smith's Zool. of S. Africa, pls. 108 & 109.
Erismatura maccoa, Layard's Cat. No. 661.

[Mr. Andersson does not appear to have met with this species, which, however, must not be omitted from the present work, as a female specimen is recorded (*loc. cit.*) as having been obtained at the mouth of the Orange River by Sir Andrew Smith.—ED.]

PODICIPIDÆ *.

398. Podiceps cristatus, Linn. Great Crested Grebe.

Podiceps cristatus, Gould's Birds of Europe, pl. 388.
" " Layard's Cat. No. 691.
" " Chapman's Travels in S. Afr., App. p. 423.
" " Gurney in Ibis, 1869, p. 303.

I have only observed this handsome species on the sea-coast, chiefly at or near Walwich Bay, and there by no means numerously.

It is seldom that more than three or four of these Grebes are seen together, and generally not so many. They are rather wary, but may nevertheless be successfully surprised if the sportsman is acquainted with their habits.

Not unfrequently they may be seen asleep on the water, when of course it does not require much art to secure them. It is, however, a bad plan to fire at them in such a position, as there is then but a small portion of the body exposed to view; it is best to startle the birds lightly, when they immediately stretch forth their

---

* [Mr. Andersson's last collection contained a male specimen in breeding-plumage of *Podica Petersi,* Hartl.; but this species is not referred to in his MS. notes, and I can find no trace of where the specimen in question was obtained. Under these circumstances, and especially as the collection which Mr. Andersson had formed contained several specimens from the Knysna, I do not feel at liberty to include *Podica Petersi* amongst the species enumerated in the present volume.

I may take this opportunity of mentioning that I cannot agree with Mr. G. R. Gray, in his 'Hand-list of Birds,' and with Drs. Finsch and Hartlaub, in their work on the birds of East Africa, in placing the family of Heliornithidæ next to that of Gallinulidæ; in my view it should follow the Podicipidæ, which it appears to me, in some respects, naturally to connect with the genus *Plotus.*—ED.]

long necks to the full extent, besides considerably raising the body.

If these birds are seen swimming in deep water, but within gunshot of the shore, the gunner should run as fast as possible straight for the birds and as far as the land will allow him, when they rarely take wing, but appear surprised and half stupified. When fired at they dive if not killed, but generally reappear within range for a second or even a third shot. In shallow water, however, they are very difficult to reach, as in such positions they take wing at once on the slightest approach of danger.

I dissected all the specimens which I obtained, and invariably found that the stomachs contained scarcely any thing but fine sea-grasses; this species does, however, feed on shrimps, sea-lice, small mollusca, &c.

The flesh of this bird is not very palatable.

The iris is reddish yellow; the under mandible of the bill is vermilion; the lower edges of the upper mandible are of the same colour but paler; the ridge of the upper mandible is dull bluish black, tinged with olive; the legs and toes are bluish black on the outer, and yellowish green on the inner side, the latter tint prevailing especially at the joints.

[The South-African specimens of this Grebe which have come under my notice have all been slightly smaller than European examples; some details of their comparative measurements are given in 'The Ibis' (*loc. cit.*).—ED.]

399. Podiceps nigricollis, Sund. Eared Grebe.

Podiceps auritus, Gould's Birds of Europe, pl. 301.
,, ,, Layard's Cat. No. 692.

Podiceps nigricollis, Gurney, in Ibis, 1868, p. 263, & 1869, p. 303.
Proctopus nigricollis, Gray's Hand-list of Birds, No. 10753.

The only locality in Damara Land in which I have met with this species is Walwich Bay, where, however, it is a rare bird.

[I have not seen a Damara-Land example of this Grebe; but Mr. Andersson's portfolio contained an excellent drawing, by Mr. Baines, of a specimen, partially in breeding-dress, which was obtained at Walwich Bay on November 8th.

In this, as in the preceding species, South-African examples are smaller than most of those from various localities north of the tropics; and some particulars illustrating this fact will be found in 'The Ibis' (*loc. cit.*).—ED.]

400. Podiceps minor, Linn. British Little Grebe.

Podiceps minor, Gould's Birds of Europe, pl. 392.
 „   „   Strickland & Sclater, Birds Damar., Contr. Orn. 1852, p. 160.
 „   „   Layard's Cat. No. 693.
 „   „   Finsch & Hartlaub's Vögel Ost-Afrika's, p. 811.

I have repeatedly shot this diminutive Grebe at Lake Ngami, Otjikoto, Omanbondé, and Walwich Bay, but have nowhere found it abundant except in the vleys of the Ondonga country, where it breeds in vast numbers.

Its eggs are from four to six in number and of a dirty white. The nest is a mass of weeds and grasses, and lies on the water. The eggs are always found covered over, often several inches deep. Out of the numerous nests I have taken and seen, in no one instance (except where the nest contained only one or two eggs) did I find the eggs uncovered; and the covering is so complete and regular that it is not possible that it can be the work of an instant or performed whilst the bird was making a hurried retreat from its nest.

The iris is dirty brown; the bill black except the tip, which is whitish horn-colour; the angle of the mouth is pale greenish yellow; the legs and toes are dusky livid, with a greenish tinge near the tarsal joint.

Measurements of a female:—

	in.	lin.
Entire length	8	0
Length of folded wing	3	8
,, tarsus	1	2
,, middle toe	1	8
,, tail	1	3
,, bill	0	11

SPHENISCIDÆ.

401. Spheniscus demersus, Linn. South-African Penguin.

Le Manchot du Cap de Bonne-Espérance, Buffon's Planches Enl. pl. 382, vol. x. p. 219.
Spheniscus demersa, Layard's Cat. No. 695.
Spheniscus demersus, Gray's Hand-list of Birds, No. 10790.

This is the only description of Penguin found on the south-west coast of Africa; but if this coast-line lacks variety in this genus, the deficiency is in some measure compensated by the great abundance of this particular species, which is found on almost all parts of the coast from the Cape of Good Hope to Walwich Bay. How much further it may extend to the north I cannot say.

This species is known to sea-faring men as the "Jackass Penguin;" and its most favourite resorts are the Ichaboe, Mercury, Hollanis-bird, and Possession islands.

Next to the Gannet it is the most valuable depositor

of African guano; but its value is somewhat diminished by its habit of scratching the ground which it occupies, in consequence of which the real guano becomes much mingled with sand.

From the latter end of March to late in May these Penguins go to sea in a body, and may then be seen forty or fifty miles away from land.

The egg of this species is bluish white, very large, round at one end, but tapering towards the opposite extremity. In the breeding-places frequented by these birds they always occupy the highest parts, the Gannets coming next, and nearest to the sea the Cormorants.

This Penguin, if undisturbed, breeds about September and October—but if judiciously deprived of its eggs, will continue to lay for seven or eight consecutive months. It makes no particular nest; but sometimes it surrounds its sitting-place with a few seaweeds, remnants of decayed birds, scraps of bones, and even broken bottles. The number of eggs is one or two, and occasionally, though rarely, three. The bird sits quite upright on its egg or eggs, which it keeps close between its heels and tail; and if compelled to move off, it frequently shuffles away its eggs unperceived.

These Penguins sit very close together, and when incubating are not only perfectly fearless but somewhat vicious; and a man cannot then move amongst them without the probability of being bitten, which is no laughing matter, for the bite of this Penguin will not unfrequently take a piece of flesh clean out of a man's leg. The guano-collectors always use a stout stick when moving about among the birds; and they by no means spare it.

I have been assured on good authority that the Penguins will remain one or two months on islands where they breed without going in search of food; they are very fat when they arrive, but very much reduced when they leave.

The young birds remain on the land until they are well grown, and do not follow their parents until they are able to provide for themselves.

The iris in this species is dark brown, and the bill blue-black, with a transverse bar of livid flesh-colour across both mandibles.

[A striking view of a group of these Penguins on the island of Ichaboe is contained in an engraving at page 349 of Mr. Andersson's second work, 'The Okavango River.'—ED.]

PROCELLARIIDÆ.

402. Puffinus major, Fab. Greater Shearwater.

Puffinus cinereus, Smith's Zool. of S. Africa, pl. 56.
„ „ Layard's Cat. No. 662.
Puffinus major, Gray's Hand-list of Birds, No. 10828.

This species is common in the Cape seas; and I have reason to think that it is not unfrequently met with off the coast to considerably north of the Orange River. It is generally observed in the Cape seas from May till September, when it retires to its breeding-grounds.

[I have not had the opportunity of examining a South or

South-west African example of this species, and am therefore unable to confirm Mr. Andersson's identification of it.—ED.]

403. Procellaria pelagica, Linn. Stormy Petrel.

Thalassidroma pelagica, Gould's Birds of Europe, pl. 447. fig. 2.
Procellaria pelagica, Gray's Hand-list of Birds, No. 10851.

This Petrel is occasionally seen rather numerously at Walwich Bay, and is quite common off the rest of the south-west coast of Africa.

[This species is included in Mr. Andersson's MS. list of birds which he had sent to England for identification, and the specific names of which he had thus accurately ascertained; but I did not find any specimens of it in his collection.—ED.]

404. Procellaria oceanica, Kuhl. Wilson's Petrel.

Thalassidroma Wilsonii, Gould's Birds of Australia, vol. vii. pl. 65.
Thalassidroma Wilsoni, Layard's Cat. No. 665.
Oceanites oceanicus, Gray's Hand-list of Birds, No. 10860.

This bird is not unfrequently met with off the south-west coast of Africa, as well as in many of the bays and inlets. I have occasionally seen it very abundant about the fisheries at Walwich Bay and Sandwich Harbour, where these birds would approach within a few feet of the fishermen, eagerly picking up the smaller particles of refuse thrown away by them whilst cleaning their fish on the shore.

This species is distinguished from the preceding one by its greater size, and by the yellowish spot on the webs between the toes.

405. Pseudoprion turtur (Smith). Dove-coloured Petrel.

Procellaria turtur, Smith's Zool. of S. Africa, pl. 54.
    ,,   ,, Gould's Birds of Australia, vol. vii. pl. 54.
    ,,   ,, Layard's Cat. No. 672.
Pseudoprion turtur, Gray's Hand-list of Birds, No. 10921.

This species is occasionally seen in the bays and inlets of the south-west coast of Africa, especially after a storm; but the open sea is its favourite resort. It sometimes settles on the water, but rarely remains longer than is necessary to enable it to fish up its prey. It will follow in a vessel's course for hours together, but evidently not with a view of picking up the refuse that may chance to be thrown overboard, as is the case with some other species, nor does it approach very close. It is a very powerful flier, and cleaves the air with astonishing velocity—now rising suddenly, then abruptly precipitating itself to the very crest of the foaming waves, and skimming gracefully over the intervening troughs—its varied evolutions thus affording to the voyager a constant object of never-ceasing interest.

[According to Mr. Andersson's MS. memorandum, already alluded to under the head of *Procellaria pelagica*, it appears that this species was also satisfactorily identified from a specimen which he forwarded to England some years since; but it did not occur in his last collection. Mr. Andersson records (in his MS. notes) that this Petrel, and also two or three nearly allied congeners of similar habits which frequent the southern and southwestern coast of Africa, but which he does not appear to have succeeded in positively identifying, are called by the sailors " whale-birds."

Writing of the Petrels, to which this term is thus promiscuously applied, Mr. Andersson observes:—" I have frequently seen the ' Whale-birds ' settle on the water in the neighbourhood of Table Bay in large flocks; but I have never seen them remain sufficiently long to allow a vessel to pass them, or even to come abreast of them, as they invariably rise first; other Petrels will frequently allow a vessel to pass them, often within a stone's throw. I have seen this denied; but I state a fact. A large number of Whale-birds may frequently be seen hovering over a whale as he disports himself; they fly very low on such occasions, traversing rapidly the spray and foam ejected from the whale's nostrils, as if it afforded them some kind of food; and probably such is the case. I have also seen them hover about spots where a quantity of seaweed is floating, no doubt with the object of picking up such insects &c. as may be found attached to the weeds."—ED.]

406. Daption capensis (Linn.). Cape-Petrel.

Daption capensis, Gould's Birds of Australia, vol. vii. pl. 53.
„   „   Layard's Cat. No. 671.

This is the most common Petrel off the south and south-west coast of Africa, but is rarely seen much to the north of 27° lat. It is pretty generally known as the "Cape-pigeon;" why, it is hard to say, except it be from its comparative fearlessness of man.

The first appearance of these birds after a long voyage is hailed with universal pleasure; for, whether on the wing and sailing round and round the vessel, or quietly resting on the surface of the water, they equally present a picture which never fails to attract and delight the eye.

The same birds have been known to follow a vessel for several consecutive days.

Their food is various, consisting chiefly of mollusca, but including such refuse as is thrown overboard from vessels, and sometimes also the carcass of a putrid whale, which, when it occurs, affords to these birds a welcome feast. They are voracious feeders, and will rush at a bait with the utmost avidity, disputing fiercely with each other for the treacherous morsel; they are thus easily captured with a hook and line, and sometimes by a line only, if suddenly thrown over them as they pass close under a vessel's stern.

Sometimes also they become entangled by flying against lines which have been set by the sailors for that purpose.

When caught, they are exceedingly awkward on the deck of a vessel, as they cannot rise, but attempt to make their escape by running along with outstretched wings; on such occasions they frequently vomit large quantities of a reddish-coloured offensive oil. This they seem to do from sheer fright only.

Sealers declare that the only locality known as a breeding-place of this species is the island of South Georgia.

[Mr. Andersson's collection did not contain a specimen of this well-known Petrel; but its plumage is accurately described in his MS. notes, showing that the species was correctly identified.—ED.]

407. Ossifraga gigantea (Gmel.). Giant Petrel.

Procellaria gigantea, Gould's Birds of Australia, vol. vii. pl. 45.
   „   „ Layard's Cat. No. 667.
Ossifraga gigantea, Gray's Hand-list of Birds, No. 10872.

The Giant Petrel is not unfrequently met with on the

south-west coast of Africa; and I have observed it between the 26th and 35th degrees of south latitude.

It is a most voracious feeder, and hardly any thing comes amiss to it; but it seems to be especially fond of the abandoned carcasses of whales and seals.

This species pursues the Greater Shearwater, probably with the view of compelling it to disgorge any prey that it has captured.

[Mr. Andersson's last collection did not contain a specimen of this Petrel; but Mr. Layard (*loc. cit.*) mentions a white variety of this species as being commonly found in the neighbourhood of Walwich Bay.—ED.]

408. Diomedea exulans, Linn. Wandering Albatross.

Diomedea exulans, Gould's Birds of Australia, vol. vii. pl. 38.
 „   „   Layard's Cat. No. 675.

The thirtieth degree of south latitude has been assigned as the northern limit of the Great Wandering Albatross; but I have frequently met with them off the coast two or three degrees further north. They are generally observed at sea from April to October; during the rest of the year few or none are to be seen—certainly none of the old birds, which retire at that season to their breeding-places at the island of Tristan d'Acunha, Prince Edward's Island, and the south-eastern part of Kerguelen's Land.

The iris is brown; the bill yellowish horn-colour, with a pinkish tint here and there, which gradually fades after death; the legs and toes bluish.

[Mr. Andersson's collection did not contain this well-known species; but there can be no doubt of the correctness of his identification of it.—ED.]

409. Diomedea melanophrys, Temm. Black-browed Albatross.

Diomedea melanophrys, Temminck's Pl. Col. pl. 456.
      „   „   Gould's Birds of Australia, vol. vii. pl. 43.
      „   „   Layard's Cat. No. 676.

This bird is very common on the south-west coast of Africa, and may even be occasionally seen close inshore. I have not unfrequently found them in Walwich Bay, as well as in some other bays and inlets to the southward of that locality. It is known to sealers and sailors by the name of "Mollymawk," and is very little inferior in grace of flight and strength of wing to the Great Wandering Albatross.

As far as I have observed, it is more difficult to capture in the open sea than that species; but it is frequently taken by the fishermen in the vicinity of the southern coast, and exposed for sale in Cape-Town market.

This species sometimes captures by diving the moluscous animals on which it preys, but it seems to do so rather from necessity than from inclination; and it will frequently pursue some of the smaller allied birds, especially the Greater Shearwater, when these have captured any thing in sight of their stronger kinsman, and compel them to disgorge their prey for its benefit.

[Mr. Andersson's collection did not contain this Albatross; and the identification of the species therefore rests on his authority.—ED.]

LARIDÆ.

410. Stercorarius pomarinus, Temm.   Pomarine Skua.

Lestris pomarinus, Gould's Birds of Europe, pl. 440.
Coprotheres pomarinus, Gray's Hand-list of Birds, No. 10941.

[Mr. Andersson's last collection contained two skins of this species, one of which was ticketed as having been obtained in Walwich Bay.—ED.]

411. Stercorarius parasiticus (Linn.).   Richardson's Skua.

Lestris Richardsonii, Gould's Birds of Europe, pl. 441.
Stercorarius spinicauda, Layard's Cat. No. 681.
Lestris Richardsonii, Chapman's Travels in S. Afr., App. p. 425.
Stercorarius parasiticus, Gray's Hand-list of Birds, No. 10937.

This bird is not uncommon on many parts of the south-west coast of Africa; and I have often killed it at Walwich Bay.

It frequents the innermost shallows and lagoons on the coast, but is not equally abundant throughout the the year, apparently retiring from that part of the coast during the breeding-season.

This species rarely fishes for itself, but compels the timid Gulls and Terns to disgorge their captures for its benefit.

412. Larus vetula, Bruch.   South-African Black-backed Gull.

Larus vetula, Bruch, in Journ. für Orn. 1853, p. 100, pl. 2. fig. 4 (head).
Larus dominicanus, Layard's Cat. No. 682.
Larus fuscus, Chapman's Travels in S. Afr., App. p. 425.

This is a very common Gull all along the south-west coast of Africa, from Walwich Bay to Table Bay. It is a most voracious feeder, preying on the dead carcasses of whales and seals, and also devouring freely dead rats, birds, or fish, as well as worms, insects, and shell-fish; besides which it is very destructive to the eggs of other sea-fowl. I have been assured on excellent authority that it carries off whole, and with perfect facility, the eggs of Penguins and Gannets to some distant rock, where it devours them at leisure. It is also said that this Gull has the singular habit of destroying its own eggs if it finds that its nest is in danger of being robbed of them by any person who is seeking to gather them.

The breeding-places of this species are the rocky islets off the south-west coast, to which it resorts for the purpose of incubation about the month of December. Its eggs vary much in colour, being of various shades of green, drab, or brown, profusely blotched and spotted with dark brown, especially at the larger end.

This Gull is easily caught with a hook and line; and specimens so captured are frequently offered for sale in Cape-Town market.

When kept in captivity it becomes quite tame and perfectly fearless.

[I am not aware that any complete figure of this Gull has been published.—ED.]

413. Cirrhocephalus poiocephalus (Swains.). African Grey-headed Gull.

Larus poiocephalus, Swainson's Birds of W. Afr. vol. ii. pl. 29.
Xema phæocephala, Strickland & Sclater, Birds Damar., Contr. Orn. 1852, p. 160.
Larus poiocephalus, Layard's Cat. No. 683.

Larus phæocephalus, Chapman's Travels in S. Afr., App. p. 425.
„   „   Finsch & Hartlaub's Vögel Ost-Afrika's, p. 825.
Cirrhocephalus poiocephalus, Gray's Hand-list of Birds, No. 10996.

This species is found both on the sea-coast and on the inland watersheds. It is rather a scarce species at Walwich Bay, but more common at Lake Ngami. It generally occurs singly.

414. Sterna caspia. Caspian Tern.

Sterna caspia, Gould's Birds of Europe, pl. 414.
„   „   Layard's Cat. No. 684.
„   „   Chapman's Travels in S. Afr., App. p. 425.

This splendid and powerful Tern is not uncommon at Walwich Bay, and on the south-west coast of Africa generally.

When on the wing it usually pursues a steady flight, and at a distance bears considerable resemblance to a Gull. It utters at intervals, especially when fishing, exceedingly harsh and discordant notes; and when about to do so it arrests its course, and, rising suddenly at an angle of about 45° to its line of flight, gives vent to its cries, which are repeated two or three times, and in uttering which it greatly depresses its lower mandible whenever it opens its bill for this purpose.

So far as I have seen, it lives entirely on fish.

Where not previously disturbed it is not particularly shy or difficult to approach; but when once it knows a gun it becomes exceedingly cunning.

At Walwich Bay several pairs of these Terns were in the habit of returning at dusk from their fishing-excursions within easy gunshot of the shore. At the accustomed hour I would station myself in some convenient spot, carefully concealing my gun, but I could not deceive these birds; when within some hundred yards of me they would suddenly diverge, making a circuit of from a quarter to half a mile; after which they would return to their accustomed line of flight along the shore.

The iris is reddish brown; the bill vermilion-red, clouded with horn-colour towards the points of the mandibles; the legs and toes are reddish black.

415. Sterna Bergii, Licht. Swift Tern.

Sterna velox, Rüppell's Atlas, pl. 13.
Sterna Bergii, Layard's Cat. No. 687.
Sterna velox, Chapman's Travels in S. Afr., App. p. 424.
Sterna Bergii, Finsch & Hartlaub's Vögel Ost-Afrika's, p. 828.

Next to *Sterna caspia* this is the finest Tern on the south-west coast of Africa, where it is not uncommon. I have shot large numbers of it at Walwich Bay, where I have found it frequenting the innermost parts of the bay, as well as some other inlets and bays to the southward.

It lives entirely on fish.

The iris is dark brown; the bill a rich warm yellow; the legs and toes black, but with yellow spots along and under the toes. In immature birds the feet and legs are yellowish, with irregular black markings; and the yellow of the bill is tinged with greenish.

416. Sterna cantiaca, Gmel.   Sandwich Tern.

Sterna cantiaca, Gould's Birds of Europe, pl. 415.
„   „   Layard's Cat. No. 686.
Sterna Boysii, Chapman's Travels in S. Afr., App. p. 424.

This is a common species at Walwich Bay, and occurs in all parts of the coast between that locality and Table Bay.

It is easy to obtain, and is sometimes seen on the lagoons of the south-west coast in immense flocks, mingling with *Sterna fluviatilis* and other species.

It utters short, harsh cries, especially when occupied in fishing.

The iris is dark; the points of the mandibles are pale yellow; the remainder of the bill and also the legs and toes are a shining black.

417. Sterna fluviatilis, Naum.   British Common Tern.

Sterna hirundo, Gould's Birds of Europe, pl. 417.
Sterna Dougalli, Layard's Cat. No. 685.
„   „   Chapman's Travels in S. Afr., App. p. 424.
Sterna fluviatilis, Sharpe & Dresser's Birds of Europe, pl. 45.

This Tern is pretty common on many parts of the south-west coast of Africa, though not so numerous as the preceding species at Walwich Bay.

The iris is dark brown; the bill dark horn-colour, but reddish brown near the base, and with the tips of the mandibles light horn-colour *; the legs, toes, and webs

---

* [The colour of the bill, as here described, is no doubt taken from a specimen which was undergoing the seasonal change, the bill of this species being black in winter, with light horn-coloured tips, and red during the breeding-season, with black tips.—ED.]

red; the claws dark horn-colour, tipped with yellow horn-colour.

[Messrs. Sharpe and Dresser (*loc. cit.*) describe a specimen of this Tern in winter plumage, which was procured by Mr. Andersson at Walwich Bay on October 6th, 1863; and another, in similar plumage, obtained there on the 4th of November following, was also contained in Mr. Andersson's last collection. —ED.]

418. Sternula balænarum, Strickl. Damara Tern.

Sternula balænarum, Strickland & Sclater, Birds Damar., Contr. Orn. 1852, p. 160.
  „   „   Chapman's Travels in S. Afr., App. p. 425.

This exquisite little Tern is very abundant at Walwich Bay, and on some other parts of the south-west coast of Africa. It flies in pairs, or in small flocks, uttering harsh and rapid cries. It feeds on small fish and crustacea, in search of which it explores the creeks and shallows which are left by the receding tide. It is swift of flight, and very rapid in its movements.

This species occasionally breeds at Walwich Bay, being the only Tern which does so; its eggs are deposited in a small hole which it scoops in the sand.

The iris is brown; the bill black, except close to the base, where it is of a dusky yellow; the legs and toes are yellow.

[This Tern has not been figured.—ED.]

419. Pelodes hybrida (Pall.). Whiskered Tern.

Sterna leucopareia, Gould's Birds of Europe, pl. 424.
Pelodes hybrida, Gray's Hand-list of Birds, No. 11071.

Ondonga, February 6th.—Obtained a specimen of

this Tern; it was circling for some time round a vley in company with another; these and another seen on the Great Flat a few days since are the only individuals I have yet met with.

March 6th.—Axel obtained one to-day; and since the above date I have twice observed it.

Measurements of a female:—

		in.	lin.
Entire length		10	10
Length of folded wing		9	0
,,	tarsus	0	11
,,	middle toe	0	10
,,	tail	3	6
,,	bill	1	7

[The Norwich Museum possesses a female specimen of this Tern in full breeding-dress, obtained by Mr. Andersson in Ondonga on the 6th April.—ED.]

420. **Hydrochelidon nigra** (Linn.). White-winged Black Tern.

Sterna leucoptera, Gould's Birds of Europe, pl. 423.
   ,,   ,,   Naumann's Nat. Vögel Deutsch. pl. 257.
   ,,   ,,   Ayres, in Ibis, 1871, p. 267.
Hydrochelidon nigra, Gray's Hand-list of Birds, No. 11070.

I have never seen this Tern on the sea-shore, but it is common on many inland freshwater lakes, and during the wet season hunts over the temporary rain-pools.

It feeds on fry, frogs' spawn, snails, &c.

It flies slowly and heavily, examining the ground carefully as it progresses.

Colour.—Forehead, cere, chin and throat, and a space on each side of the nape, lower part of the neck, all the underpart, including axillary plume and under wing-coverts, pure white; top of head, at the back of ears and nape, brown margined with white.

Upper part of back and shoulders brownish grey, edged with lighter grey; the rest of the back and scapulars, tertials, and wings grey, shaded and spotted with brown; outer vanes of the quills dusky silver-grey; nearly half of the inner vanes grey-white, shading lighter on the inner vanes; all shaded with brown on the extremities; shafts light wood-brown; inner side of wings grey, shading lighter on the inner vanes; upper tail-coverts pale grey; tail also grey, tinged slightly with brown; bill jet-black *, except the very extremities, which are light horn-colour; legs and toes bright red; claws dark horn-colour.

Form.—Body slender; bill rather short but strong, anteriorly slightly curved, points awl-like; cutting-edges of both mandibles bent inwards; nostrils basal, linear, oblong, and situated nearer the cutting-edges than the superior ridge; wings very long and pointed, reaching when folded far over the end of the tail; first quill the longest.

Tail moderately long and slightly forked; tarsi short and weak; toes moderately long and slender and united by a membrane to the second articulation; claws longish, nearly straight, middle the longest and dilated; hind toe and nail very short.

[As this Tern was not comprised in Mr. Andersson's last collection, and as the species was not identified by him, I have transcribed in full his description of it, which appears to me

* [The bill of this species, though black in winter, assumes more or less of a dark red tinge when the bird attains its full nuptial dress.—ED.]

to refer without doubt to the winter plumage of the White-winged Black Tern, a species which has been met with in Vaal by Mr. Ayres, as recorded in 'The Ibis' (*loc. cit.*).—ED.]

421. Rhynchops flavirostris, Vieill. African Scissor-bill.

Rhynchops flavirostris, Vieillot's Gal. des Ois. pl. 291.
Rhynchops orientalis, Rüppell's Atlas, pl. 24.
Rhynchops flavirostris, Shelley's Birds of Egypt, pl. 14.

This species occurs in Ondonga and also at Lake Ngami.

The iris is dark brown; the bill red, but yellowish at the extremity; the tarsus is also red.

PELECANIDÆ.

422. Sula capensis, Licht. South-African Gannet.

Sula capensis, Layard's Cat. No. 697.
Sula australis, Chapman's Travels in S. Afr., App. p. 424.
Sula capensis, Gray's Hand-list of Birds, No. 11104.

This is a very common bird on the south-west coast of Africa, and is found far to the north of Walwich Bay and from thence southward to the Cape of Good Hope.

Its flight is very powerful and continuous, and it also presents a very pretty picture as it rides carefully on the foaming and crested waves. It is a most expert diver, and will precipitate itself from a considerable height with fearful velocity in pursuit of its prey, which it seizes under the surface, often at some depth. As it comes sweeping down, its wings are extended and motionless, but it tucks them in a moment or two before it touches the water. The Gannet is the most valuable depositor of guano in South Africa. A single bird produces about four inches in depth annually, or 16 to 17 lbs. weight to the square foot. The chief strongholds of this species are Ichaboe, Mercury, and Possession islands.

So valuable have the deposits of this species been found, that the lessees of Ichaboe have considered it worth while to enclose the middle of the island for a distance of 300 feet by 180 feet, exclusively for the convenience of this bird. The chief object is to prevent the Penguins from mingling their adulterated deposits with those of the Gannets; and the proprietors have succeeded admirably. The former are of course kept out from want of wings available for flight.

The Gannet incubates about October and November. It makes no nest, and lays only one egg, of a pure white colour, tapering at one end, and of an oval form. Average of three eggs: length 3″ 10‴, breadth 1″ 11‴. The young, which are at first covered with white down, remain on the island where they are hatched until well able to fly, when they all suddenly disappear and are never seen again in that locality in their first plumage; when they reappear it is in the adult dress. In the interval I have frequently seen and shot them about Walwich Bay. The old birds leave the islands about the end of April and beginning of May, and at this time are generally observed in the open sea, often at a considerable distance from land; they return late in July.

[I believe that this Gannet has not been figured, except incidentally in the view of the island of Ichaboe, published in Mr. Andersson's work 'The River Okavango,' p. 349.—ED.]

423. Plotus Levaillantii, Licht.* African Darter.

Plotus Levaillantii, Temminck's Pl. Col. pl. 380.
Plotus congensis, Layard's Cat. No. 696.
Plotus Levaillantii, Gray's Hand-list of Birds, No. 11101.

During my visit to Lake Ngami I saw there a bird which I believe to have been of this species; and my friend Mr. E. Layard informs me that specimens of it were obtained in the Lake-country by Mr. James Chapman, jun., by whom they were presented to the Cape Museum.

424. Graculus carbo (Linn.). Common Cormorant.

Phalacrocorax carbo, Gould's Birds of Europe, pl. 407.
Graculus carbo, Finsch & Hartlaub's Vögel Ost-Afrika's, p. 844.

At one season of the year this species is not uncommon at Walwich Bay, and from thence southward to Table Bay, all along the coast and on the adjacent islands; it is, however, by no means so numerous as *G. capensis*.

The iris is bright sea-green, the bare skin round the eyes and the base of the lower mandible yellow; the lower mandible and the edges of the upper mandible light horn-colour, the ridge of the latter dark horn-colour; the legs and toes inky black.

* [Mr. Gray, in his 'Hand-list of Birds,' vol. iii. p. 125, adopts the view that the genus *Plotus* should be regarded as forming a distinct family; but it seems to me that the Darters are so closely allied to the Cormorants that they may more properly be treated as forming merely a subfamily of the Pelecanidæ.—ED.]

[A specimen of this Cormorant in winter plumage was contained in Mr. Andersson's last collection, and is now in that of Mr. J. H. Gurney, jun.; it is marked as having been obtained in Walwich Bay on December 16th, 1863, and is precisely similar to English examples in the winter dress.

Mr. Layard informs me that he believes he has seen the true *Graculus carbo* in Simon's Bay in breeding-plumage, but that the bird which was inserted in his 'Birds of South Africa,' p. 380, under the name of *Graculus carbo*, proved on further examination to be a distinct species, *Graculus lucidus*, Licht.—ED.]

425. Graculus capensis, Sparrm. Cape-Cormorant.

Pelecanus capensis, Sparrman's Mus. Carlson. pl. 61.
Graculus capensis, Layard's Cat. No. 699.

This is the most abundant species of Cormorant along the whole south-west coast of Africa; indeed at some seasons of the year they may be counted not merely by tens or even by hundreds of thousands, but by millions: their numbers, in fact, exceed all computation; for it is no unusual thing to see a deep unbroken line of these birds winging their way for two, or even three, consecutive hours to or from their feeding-grounds.

This Cormorant leaves Walwich Bay for its breeding-places in December; and during the nesting-season large numbers are to be found on almost every suitable rock and islet from the river Cunéné to Table Bay, in which situations, next to the Gannet and Penguin, this species is the principal depositor of guano. Its nest is composed of seaweed; its eggs are either one or two in number, white speckled with pearl-grey and pale sea-green, and about two inches in length.

About June and July these Cormorants are again to be seen in Walwich Bay, where they frequent the low sands of the peninsula which protects the entrance of the bay; but no accumulation of guano takes place in that locality, as their usual resting-places are flooded when the tide is high.

This species feeds on fish, and sometimes on mussels.

The iris is sea-green, the eyelids studded with small beads of cobalt-blue; the beak more or less brown, the bare skin under the bill and at the gape orange; the legs, webs, and toes black.

426. Graculus neglectus, Wahl. Wahlberg's Cormorant.

Graculus neglectus, Wahlberg, in Journ. für Orn. 1857, p. 4.
  „   „   Gray's Hand-list of Birds, No. 11121.

[As this Cormorant was discovered by the late Professor Wahlberg on the rocky islands adjacent to the coast of Great Namaqua Land, it may rightly claim to be included in the present volume.

Mr. Andersson does not appear to have been personally acquainted with this species, although he refers in his MS. notes to Wahlberg's description of it; and no examples of it have come under my own notice; but I think it probable that it may be the Cormorant with twelve rectrices mentioned by Mr. Layard in 'The Ibis' for 1868, p. 121, as having been obtained in two instances in the vicinity of Cape Town.

These specimens were supposed by Mr. Layard to be referable to *G. carbo*; but that species has fourteen rectrices, a peculiarity which, according to Professor Schlegel (Muséum des Pays-Bas, *Pelecani*, p. 4), is only shared by *G. lucidus* and *G. capensis*, the number of rectrices in all other Cormorants being but twelve.

The following is a translation of the account given by Professor Wahlberg (*loc. cit.*) of the present species:—

"Greenish black, but cinereous brown on the back, with

bronzy reflections; the feathers narrowly (1-1½ millim.) edged with greenish black, rounded at the tips in adults, but slightly pointed in younger specimens; throat almost bare, but the space extending from the angle of the mouth to below the nostrils feathered; the sides of the head much feathered, but a black bare ring round the eyes (2 millims. wide). Rectrices twelve. Iris ochre-yellow in adult birds, but green on the lower moiety, in younger specimens entirely a cinereous brown; bill blackish horn-colour; feet black.

"*Male.*—Entire length 715 millims.; expanse of wings 1160; bill from forehead 60, height at base 13·5; wing 273; tail 130 to 138; tarsus 55; middle toe, with claw, 82.

"*Female.*—Bill from forehead 56 millims., height 13; wing 269; tail 137; tarsus 55.

"Tolerably frequent on the islands off the coast of Western South Africa, such as Possession, Halifax, Ichaboe, &c."

This species has not been figured—ED.]

427. Graculus africanus (Gmel.). Long-tailed Cormorant.

Carbo longicauda, Swainson's Birds of W. Afr. vol. ii. pl. 31.
Graculus coronatus, Wahlberg, in Journ. für Orn. 1857, p. 4.
Graculus africanus, Layard's Cat. No. 700.
Phalacrocorax africanus, Chapman's Travels in S. Afr., App. p. 423.
Graculus africanus, Finsch & Hartlaub's Vögel Ost-Afrika's, p. 847.

This Cormorant occurs on Lake Ngami and its watersheds; but I have never met with it, except on inland waters. It feeds on fish, and is a most expert diver. Its flight is strong and rapid; and it perches on trees both during the day and at night.

This species feeds chiefly at night; as the sun declines it is seen in flocks flying from its roosting-places to its fishing-grounds. During the day it remains in great measure stationary, either lazily sunning itself on some branch overhanging the water, or on a bunch of reed; or it may be seen standing erect on a sandbank, with outstretched wings. When in the water, it has the habit of submerging its body to such an extent as to leave little more than the neck exposed.

[Although this Cormorant appears for the most part to frequent inland waters, it was found breeding, by the late Professor Wahlberg, on the small islands off the coast of Great Namaqua Land; and specimens which he obtained in nuptial dress from these localities were described by him under the name of "*Graculus coronatus*" in the 'Journ. für Orn.' *loc. cit.* One of the specimens so named by Wahlberg, and obtained by him from the island of Ichaboe, is recorded by Drs. Finsch and Hartlaub (*loc. cit.*) as being now preserved in the Bremen Museum.

The following is a translation of Professor Wahlberg's remarks on the nidification of this species:—

"Tolerably common on Possession and other islands; makes a nest of seaweed without much substructure; lays one white egg."—ED.]

428. Pelecanus minor (Rüpp.). Rüppell's Pelican.

Onocrotalus minor, Rüppell's System. Uebers. pl. 49.
Pelecanus mitratus, Gurney, in Ibis, 1861, p. 135.
Pelicanus onocrotalus, Layard's Cat. No. 701.
Pelecanus onocrotalus, Chapman's Travels in S. Afr., App. p. 424.
Pelecanus mitratus, Sclater, in Proc. Zool. Soc. 1868, p. 266, fig. 3 (head).
Pelecanus minor, Elliot, ibid. 1869, p. 580.

This species is very numerous at Walwich Bay, and is also tolerably common at Sandwich Harbour; but south of this port its numbers rapidly diminish. On approaching Table Bay the Pelican reappears and is abundant in many parts of the Cape seas nearly inshore; it is also found at Lake Ngami, Lake Onondava, and a few other localities in the interior.

Where not previously disturbed the Pelican is not of a particularly shy nature; but at Walwich Bay it has now become nearly impossible for the European to approach it, even within rifle-shot. It, however, shows little or no fear of the natives; and it is no unusual thing to see Pelicans watching the natives at a short distance whilst they are spearing fish in the shallows, and even anticipating them in the capture and thus cheating them of their quarry.

From January to May, Walwich Bay is very bare of these as well as of other birds, as they then resort to the interior for the purpose of incubation. At Lake Ngami the Pelican makes its nest among the bushes, and deposits a single white egg of a beautiful oval shape.

[I have not had an opportunity of examining a Pelican from South-western Africa; but Mr. Andersson, though he described the species found in Damara Land under the name of *P. onocrotalus*, seems from his MS. notes to have considered it identical with that found at the Cape, which, according to Mr. Elliot (*loc. cit.*), appears to be referable to the smaller species *P. minor*, to which it is therefore most probable that Mr. Andersson's observations also refer.

Mr. Chapman, in his 'Travels in South Africa,' vol. i. p. 346, speaking of the Pelicans observed by him at Walwich Bay, remarks that they "pursue their prey nightly in the shallow lagoons, driving the fish, by flapping their wings on the water, near the shore, and catching them while floundering."—ED.]

ALPHABETICAL INDEX

OF GENERA AND SPECIES.

	Page		Page
Abdimii, Ciconia	280	afra, Cinnyris	70
——, Sphenorrhynchus	280	——, Columba	236
Abingoni, Campethera	221	——, Eupodotis	260
——, Chrysopicus	222	——, Nectarinia	70
abyssinica, Coracias	53	——, Peristera	236
abyssinicus, Bucorvus	205	African Jacana, Greater	328
——, Ploceus	171	——, Lesser	330
——, Tmetoceros	206	African Kingfisher, Great	59
ACCIPITER	28	African Swift, Little	48
ACTITIS	303	africana, Buphaga	163
ACTODROMAS	308	——, Parra	328
Adjutant	282	——, Sarkidiornis	335
adspersa, Scleroptera	247	——, Strix	42
adspersus, Francolinus	247	——, Upupa	64
AËDON	90	africanoides, Megalophonus	198
Aëdon, Fasciolated	90	africanus, Graculus	370
——, Kamiesberg	90	——, Phalacrocorax	370
——, Smith's	92	afroides, Eupodotis	260
——, White-browed	92	——, Otis	260
ÆGIALITES	272	AGAPORNIS	216
ÆGITHALUS	79	Aguimp	112
ægyptiacus, Chenalopex	335	aguimp, Motacilla	112
ægyptica, Anas	335	ALARIO	175
ægyptius, Merops	61	alario, Alario	175
——, Milvus	22	——, Amadina	175
——, Nycticorax	293	ALAUDA	192
æquatorialis, Gallinago	312	alaudarius, Tinnunculus	18
æthiopica, Ibis	297	alba, Ardea	289
æthiopicus, Geronticus	297	——, Ciconia	280
afer, Oxylophus	225	——, Herodias	289
——, Parus	81	Albatross, Black-browed	356
affinis, Drymœca	83	——, Wandering	355
——, Drymoica	83	albicans, Saxicola	105
——, Platysteira	131	albicauda, Platystira	132
——, Platystira	131	albiceps, Charadrius	267
——, Strix	36	——, Hoplopterus	268
afra, Chalcopeleia	236	albicollis, Archicorax	153
——, Chalcopelia	236	——, Corvultur	153

ALPHABETICAL INDEX

albicollis, Corvus 153
ALCEDO 58
alcinus, Machærhamphus...... 23
——, Macheirhamphus........ 23
ALECTHELIA................ 320
alexandrinus, Ægialites 272
——, Charadrius 272
Alleni, Gallinula 327
——, Porphyrio 327
Alouette à dos roux 113
—— à gros bec.............. 195
—— petite à tête rousse 197
alpina, Saxicola.............. 109
AMADINA 174
ambrosiacus, Cypselus........ 48
amurensis, Erythropus........ 17
——, Falco 17
AMYDRUS 162
ANAS 335, 339
ANASTOMUS 283
anchietæ, Chætops 117
——, Drymœca.............. 117
Anderssoni, Alauda 197
——, Calandrella 197
——, Campephaga 134
——, Enneoctonus 135
——, Machærhamphus 23
——, Megalophonus.......... 197
——, Nectarinia 72
——, Oriolus................ 125
——, Stringonyx 23
Anderssonii, Cinnyris 72
angolensis, Crithagra 183
——, Fringilla 183
——, Poliospiza 183
anguitimens, Eurocephalus.... 140
angulata, Gallinula 321
ANSER 334
Ant-eating Wheatear, Southern 110
ANTHOBAPHES.............. 74
ANTHOSCOPUS 79
ANTHROPOIDES 278
ANTHUS.................... 113
apiaster, Merops 60
apus, Cypselus 47
aquaticus, Rallus 316
AQUILA 5
ARCHICORAX 153
arcuatus, Passer 185
ARDEA..................279, 284
ARDEIRALLA................ 291
ARDEOLA 288
ARDETTA 291

arenaria, Calidris 311
argala, Ciconia 282
armatus, Charadrius.......... 268
——, Hoplopterus............ 267
——, Pluvianus 267
arquata, Numenius 299
arquatus, Numenius 299
arundinacea, Calamodyta 99
asiaticus, Charadrius......xxx, 271
——, Eudromias 271
ASTER 26
astrild, Estrelda 178
——, Estrilda 178
——, Fringilla 178
——, Habropyga 178
ASTUR 26
ater, Circus 33
——, Melænornis 128
——, Milvus................ 21
aterrimus, Irrisor 68
——, Scopteltus.............. 68
ATHENE.................... 37
Atmorei, Saxicola............ 110
Atmorii, Saxicola............ 110
atricapilla, Ardea 292
——, Butorides.............. 292
atricollis, Ardea 284
atrococcineus, Laniarius 144
——, Lanius 144
——, Malaconotus 144
atrovarius, Caprimulgus 45
aurautia, Alario 175
aurata, Juida................ 160
aurautus, Chalcites............ 228
——, Oriolus................ 124
auricularis, Otogyps.......... 2
auritus, Nettapus 336
——, Podiceps.............. 346
australis, Juida.............. 158
——, Lamprotornis 158
——, Pyrrhulauda 191
——, Struthio 251
——, Sula 365
——, Treron 230
autumnalis, Himantopus 315
Avocet, European............ 314
avocetta, Recurvirostra 314
Ayresii, Spizaëtus............ 7
AYTHIA 342

Babbler, Dark-faced.......... 123
——, Hartlaub's 124
——, Jardine's.............. 123

OF GENERA AND SPECIES. 375

	Page		Page
Babbler, Southern Black-and-White	121	Bishop bird, Yellow-Fink	170
Bacbakiri	147	bispecularis, Spreo	160
bacbakiri, Telophonus	147	Bittern, European Little	292
backbakiri, Lanius	147	Black Hawk, African	29
badius, Nisus	30	—— Swift, South-African	47
bæticata, Calamodyta	99	—— Tern, White-winged	363
——, Calamoherpe	99	Black-and-White Babbler, Southern	121
——, Sylvia	99	Black-and-White Tit, Southern	81
bæticula, Calamodyta	99	Black-backed Gull, South-African	357
Bailloni, Ortygometra	317		
Baillonii, Crex	317	Blossom-pecker, Andersson's	80
——, Zaporna	317	——, Dwarf	79
Bairdi, Tringa	308	Bocagei, Cinnyricinclus	156
Bairdii, Actodromas	308	Bonapartei, Actodromas	309
——, Tringa	308	——, Tringa	308
balænarum, Sternula	362	Bonellii, Aquila	7
BALEARICA	279	Bouvreuil du Cap de Bonne Espérance	175
barbatus, Cypselus	47		
Barbican, Black-throated	217	Boysii, Sterna	361
Barred-tail Owl, African	38	BRACHYPTERUS	93
Bateleur	10	brachyrhynchos, Herodias	289
Bateleur Eagle, Rufous-backed	10	brachyura, Drymoica	97
Becquefleur	79	——, Sylvietta	77
Bee-eater, Carmine-throated	62	BRADORNIS	128
——, Blue-cheeked	61	BRADYORNIS	128
——, European	60	brevicaudata, Camaroptera	94
——, Rufous-winged	62	brevipes, Monticola	116
——, Swallow-tailed	63	Brubru	139
belisarius, Falco	6	brubru, Lanius	139
bellicosa, Aquila	8	——, Nilaus	139
bellicosus, Pseudaëtus	8	Brucei, Chrysopicus	221
——, Spizaëtus	8	——, Ipagrus	222
benghala, Estrelda	179	Brucii, Campethera	222
Benghala Finch, Southern	179	brunnea, Nyroca	342
Bennetti, Campethera	222	Brunoir	119
——, Chrysoptilus	222	BUBALORNIS	165
——, Picus	222	BUBO	40
Bergii, Sterna	360	BUBULCUS	288
BESSONORNIS	118	bubulcus, Ardea	288
biarmicus, Dendropicus	219	BUCEROS	206
——, Falco	13	BUCORVUS	205
bicincta, Ceryle	59	BUDYTES	112
bicinctus, Cursorius	261	BUGERANUS	278
——, Pterocles	241	Bulbul, Angola	120
bicolor, Cossypha	119	——, Brunoir	119
——, Crateropus	121	Bullfinch, Black-throated	182
——, Juida	161	——, Decken's	183
——, Spreo	161	——, Golden-rumped	182
bifasciata, Cinnyris	70	Bunting, Lark-like	189
——, Nectarinia	70	——, Southern Yellow-bellied	188
Bishop bird, Red-Fink	172	BUPHAGA	163
——, Taha	171	Burchelli, Lamprotornis	158

	Page		Page
Burchellii, Centropus	224	candidus, Himantopus	315
——, Lamprotornis	158	canescens, Glottis	301
Bush-chirper, Brown-throated	98	——, Totanus	301
——, Yellow-bellied	97	canorus, Cuculus	227
Bush-creeper, Coriphée	93	cantiaca, Sterna	361
Bush-Shrike, Rufous-winged	149	cantianus, Ægialites	272
——, Three-streaked	151	——, Charadrius	272
Bustard, Black-and-white-winged	260	canutus, Calidris	306
——, Cape Knorhaan	260	——, Tringa	306
——, Kori	258	capensis, Ægithalus	79
——, Rufous-crested	259	——, Anas	339
——, Rüppell's	259	——, Athene	38
——, White-eared	260	——, Aythia	242
BUTEO	11	——, Colius	202, 203
BUTORIDES	292	——, Columba	235
butyracea, Crithagra	183	——, Corvus	155
Buzzard, Desert	12	——, Daption	353
——, Jackal	11	——, Drymoica	82
		——, Ephialtes	38
		——, Euplectes	170
		——, Graculus	368
Cabanisi, Hyphantornis	169	——, Hirundo	51
caffer, Amydrus	162	——, Mareca	339
——, Anthus	113	——, Motacilla	111
——, Coccystes	226	——, Nilaus	139
——, Nabouroupus	162	——, Œdicnemus	266
——, Oxylophus	225	——, Œna	235
caffra, Bessonornis	118	——, Otus	43
——, Cossypha	118	——, Paroides	79
caffrensis, Hagedashia	298	——, Parra	330
CALAMODUS	100	——, Pelecanus	368
CALAMODYTA	94, 99	——, Phasmaptynx	43
CALAMOHERPE	99	——, Phyllastrephus	120
CALANDRELLA	198	——, Ploceus	170
Calao caronculé	205	——, Pycnonotus	120
—— couronné	208	——, Querquedula	339
—— nasique	206	——, Rhynchæa	313
—— Toc	211	——, Rhynchaspis	341
CALENDULA	195	——, Scops	38
CALIDRIS	306, 311	——, Spatula	341
calidris, Arenaria	311	——, Sula	365
——, Totanus	300	——, Tænioglaux	38
——, Tringa	311	——, Zosterops	75
calva, Phalacrotreron	230	capitalis, Hyphantornis	169
calvus, Geronticus	297	capricorni, Campethera	221
CAMAROPTERA	94	——, Ipagrus	221
Camaroptera, Olivaceous	94	CAPRIMULGUS	44
cambayensis, Columba	232	CARBO	370
camelus, Struthio	251	carbo, Graculus	367
CAMPEPHAGA	133	——, Phalacrocorax	367
campestris, Anthus	114	cardinalis, Dendropicus	210
CAMPETHERA	221	carmelita, Estrelda	181
Canard du Maragnan	338	Caroli, Ægithalus	80

OF GENERA AND SPECIES. 377

	Page		Page
Caroli, Anthoscopus	80	cinctus, Cursorius	262
carunculata, Gracula	162	——, Hemerodromus	262
——, Grus	278	cineraceus, Circus	33
carunculatus, Bugeranus	278	cinerarius, Circus	33
——, Dilophus	162	cinerascens, Circus	33
caspia, Sterna	359	——, Parus	81
caspius, Charadrius	xxx, 271	cinerea, Alauda	197
castanonotus, Hypotriorchis	19	——, Ardea	284
Caterpillar-eater, Black	133	——, Limosa	304
——, Pectoral	134	——, Squatarola	270
CATHARTES	2	——, Terekia	304
caudata, Coracias	53	cinereus, Megalophonus	197
CEBLEPYRIS	134	——, Puffinus	350
Cenchris, Falco	17	cinnamomeus, Anthus	114
——, Tinnunculus	17	CINNYRICINCLUS	156
CENTROPUS	224	CINNYRIS	69
CERTHILAUDA	201	CIRCAËTUS	10
Cervicalis, Falco	13	CIRCUS	32
CERYLE	59	CIRRHOCEPHALUS	358
CHÆTOPS	117	cirtensis, Falco	12
Chætops, Damara	117	cissoides, Lanius	138
CHALCITES	228	——, Urolestes	138
CHALCOMITRA	73	CISTICOLA	88
CHALCOPELEIA	236	Cisticola, Ground	88
CHALCOPELIA	236	Citrin	84
chalcoptera, Ibis	298	clamosus, Cuculus	226
chalcopterus, Cursorius	263	CLANGULA	343
chalcospilos, Chalcopelia	236	COCCYSTES	225
chalybea, Cinnyris	69	COCCYZUS	225
——, Nectarinia	69	cærulescens, Rallus	316
CHARADRIUS	267, 315	cæruleus, Elanus	20
Chassefiente	5	Colei, Eupodotis	260
Chat-Thrush, Jan-frédric	118	Coliou à dos blanc	202
——, Vociferous	119	——, Quiriwa	203
chelicutensis, Halcyon	57	COLIUS	202
CHENALOPEX	335	collaris, Fiscus	136
CHETTUSIA	267, 268	——, Lanius	136
Chevêchette perlée	37	——, Strepsilas	276
CHICQUERA	14	COLLURIO	134
chiniana, Drymœca	86	collurio, Enneoctonus	135
——, Drymoica	86, 87	——, Enneoctornis	135
CHIZÆRHIS	204	——, Lanius	135
CHLOROPHONEUS	148	COLUMBA	231
chloropsis, Crithagra	183	Coly, Quiriwa	203
chloropus, Gallinula	323	——, White-backed	202
CHRYSOCOCCYX	228	comata, Ardea	288
CHRYSOPICUS	221	——, Ardeola	288
CHRYSOPTILUS	222	Common Tern, British	361
chrysopyga, Crithagra	182	communis, Coturnix	248
chrysura, Campethera	222	concolor, Chizærhis	204
CICONIA	280	——, Schizorhis	204
ciconia, Ardea	280	congensis, Plotus	367
CINCLUS	276	conirostris, Alauda	192

ALPHABETICAL INDEX

	Page
Cooperi, Actodromas	309
Coot, Rufous-knobbed	327
COPROTHERES	357
CORACIAS	53
CORAPHITES	190
Corbivau	153
——, Southern	153
CORETHRURA	320
coriphæus, Brachypterus	93
Coriphée	93
Cormorant, Cape	368
——, Common	367
——, Long-tailed	370
——, Wahlberg's	369
Corneille à scapulaire blanc	154
—— du Cap	155
cornuta, Numida	238
Coromandelianus, Nettapus	337
coronata, Chettusia	268
coronatus, Buceros	208
——, Graculus	370
——, Hoplopterus	268
——, Spizaëtus	9
CORVULTUR	153
CORVUS	153
coryphæus, Thamnobia	93
CORYTHORNIS	60
COSMETORNIS	45
COSSYPHA	118
COTURNIX	248
COTYLE	52
Coucal, Houhou	224
Coucou criard	226
—— didric	228
—— edolio (femelle)	225
—— —— (mâle)	226
—— —— (variété)	225
—— de Klaas	229
—— vulgaire d'Afrique	228
Courly à tête nue	297
Courser, Heuglin's	262
——, Senegal	261
——, South-African Double-collared	261
——, Violet-winged	263
Crake, Baillon's	317
——, Black	321
——, Olive-margined	318
——, Temminck's	320
——, Southern Crowned	279
——, Stanley	278
——, Wattled	278
crassirostris, Alauda	195

	Page
crassirostris, Calendula	195
CRATEROPUS	121
Crested Cuckoo, Great	225
Crested Grebe, Great	345
CREX	317
Crimson-breasted Shrike, Southern	144
CRINIGER	121
Criniger, Yellow-bellied	121
cristata, Alcedo	60
——, Corythornis	60
——, Eupodotis	258
——, Fulica	327
——, Muscipeta	130
——, Tchitrea	130
——, Terpsiphone	130
cristatella, Upupa	64
cristatus, Podiceps	345
CRITHAGRA	182
crocopygia, Poliospiza	184
Crombec	77
——, South-African	77
Crow, Scapulary	154
Crowned Crane, Southern	279
crumeniferus, Leptoptilos	282
——, Leptoptilus	282
Cubla	146
cubla, Dryoscopus	146
——, Laniarius	146
Cuckoo, Black-and-White	225
——, Didric	228
——, Edolio	226
——, European	227
——, Great Crested	225
——, Klaas's	229
——, Levaillant's	225
——, Lineated	228
——, Noisy	226
——, Senegal Spur-heeled	224
cucullata, Hirundo	50
cucullatus, Lanius	149
——, Pomatorhynchus	149
CUCULUS	226
cupreus, Chrysococcyx	228
curlew, Common	299
CURRUCA	100
cursoria, Saxicola	109
CURSORIUS	261
cyaneus, Circus	33
cyanogastra, Estrelda	179
——, Loxia	179
cyanoleuca, Halcyon	56
cyanomelas, Irrisor	67

OF GENERA AND SPECIES. 379

	Page
cyanomelas, Rhinopomastus	67
cyanostigma, Alcedo	60
——, Corythornis	60
CYPSELUS	46
dactylisonans, Coturnix	248
damarensis, Caprimulgus	44
——, Charadrius	xxx, 271
——, Dryodromas	95
——, Eremomela	95
——, Eudromias	xxx
——, Halcyon	57
——, Streptopelia	233
——, Turtur	233
DAPTION	353
Darter, African	367
Delegorguei, Coturnix	249
demersa, Spheniscus	348
demersus, Spheniscus	348
DENDROBATES	219
DENDROCYGNA	338
DENDROPICUS	219
desertorum, Buteo	12
DICÆUM	77
DICROCERCUS	63
DICRURUS	125
diffusus, Passer	187
DILOPHUS	162
dimidiata, Alecthelia	320
——, Corethrura	320
——, Gallinula	320
——, Hirundo	50
DIOMEDEA	355
divaricatus, Dicrurus	125
Diving Duck, Maccoa	344
——, Yellow-throated	343
dominicanus, Larus	357
dorsalis, Tringa	309
Double-collared Courser, South-African	261
Double-collared Sun-bird, Greater	70
—— ——, Lesser	69
Dougalli, Sterna	361
Dove, Black-throated	235
——, Damara	233
——, Emerald-spotted	236
——, Red-eyed	234
——, Senegal	232
Drongeur	125
Drongo, Musical	125
DRYMŒCA	82, 97, 116
DRYMOICA	82
Drymoica, Allied	83

	Page
Drymoica, Andersson's	87
——, Cape	82
——, Kurichane	86
——, Levaillant's	87
——, Pectoral	84
——, Rufous-cheeked	85
——, Smith's	86
——, Tawny-headed	87
DRYODROMAS	95
Dryodrome, Black-breasted	96
——, Damara	95
DRYOSCOPUS	145, 146
Duck, Maccoa Diving	344
——, White-spotted	341
——, Widow Tree-	338
——, Yellow-billed	342
——, Yellow-throated Diving	343
Dwarf Falcon, African	19
Dwarf Goose, African	336
Dwarf Heron, Black-headed	292
——, Sturm's	291
Eagle, African Sea-	9
——, Black-breasted Harrier	10
——, Booted	7
——, Martial Hawk	8
——, Rufous-backed Bateleur	10
——, Spotted-breasted Hawk	7
——, Tawny	6
——, Verreaux's	5
Eagle-Owl, Spotted	42
——, Verreaux's	41
ecaudatus, Helotarsus	10
Echenilleur jaune	133
—— noir	133
Ecorcheur	135
edolius, Oxylophus	226
Egret, European Greater	289
——, European Lesser	290
——, Short-billed	289
EGRETTA	289
egretta, Ardea	289
egyptica, Anas	335
ELANUS	20
elegans, Fringilla	176
——, Tockus	209
EMBERIZA	181
Engoule-vent à collier	45
ENNEOCTONUS	135
ENNEOCTORNIS	135
EPHIALTES	38
ephippiorhyncha, Ciconia	281
EPHIPPIORHYNCHUS	281

ALPHABETICAL INDEX

	Page
epirhinus, Buceros	206
EREMOMELA	95, 97
ERISMATURA	344
ERITHACUS	104
erythræus, Phœnicopterus	331
erythrocephala, Amadina	174
——, Loxia	174
erythrochlamys, Alauda	194
erythromelas, Colius	203
erythromelon, Colius	203
erythronota, Alauda	114
——, Estrelda	178
erythronotus, Anthus	114
erythrophrys, Turtur	234
erythropterus, Lanius	149
——, Merops	62
——, Pomatorhynchus	149
——, Telephonus	149, 151
——, Telophonus	149
ERYTHROPUS	15
erythropus, Colius	202
——, Porphyrio	325
ERYTHROPYGIA	92, 103
erythrorhyncha, Anas	339
——, Nettion	339
——, Pœcilonetta	339
——, Pœcilonitta	339
——, Vidua	181
erythrorhynchos, Irrisor	65
erythrorhynchus, Bubalornis	165
——, Buceros	211
——, Irrisor	65
——, Textor	165
——, Tockus	211
erythrorhyncus, Irrisor	65
ESTRELDA	176, 178, 181
ESTRILDA	178
EUDROMIAS	271
eulophus, Vultur	4
EUPLECTES	167, 170
EUPODOTIS	258
EUROCEPHALUS	140
excubitor, Lanius	135
explorator, Petrociucla	116
exulans, Diomedea	355
FALCO	6, 12
Falcon, African Dwarf	19
——, Rufous-necked	14
——, South-African Lanneroid	13
——, South-African Peregrinoid	12
familiaris, Aëdon	103
——, Saxicola	103

	Page
famosa, Nectarinia	68
fasciata, Aquila	7
fasciolata, Aëdon	90
——, Drymœca	90
——, Drymoica	90
Faucon chanteur	26
FICEDULA	100
Finch, Amadavat	176
——, Berg Canarie	175
——, Black-cheeked	178
——, Grenadier	180
——, Latham's	173
——, Red-headed	174
——, Scutellated	177
——, Southern Benghala	179
——, Southern Red-faced	176
——, Ultramarine	175
——, Waxbill	178
Finch Lark, Dark-naped	191
——, Grey-backed	190
——, Smith's	189
Fiscal	136
FISCUS	136
Flamingo, Greater South-African	331
——, Lesser South-African	333
flammea, Strix	36
flava, Budytes	112
——, Motacilla	112
——, Zosterops	76
flavicans, Drymœca	84
——, Drymoica	84
——, Sylvia	84
flavida, Drymœca	96
——, Dryodromas	96
——, Eremomela	96
flavigaster, Emberiza	
flavigula, Xanthodira	185
flavirostra, Galliuula	321
flavirostris, Anas	342
——, Buceros	210
——, Egretta	289
——, Rhynchops	365
——, Tockus	210
flaviventris, Criniger	121
——, Emberiza	188
——, Eremomela	97
——, Fringillaria	188
——, Sylvia	97
flavus, Budytes	112
fluviatilis, Sterna	361
Flycatcher, Black	128
——, European Spotted	129
——, Mariqua	128

OF GENERA AND SPECIES. 381

	Page
Flycatcher, Pririt	131
—, Tchitrec	130
—, Wahlberg's	131
—, White-tailed	132
formicivora, Myrmecocichla	110
—, Saxicola	110
Forskahli, Milvus	22
Forskali, Milvus	22
Francolin, Coqui	246
—, Crimson-throated	244
—, Orange-River	245
—, Pileated	247
—, Red-billed	247
—, Swainson's	244
FRANCOLINUS	244
FRINGILLA	173, 175, 180, 182
FRINGILLARIA	188
FULICA	325, 327
FULIGULA	342
fuligula, Cotyle	52
fulvipennis, Juida	162
—, Lamprotornis	162
fulviscapus, Dendropicus	220
—, Picus	220
fulvus, Gyps	5
—, Vultur	5
fusca, Cinnyris	71
—, Nectarinia	71
fuscescens, Dendrobates	220
Fuscous Wheatear, Great	107
fuscus, Larus	357
Gabar	28
gabar, Accipiter	28
—, Melierax	28
—, Nisus	28
galbula, Oriolus	124
GALLINAGO	312
GALLINULA	320
Galtoni, Aëdon	103
—, Erythropygia	103
gambensis, Anser	334
—, Plectopterus	334
Gannet, South-African	365
Garden-Warbler, British	100
gariepensis, Francolinus	245
—, Scleroptera	245
garrula, Certhilauda	201
—, Coracias	56
garrulus, Coracias	56
garzetta, Ardea	290
—, Herodias	290
GERONTICUS	297

	Page
gigantea, Alcedo	59
—, Ossifraga	354
—, Procellaria	354
glandarius, Coccystes	225
—, Oxylophus	225
GLAREOLA	264
glareola, Totanus	302
Glossy Starling, Burchell's	158
— —, Caffre	162
— —, Meve's	159
— —, Nabirop	160
— —, Spreo	161
— —, Verreaux's	156
GLOTTIS	301
glottis, Totanus	301
Goatsucker, Freckled	45
—, Pectoral	45
—, Rufous-cheeked	44
—, Standard-wing	45
goliat, Ardea	285
goliath, Ardea	285
Goose, African Dwarf	336
—, Egyptian	335
—, Knob-billed	335
—, Western Spur-wing	334
Gouldii, Anthus	114
GRACULUS	367
granatina, Estrelda	180
—, Fringilla	180
granatinus, Uræginthus	180
Grande Veuve d'Angola	181
Grayi, Alauda	193
—, Chætops	117
Greater Egret, European	289
Grebe, British Little	347
—, Eared	346
—, Great Crested	345
Greenshank	301
Grenouillard	34
Grey Shrike, European Lesser	134
Grey-headed Gull, African	358
— — Sparrow, Southern	187
Grey-winged Kestrel, Western	17
Griffon-Vulture, Rüppell's	5
—, South-African	5
Grignet	77
—, Layard's	78
—, Rufous-vented	77
griseola, Muscicapa	129
griseus, Nycticorax	293
grisola, Butalis	129
—, Muscicapa	129
Grivron	116

	Page		Page
Grosbeak, Damara Yellow-rumped	184	Hawk, Minulle Sparrow-	31
——, Southern Yellow-throated	185	——, One-streaked	26
——, Streaky-headed	183	——, Rufous-bellied Sparrow-	32
Grosbec du Coromandel	170	——, Rüppell's Many-zoned	27
—— tacheté du Cap	170	——, Tachiro Sparrow-	29
Grouse, Double-banded Sand-	241	Hawk Eagle, Martial	8
——, Namaqua Sand-	242	——, Spotted-breasted	7
——, Variegated Sand-	242	Helmet-Shrike, Retz's	142
GRUS	278	——, Smith's	141
Guépier à collier bleu	62	HELOTARSUS	10
—— minulle	62	helvetica, Squatarola	270
Guinea, Columba	231	HEMERODROMUS	262
Guineæ, Columba	231	Hemipode, Kurichane	249
Guinea-fowl, Cape	238	HEMIPODIUS	249
guineensis, Columba	231	HERODIAS	289
gularis, Crithagra	182	Heron, Black-headed Dwarf	292
——, Cuculus	228	——, Black-throated	284
——, Fringilla	183	——, Buff-backed	288
——, Linaria	183	——, Common	284
——, Poliospiza	183	——, European Night	293
Gull, African Grey-headed	358	——, Goliath	285
——, South-African Black-backed	357	——, Purple	286
gutturalis, Ardea	291	——, Rufous-bellied	287
——, Chalcomitra	73	——, Squacco	288
——, Cypselus	46	——, Sturm's Dwarf	291
——, Laniarius	147	Heywoodi, Charadrius	272
——, Nectarinia	73	hiaticula, Ægialites	276
——, Telophorus	147	——, Charadrius	276
GYPS	5	HIERAËTUS	7
		HIMANTOPUS	315
		himantopus, Charadrius	315
		Hirondelle fauve	52
HABROPYGA	178	—— rouselline	50
HÆMATOPUS	277	hirundinaceus, Dicrocercus	63
Hagedash, Geronticus	298	——, Merops	63
——, Hagedashia	298	hirundineus, Melittophagus	63
HAGEDASHIA	298	——, Merops	63
HALCYON	56	HIRUNDO	49
HALIAËTUS	9	hirundo, Sterna	361
Harrier, Fuliginous	33	histrionica, Coturnix	249
——, Levaillant's	34	Hobby, British	14
——, Montagu's	33	——, Eastern Red-footed	17
——, Swainson's	32	——, Western Red-footed	15
Harrier-Eagle, Black-breasted	10	Honey-Guide, Little	223
Hartlaubi, Crateropus	124	Hoopoe, South-African	64
——, Picus	219	HOPLOPTERUS	267
Hartlaubii, Crithagra	182	Hornbill, Crowned	208
——, Dendropicus	219	——, Ground	205
hastatus, Buceros	207	——, Monteiro's	208
Hawk, African Black	29	——, Red-billed	210
——, Chanting	26	——, Tock	206
——, Gabar	28	——, Yellow-billed	210
——, Many-banded Sparrow-	30	hortensis, Curruca	100

OF GENERA AND SPECIES. 383

	Page
hortensis, Sylvia	100
hottentota, Nettion	340
——, Querquedula	340
hottentotta, Querquedula	340
——, Saxicola	108
HUHUA	41
hybrida, Pelodes	363
HYDROCHELIDON	363
HYPHANTORNIS	169
HYPOCHERA	175
HYPOLAIS	100
hypolais, Ficedula	100
——, Phyllopseuste	100
——, Sylvia	100
hypoleucus, Actitis	303
——, Totanus	303
——, Tringoides	303
HYPOTRIORCHIS	14
IBIS	297
——, African Roseate	296
——, Bald	297
——, Bubulcus	288
——, Hagedash	298
——, Sacred	297
——, Tantalus	296
—— blanc d'Egypte	296
impetuani, Fringillaria	180
Indicateur, petit	223
INDICATOR	223
infuscata, Saxicola	107
intermedia, Ardea	289
——, Herodias	289
interpres, Cinclus	276
——, Strepsilas	276
——, Tringa	276
IPAGRUS	221
IRRISOR	65
Irrisor, Black	68
——, Namaqua	67
——, Red-billed	65
Isabelle	99
Ixos	120
Jaboteur	120
——, Cape	120
Jacana, Greater African	328
——, Lesser African	330
jackal, Buteo	11
jacobinus, Coccystes	225
——, Oxylophus	225
Jan frédric	118
Jardinii, Crateropus	123

	Page
JUIDA	156
KAUPIFALCO	26
Kestrel, British	18
——, Greater South-African	19
——, Lesser South-African	18
——, Western Grey-winged	17
Kingfisher, African Malachite-crested	60
——, African White-headed	57
——, Angola	56
——, Black-and-White	59
——, Great African	59
——, Half-collared	58
——, Striped	57
Kite, Black	21
——, Sonnini's	20
——, Yellow-billed	22
Kittlitzi, Ægialites	274
——, Charadrius	274
Klaasi, Chrysoccyx	229
Klaasii, Chalcites	229
Knot	306
Kolbii, Gyps	5
Kori, Eupodotis	258
——, Otis	258
lactea, Strix	41
lacteus, Bubo	41
lagepa, Alauda	200
——, Megalophonus	200
LAGONOSTICTA	176
LAIMODON	217
Lamelligerus, Anastomus	283
LAMPROCOLIUS	160
LAMPROTORNIS	156
LANIARIUS	144
LANICTERUS	134
LANIOTURDUS	132
LANIUS	134
Lanneroid Falcon, South-African	13
Lark, Andersson's	198
——, Dark-lined	199
——, Dark-naped Finch	191
——, Garrulous	201
——, Grey-backed Finch	190
——, Gray's	193
——, Grey-collared	201
——, Lagepa	200
——, Objimbinque	198
——, Pink-billed	192
——, Rufous-mantled	194
——, Sabota	195

ALPHABETICAL INDEX

Lark, Smith's Finch 189
——, South-African Rufous-capped 197
——, South-African Thick-billed 195
LARUS 357
lateralis, Chettusia 267
——, Lobivanellus 267
——, Vanellus 267
Lathamii, Ploceus............. 173
latipennis, Scops 38
Lavandière brune............. 111
Layardi, Parisoma 78
lentiginosus, Caprimulgus 45
lepidus, Euplectes............ 167
——, Philetærus 167
LEPTOPTILOS 282
LEPTOPTILUS 282
lepurana, Hemipodius 249
——, Turnix 249
Lesser Egret, European 290
—— Waterhen, South-African 321
LESTRIS 357
letsitsirupa, Turdus 114
leucogaster, Juida............. 156
——, Lamprotornis 156
——, Pholidauges............. 156
leucomelæna, Saxicola........ 109
leucomelas, Laimodon 217
——, Pogonias 217
——, Pogonorhynchus........ 217
——, Tockus................. 211
leuconota, Clangula 343
——, Thalassornis............. 343
leucopareia, Sterna 363
leucophrys, Ædon 92
——, Aëdon 92
——, Anthus................. 114
leucopolius, Charadrius 272
leucoptera, Ardeola 288
——, Sterna 363
leucopterus, Parus 82
LEUCORODIA 295
leucorodia, Platalea 295
leucostigma, Anas 341
leucotis, Bubo 40
——, Ephialtes............... 40
——, Pyrrhulauda 189
——, Scops 40
——, Strix................... 40
Levaillantii, Drymœca........ 87
——, Drymoica............... 87
——, Plotus 367
——, Psittacus 213

libonyana, Turdus............ 115
libonyanus, Turdus 115
licua, Athene 37
LIMNOCORAX................. 321
LIMOSA 304
LINARIA..................... 183
lineatus, Cuculus 228
Linnets from Angola 182
lipiniana, Estrelda 178
Little Bittern, European 292
Little Grebe, British.......... 347
littoralis, Charadrius 272
LOBIVANELLUS............... 267
longicauda, Carbo........... 370
Long-tailed Shrike, South-African 138
Love-bird, Rosy-necked 216
LOXIA 172, 174
lucidus, Graculus 368, 369
Ludwigii, Eupodotis.......... 260
lugubris, Muscicapa 128

maccoa, Erismatura 344
——, Oxyura................. 344
MACHÆRHAMPHUS 23
MACHEIRHAMPHUS 23
MACHETES................... 304
macrodactyla, Gallinago 312
macrurus, Colius 203
maculata, Actodromas 309
——, Tringa 308
maculosa, Drymœca... 82
——, Drymoica............... 82
maculosus, Bubo 42
——, Œdicnemus............. 266
madagascariensis, Nettapus.... 336
mahali, Plocepasser 166
major, Dryoscopus........... 145
——, Gallinago............... 312
——, Laniarius............... 145
——, Puffinus 350
——, Scolopax 312
——, Tchagra 145
Malachite-crested Kingfisher, African 60
MALACONOTUS 144, 148
Manchot du Cap de Bonne Espérance 348
Mangeur de Serpents 34
Many-zoned Hawk, Rüppell's.. 27
marabou, Ciconia 282
Marabou Stork, African 282
MARECA..................... 339

OF GENERA AND SPECIES. 385

	Page
marginalis, Ortygometra	318
——, Porzana	318
marginatus, Ægialites	272
——, Charadrius	272
MARIPOSA	179
mariquensis, Bradornis	128
——, Bradyornis	128
——, Hyphantornis	169
——, Ploceus	169
——, Saxicola	129
Martin, Fawn-breasted	52
Martinet à gorge blanche	46
maurus, Circus	xxv, 33
——, Falco	33
maxima, Ceryle	59
MEGALOPHONUS	195
MELÆNORNIS	148
MELÆNORNIS	128
melanocephala, Ardea	284
melanoleucus, Buceros	208
——, Lanius	138
——, Oxylophus	225
——, Tockus	208
——, Urolestes	138
MELANOPEPLA	128
melanophrys, Diomedea	356
melanops, Crateropus	123
melanoptera, Glareola	264
melanopterus, Elanus	20
——, Falco	20
——, Himantopus	315
melanorhyncha, Herodias	289
melanota, Tringa	309
melanotus, Sarcidiornis	335
——, Sarkidiornis	335
melba, Cypselus	46
——, Estrelda	176
——, Pytelia	176
MELIERAX	26
MELITTOPHAGUS	62
Melodious Warbler, European	100
MEROPS	60
Mésange grisette	81
—— noire	81
Mevesi, Lamprotornis	159
Mevesii, Juida	159
Meyeri, Pionias	213
——, Pœocephalus	213
——, Poicephalus	213
——, Psittacus	213
migrans, Milvus	21
MILVUS	21
minima, Estrelda	176

	Page
minima, Fringilla	176
——, Lagonosticta	176
——, Pytelia	176
minor, Collurio	134
——, Falco	12
——, Indicator	223
——, Lanius	134
——, Onocrotalus	371
——, Pelecanus	371
——, Phœnicopterus	333
——, Podiceps	347
——, Upupa	64
Minulle	31
minullus, Accipiter	31
minuta, Ardea	292
——, Ardetta	292
——, Ortygometra	317
——, Pelidna	310
——, Tringa	310
minutella, Actodromas	309
minutus, Ægithalus	80
——, Anthoscopus	79
MIRAFRA	195, 198
mitrata, Numida	238
mitratus, Pelecanus	371
modesta, Saxicola	104
mogilnik, Aquila	6
Moineau du Cap de Bonne Espérance	185
monachus, Cathartes	2
monogrammicus, Aster	26
——, Astur	26
——, Kaupifalco	26
——, Melierax	26
Montagnard	18
Monteiri, Hirundo	49
——, Toccus	208
——, Tockus	208
MONTICOLA	116
Moquini, Hæmatopus	277
MOTACILLA	111
motitensis, Passer	186
MUSCICAPA	128
MUSCIPETA	130
musicus, Dicrurus	125
——, Falco	26
——, Melierax	26
MYCTERIA	281
MYRMECOCICHLA	110
Nabirop	160
Nabouroup	162
nabouroup, Spreo	162

2 c

	Page		Page
NABOUROUPUS	162	nudicollis, Francolinus	244
nævia, Alauda	193, 199	——, Pternises	244, 254
——, Coracias	54	NUMENIUS	299, 306
nævioides, Aquila	6	NUMIDA	238
nævius, Megalophonus	199	NYCTICORAX	293
namaqua, Pteroclurus	242	nycticorax, Ardea	293
namaquus, Dendrobates	219	NYROCA	342
——, Dendropicus	219	nyroca, Fuligula	342
——, Picus	219		
——, Thripias	219		
nasutus, Buceros	206	obscura, Sylvia	100
——, Tockus	206	occipitalis, Vultur	4
natalensis, Merops	62	oceanica, Oceanites	351
——, Nectarinia	73	——, Procellaria	351
NECTARINIA	68	OCEANITES	351
neglecta, Motacilla	112	ocularius, Drymœca	85
neglectus, Graculus	369	——, Drymoica	85
NEOPHRON	1	ŒDICNEMUS	266
NETTAPUS	336	ŒNA	235
NETTION	339	olivacea, Calamodyta	94
niger, Accipiter	29	——, Camaroptera	94
——, Campephaga	134	olivaceus, Turdus	116
——, Gallinula	321	Ombrette du Sénégal	294
——, Hæmatopus	277	ONOCROTALUS	371
——, Lanicterus	134	onocrotalus, Pelecanus	371
——, Limnocorax	321	——, Pelicanus	371
——, Melierax	29	Oricou	2
——, Nisus	29	orientalis, Rhynchops	365
——, Parus	81	Oriole, Andersson's	124
——, Promerops	67	——, Golden	124
——, Rallus	321	ORIOLUS	124
——, Sparvius	29	orix, Loxia	172
——, Textor	165	Ortleppi, Drymœca	84
Night Heron, European	293	ORTYGOMETRA	317, 321
nigra, Campephaga	133	oryx, Euplectes	122
——, Hydrochelidon	363	——, Ploceus	172
——, Limnocorax	321	——, Pyromelana	172
——, Ortygometra	321	oscitans, Anastomus	283
nigricans, Pycnonotus	119	OSSIFRAGA	354
nigricollis, Podiceps	346	Ostrich	251
——, Proctopus	346	——, South-African	251
nigripennis, Gallinago	312	OTIS	258
nigrogularis, Dendrobates	221	OTOGYPS	2
NILAUS	139	OTUS	43
NISUS	27	Owl, African Barred-tail	38
nitens, Amadina	175	——, African Pearl-spotted	37
——, Fringilla	175	——, African Short-eared	43
——, Hypochera	175	——, Cape Scops	38
nivifrons, Charadrius	272	——, South-African Screech	36
Nordmanni, Glareola	264	——, Spotted Eagle	42
notatus, Oriolus	124	——, Tufted	40
nubicoides, Merops	62	——, Verreaux's Eagle	41
nuchalis, Coracias	54	——, White-faced Scops	40

OF GENERA AND SPECIES.

	Page
Oxpecker, Greater	163
Oxylophus	225
OXYURA	344
Oye de la côte de Coromandel	335
Oyster-catcher, Moquin's	277
pæna, Aëdon	92
pæna, Ædon	92
paena, Erythropygia	92
pæna, Erythropygia	92
Painted Snipe, African	313
pallida, Drymoica	84
pallidus, Charadrius	272
pammelæna, Bradyornis	128
——, Melanopepla	128
paradisea, Grus	278
——, Scops	278
——, Steganura	181
——, Tetrapteryx	278
——, Vidua	181
Parasite	22
parasiticus, Milvus	22
——, Stercorarius	357
PARISOMA	77
PAROIDES	79
PARRA	328
Parrot, Levaillant's	213
——, Meyer's	213
——, Rüppell's	214
PARUS	81
parvus, Cypselus	48
——, Phœnicopterus	333
PASSER	185
pastor, Pratincola	102
pavonia, Ardea	279
Pearl-spotted Owl, African	37
pectoralis, Campephaga	134
——, Caprimulgus	45
——, Ceblepyris	134
——, Circaëtus	10
——, Drymoica	84
——, Erythopygia	92
——, Tringa	308
pecuarius, Charadrius	274
——, Ægialites	274
pelagica, Procellaria	351
——, Thalassidroma	351
PELECANUS	368, 371
Pelican, Rüppell's	371
PELICANUS	371
PELIDNA	306
PELODES	363
Penguin, South-African	348

	Page
pennata, Aquila	7
——, Hieraëtus	7
pennatus, Hieraëtus	7
percnopterus, Neophron	1
Peregrinoid Falcon, South-African	12
PERISTERA	236
perlata, Athene	37
——, Strix	37
Pern, Andersson's	23
Perroquet à franges souçi	213
Petersi, Podica	345
Petrel, Cape	353
——, Dove-coloured	352
——, Giant	353
——, Stormy	351
——, Wilson's	351
PETROCINCLA	116
petronella, Petronia	185
PETRONIA	185
phæocephala, Xema	358
PHÆOCEPHALUS	214
phæocephalus, Larus	358
phæonotus, Stictœnas	231
phæopus, Numenius	300
PHALACROCORAX	367
PHALACROTRERON	230
PHASMAPTYNX	63
PHILETÆRUS	167
PHILOMACHUS	304
phœnicoptera, Juida	160
PHŒNICOPTERUS	331
phœnicopterus, Lamprocolius	160
phœnicurus, Bessonornis	118
PHOLIDAUGES	156
phragmitis, Salicaria	100
PHYLLASTREPHUS	120
PHYLLOPNEUSTE	101
PHYLLOPSEUSTE	100
Pic à double moustache	219
Pic petit à baguettes d'or	220
picturata, Otis	259
PICUS	219, 222
Pie-grièche rousse à tête noire du Sénégal	149
Pigeon à front nu	230
——, Bald-fronted	230
——, Cape	353
——, Roussard	231
pileata, Saxicola	108
——, Scleroptera	247
pileatus, Francolinus	247
——, Neophron	2

ALPHABETICAL INDEX

pileatus, Scleroptera 247
pilosa, Coracias.............. 54
pilosus, Coracias 54
PIONIAS...................... 213
Pipit, Caffre 113
——, Cinnamon-backed 113
——, Raalten's.............. 113
——, Tawny................. 114
Pique-bœuf 163
Plantain-eater, Whole-coloured 204
PLATALEA.................. 295
PLATYSTEIRA 131
PLATYSTIRA................ 131
PLECTROPTERUS............ 334
PLOCEPASSER 166
PLOCEUS 169
PLOTUS 367
Plover, Blacksmith 267
——, Caspian 271
——, Grey.................. 270
——, Heywood's 272
——, Kentish 272
——, Kittlitz's 274
——, Ringed................ 276
——, South-African Wattled.. 267
——, Treble-collared 274
——, Wreathed 268
plumiferus, Herodias 289
PLUVIANUS 267
Pluvier du Cap de Bonne Espérance 268
Pochard, South-African 342
PODICA 345
PODICEPS 345
podiceps, Ardetta............ 293
PŒCILONETTA 339
PŒCILONITTA 339
pœna, Aëdon................ 92
poënsis, Strix 36
PŒOCEPHALUS.............. 213
POGONIAS 217
POGONORHYNCHUS 217
POIOCEPHALUS.............. 213
poiocephalus, Cirrhocephalus .. 358
——, Larus 358
POLIHIERAX................ 19
POLIOHIERAX 19
POLIOSPIZA 183
polyzonoides, Accipiter 29, 30
——, Falco 30
polyzonus, Melierax........ 27, 30
——, Nisus 27
pomarinus, Coprotheres 357

pomarinus, Lestris 357
——, Stercorarius........... 357
POMATORHYNCHUS 149
PORPHYRIO 325
Porphyrio, Allen's 327
——, Green-backed 325
porphyrio, Fulica............ 325
Porte-lambeau 162
PORZANA 318
PRATINCOLA................ 102
pratincola, Glareola 265
Pratincole, Collared 265
——, Nordmann's........... 264
principalis, Vidua............ 181
PRIONOPS 141
Pririt...................... 131
pririt, Platysteira............ 131
——, Platystira.............. 131
PROCELLARIA 351
PROCTOPUS 346
PROMEROPS 67
Promerops moqueur.......... 65
PSEUDAËTUS................ 7
PSEUDOPRION 352
PSITTACULA................ 216
PSITTACUS 213
PTERNISES 244
PTEROAËTUS................ 5
PTEROCLES 241
PTEROCLURUS 242
PUFFINUS 350
pugnax, Machetes............ 304
——, Philomachus 304
pumila, Gallinula 321
purpurea, Ardea 286
pusillus, Melittophagus 62
PYCNONOTUS................ 119
pycnopygius, Chætops........ 117
——, Sphenœacus............ 117
pygmæa, Ortygometra........ 317
——, Zapornia 317
pygmæus, Numenius 306
PYROMELANA 172
pyrrhonota, Alauda 113
pyrrhonotus, Anthus 113
PYRRHULAUDA.............. 189
PYTELIA 176

Quail, European 248
——, Harlequin 249
QUELEA.................... 173
quelea, Fringilla............ 173
QUERQUEDULA.............. 339

OF GENERA AND SPECIES. 389

	Page
Raalteni, Anthus	113
Raaltenii, Anthus	113
Raddei, Falco	17
Rail, Caffer	316
RALLUS	316, 321
Ramier roussard	231
ranivorus, Circus	xxv, 34
Réclameur	119
Recurvirostra	314
Red-faced Finch, Southern	176
Red-footed Hobby, Eastern	17
———, Western	15
Redshank, Common	300
Reed-Warbler, Isabelle	99
———, Thrush-like	99
regia, Emberiza	181
———, Vidua	181
regulorum, Balearica	279
———, Grus	279
religiosa, Ibis	297
reptilivorus, Secretarius	34
———, Serpentarius	34
Retzii, Prionops	142
RHINOPOMASTUS	67
RHYNCHÆA	313
RHYNCHASPIS	341
RHYNCHOPS	365
Richardsonii, Lestris	357
risorius ?, Columba	233
robustus, Poicephalus	213
———, Psittacus	213
Rock-Thrush, Short-footed	116
Roller, European	56
———, Green-necked	53
———, White-naped	54
Rollier varié d'Afrique	54
Rook, South-African	155
Roseate Ibis, African	296
roseicollis, Agapornis	216
———, Psittacula	216
Rounoir	11
rubicola, Saxicola	102
rudis, Ceryle	59
Rueppelli, Psittacus	214
rufescens, Anthus	114
———, Calamodyta	99
———, Calamoherpe	99
———, Dicæum	77
———, Sylvietta	77
Ruff	304
ruficapilla, Drymoica	86
ruficollis, Chicquera	14
———, Falco	14

	Page
ruficollis, Hypotriorchis	14
ruficrista, Eupodotis	259
———, Otis	259
rufigena, Caprimulgus	44, 45
rufilata, Drymoica	87
rufipes, Falco	15
rufirostris, Buceros	211
rufiventris, Accipiter	32
———, Ardea	287
———, Parisoma	78
Rufous-capped Lark, South-African	197
rufula, Certhilauda	201
rufulus, Anthus	201
rupestris, Hirundo	52
rupicoloides, Falco	19
———, Tinnunculus	19
rupicolus, Tinnunculus	18
Rüppelli, Gyps	5
———, Otis	259
———, Pœocephalus	214
———, Poicephalus	214
———, Psittacus	214
Rüppellii, Eupodotis	259
———, Gyps	5
———, Otis	259
———, Phæocephalus	214
———, Poicphalus	214
russatus, Bubulcus	288
rustica, Hirundo	50
sabota, Alauda	195
———, Megalophonus	195
———, Mirafra	195
SAGITTARIUS	34
SALICARIA	99
salicaria, Hypolais	100
Sanderling	311
Sand-Grouse, Double-banded	241
———, Namaqua	242
———, Variegated	242
Sandpiper, Baird's	308
———, Common	303
———, Curlew	306
———, Marsh	302
———, Terek	304
———, Wood	302
sanguinirostris, Ploceus	173
———, Quelea	173
Sarcelle de Madagascar	336
SARCIDIORNIS	335
SARKIDIORNIS	335
Savigni, Merops	61

ALPHABETICAL INDEX

	Page
SAXICOLA	102
scapularis, Hirundo	52
scapulatus, Corvus	154
Schinzii, Tringa	308
SCHIZORHIS	204
Schlegelii, Erithacus	104
——, Saxicola	104
schœnobænus, Calamodus	100
——, Calamodyta	100
Scissor-bill, African	365
SCLEROPTERA	245
SCOPS	38, 278
Scops Owl, Cape	38
——, White-faced	40
scops, Strix	38
SCOPTELUS	68
SCOPUS	294
Screech-Owl, South-African	36
Sea-cow Bird	275
Sea-Eagle, African	9
SECRETARIUS	34
secretarius, Sagittarius	34
Secretary bird	34
Sedge-Warbler, British	100
segetum, Corvus	155
SEMEIOPHORUS	45
semicærulea, Halcyon	57
semitorquata, Alcedo	58
——, Certhilauda	201
——, Columba	234
——, Streptopelia	234
semitorquatus, Falco	19
——, Polihierax	19
——, Poliohierax	19
——, Turtur	233, 234
senegalensis, Centropus	224
——, Colius	203
——, Cursorius	26
——, Ephialtes	38
——, Ephippiorhynchus	281
——, Halcyon	56
——, Irrisor	65
——, Mycteria	281
——, Nectarinia	73
——, Œdicnemus	266
——, Scops	38
——, Strix	37
——, Tachydromus	261
——, Telephonus	149
——, Turtur	232
——, Zosterops	76
senegalla, Aquila	6
senegalus, Telophonus	149

	Page
senex, Crateropus	124
SERPENTARIUS	34
serratus, Coccystes	226
——, Oxylophus	226
Shearwater, Greater	350
Short-eared Owl, African	43
Shoveller, South-African	341
Shrike, Bacbakiri	147
——, Brubru	139
——, Chapman's	146
——, Coronetted	138
——, Cubla	146
——, European Lesser Grey	134
——, Fiscal	136
——, Greater White-breasted	145
——, Red-backed	135
——, Retz's Helmet	142
——, Rufous-winged Bush	149
——, Smith's Helmet	141
——, South-African Long-tailed	138
——, Southern Crimson-breasted	144
——, Southern White-headed	140
——, Three-streaked Bush	151
——, Yellow-browed	148
similis, Chlorophoneus	148
——, Laniarius	148
——, Melaconotus	148
simplex, Passer	187
Skua, Pomarine	357
——, Richardson's	357
smaragdineus, Chrysococcyx	229
smaragnotus, Porphyrio	325
Smithi, Ægithalus	79
——, Coraphites	190
——, Pyrrhulauda	189
Smithii, Ægithalus	79
——, Drymœca	86
——, Drymoica	86
——, Pyrrhulauda	190
——, Rhinopomastus	67
Snail-eater, African	283
Snipe, African Painted	313
——, Solitary	312
socius, Philetærus	167
sordidus, Anthus	114
South-African Flamingo, Greater	331
——, —— Lesser	331
South-African Sparrow, Greater	186
Sparrow, Cape	185
——, Greater South-African	186
——, Southern Grey-headed	187
Sparrow-Hawk, Many-banded	30
——, Minulle	31

OF GENERA AND SPECIES. 391

	Page
Sparrow-Hawk, Rufous-bellied	32
——, Tachiro	29
sparsa, Anas	341
SPARVIUS	29
SPATULA	341
speciosus, Hoplopterus	267
sperata, Saxicola	103
SPHENISCUS	348
SPHENŒACUS	117
SPHENORRHYNCHUS	280
spilogaster, Aquila	7
——, Pseudaëtus	7
——, Spizaëtus	7
spilonotus, Hyphantornis	169
——, Ploceus	169
spinicauda, Lestris	357
——, Stercorarius	357
SPIZAËTOS	7
SPIZAËTUS	7
splendidus, Chrysococcyx	229
spleniata, Alauda	197
Spoonbill, Slender-billed	295
SPOROPIPES	177
Spotted Flycatcher, European	129
SPREO	159, 161
Spur-heeled Cuckoo, Senegal	224
Spur-wing Goose, Western	334
squamifrons, Amadina	177
——, Estrelda	177
——, Philetærus	177
——, Sporopipes	177
SQUATAROLA	270
squatarola, Tringa	270
stagnatilis, Totanus	302
stanleyanus, Anthropoides	278
Starling, Burchell's Glossy	158
——, Caffre Glossy	162
——, Meves's Glossy	159
——, Nabirop Glossy	160
——, Spreo Glossy	161
——, Verreaux's Glossy	156
——, Wattled	162
STEGANURA	181
STERCORARIUS	357
STERNA	359
STERNULA	362
STICTŒNAS	231
sticturus, Laniarius	146
——, Tchagra	146
Stilt, European	315
Stint, Little	310
Stonechat, South-African	102
Stork, Abdim's	280

	Page
Stork, African Marabou	282
——, Saddle-billed	381
——, White	280
strepitans, Platysteira	131
——, Turdus	114
STREPSILAS	276
STREPTOPELIA	233
striaticeps, Fringilla	183
stricklandii, Saxicola	105
striolata, Halcyon	57
STRIX	36
STRUTHIO	251
Sturmii, Ardea	291
——, Ardeiralla	291
——, Ardetta	291
subarcuata, Tringa	306
subarquata, Pelidna	306
——, Tringa	306
subbuteo, Falco	14
——, Hypotriorchis	14
subcæruleum, Parisoma	77
subcinnamomea, Aëdon	90
——, Drymœca	90
——, Drymoica	90
subcoronatus, Fiscus	138
——, Lanius	138
subflava, Drymoica	84
subruficapilla, Drymœca	87
——, Drymoica	86, 87
subtorquata, Scleroptera	246
subtorquatus, Francolinus	246
Sucrier malachitte	68
—— namaquois	71
—— orangé	74
—— protée	73
SULA	365
sulfureipectus, Laniarius	148
Sun-bird, Andersson's	72
——, Bifasciated	70
——, Greater Double-collared	70
——, Lesser Double-collared	69
——, Malachite	68
——, Orange-breasted	74
——, Proteus	73
——, White-vented	71
superciliosus, Merops	61
Swainsoni, Circus	32
——, Francolinus	244
——, Passer	187
——, Pternises	244
Swainsonii, Circus	32
——, Francolinus	244
——, Halcyon	57

ALPHABETICAL INDEX

	Page
Swainsonii, Pternistes	244
Swallow, Chimney	50
——, Monteiro's	49
——, Pearly-breasted	52
——, Rousseline	50
Swift, Little African	48
——, South-African Black	47
——, White-throated	46
SYLVIA	84, 97, 100
SYLVIETTA	77
tachardus, Falco	12
tachiro, Accipiter	29
——, Falco	29
——, Nisus	29
TACHYDROMUS	261
tachypetes, Pterocles	242
TÆNIOGLAUX	38
taha, Euplectes	171
——, Ploceus	171
talacoma, Prinops	141
talatala, Cinnyris	72
——, Nectarinia	72
TANTALUS	296
TCHAGRA	145
tchagra, Lanius	149
Tcheric	75
TCHITREA	130
Tchitrec	130
Teal, Hottentot	340
——, Red-billed	339
TELEPHONUS	149
TELOPHONUS	147, 149
TELOPHORUS	147
tenuirostris, Leucorodia	295
——, Platalea	295
terek, Limosa	304
TEREKIA	304
Tern, British Common	361
——, Caspian	359
——, Damara	362
——, Sandwich	361
——, Swift	360
——, Whiskered	362
——, White-winged Black	363
TERPSIPHONE	130
terrestris, Cisticola	88
——, Drymoica	88
TETRAPTERYX	278
TEXTOR	165
THALASSIDROMA	351
THALASSORNIS	343
THAMNOBIA	93

	Page
Thick-billed Lark, South-African	195
Thicknee, Spotted	266
——, Vermiculated	266
thoracicus, Circaëtus	10
THRIPIAS	219
Thrush, Ground-scraper	114
——, Jan-frédric Chat	118
——, Kurichane	115
——, Olivaceous	116
——, Short-footed Rock	116
——, Vociferous Chat	119
tinnunculoides, Falco	17
TINNUNCULUS	17
tinnunculus, Falco	18
Tit, Grisette	81
——, Southern Black-and-White	81
TMETOCEROS	206
TOCCUS	208
TOCKUS	206
torquata, Glareola	265
——, Platysteira	132
——, Pratincola	102
torquatus, Lanioturdus	132
TOTANUS	300
Tourterelle à cravatte noire	235
—— blonde à collier	233
—— émeraudine	236
—— maillée	232
Traquet familier	103
—— fourmilier	110
—— imitateur	108
—— pâtre	102
Tree-Duck, Widow	338
TRERON	230
TRICHOPHORUS	121
tricollaris, Ægialites	274
——, Charadrius	274
tricolor, Ixos	120
——, Pycnonotus	120
trigonigera, Columba	231
TRINGA	270, 276, 306
TRINGOIDES	303
trivirgatus, Lanius	151
——, Pomatorhynchus	151
——, Telephonus	151
——, Telophonus	151
trochilus, Phyllopneuste	101
——, Phyllopseuste	101
——, Sylvia	101
turdoides, Salicaria	99
TURDUS	114
TURNIX	249
Turnstone, Common	276

OF GENERA AND SPECIES.

	Page
TURTUR	232
turtur, Procellaria	352
——, Pseudoprion	352
ultramarina, Hypochera	175
umbretta, Scopus	294
Umbrette, Tufted	294
unidentatus, Laimodon	217
UPUPA	64
URÆGINTHUS	180
UROLESTES	138
usticollis, Eremomela	98
vaillantii, Coccyzus	225
——, Motacilla	112
VANELLUS	267
varia, Squatarola	270
variegatus, Pterocles	242
variolosa, Ipagrus	222
variolosus, Ipagrus	222
varius, Charadrius	270
velatus, Hyphantornis	169
——, Ploceus	169
velox, Sterna	360
veredus, Eudromias	xxx
vermiculatus, Œdicnemus	266
Verreauxi, Bubo	41
——, Cinnyricinclus	156
——, Huhua	41
——, Pholidauges	156
Verreauxii, Aquila	5
——, Nyctaëtus	41
verticalis, Pyrrhulauda	190
vespertinus, Erythropus	15
——, Falco	15
——, ——, var. amurensis	17
vetula, Larus	357
vexillarius, Cosmetornis	45
——, Semeiophorus	45
VIDUA	181
viduata, Dendrocygna	338
vinaceus, Turtur	233, 234
violacea, Anthobaphes	74
——, Nectarinia	74
viridis, Tchitrea	130
——, Terpsiphone	130
vocifer, Haliaëtus	9
vociferans, Bessonornis	119
VULTUR	4
Vulture, Egyptian	1
——, Pileated	2
——, Rüppell's Griffon	5
——, Sociable	2

	Page
Vulture, South-African Griffon	5
——, White-headed	4
vulturina, Aquila	5
——, Pteroaëtus	5
Wagtail, Blue-headed Yellow	112
——, Cape	111
——, Levaillant's	112
Warbler, British Garden	100
——, British Sedge	100
——, British Willow	101
——, European Melodious	100
——, Isabelle Reed	99
——, Thrush-like Reed	99
Waterhen, British	323
——, South-African Lesser	321
Wattled Plover, South-African	267
Weaver-bird, Buffalo	165
——, Damara	169
——, Social	167
——, Spotted-backed	169
——, White-browed	166
Whale-bird	352
Wheatear, Atmore's	110
——, Familiar	103
——, Great Fuscous	107
——, Imitative	108
——, Mountain	109
——, Schlegel's	104
——, Southern Ant-eating	110
——, Strickland's	105
Whimbrel, British	300
White-breasted Shrike, Greater	145
White-eye, Cape	75
——, Yellow	76
White-headed Kingfisher, African	57
White-headed Shrike, Southern	140
Widow bird, Dominican	181
——, Paradise	181
——, Shaft-tailed	181
Wigeon, Cape	339
Willow-Warbler, British	101
Wilsoni, Thalassidroma	351
Wilsonii, Thalassidroma	351
Woodpecker, Bearded	219
——, Bennett's	222
——, Bruce's	221
——, Capricorn	221
——, Cardinal	220
——, Hartlaub's	219
XANTHODIRA	185

2 D

INDEX OF GENERA AND SPECIES.

	Page
xanthorhyncha, Anas	342
XEMA	358
Yellow-bellied Bunting, Southern	188
Yellow-rumped Grosbeak, Damara	184
Yellow-throated Grosbeak, Southern	185
Yellow Wagtail, Blue-headed	112
ZAPORNA	317
ZAPORNIA	317
zonurus, Spizaëtos	7
ZOSTEROPS	75

THE END.

PRINTED BY TAYLOR AND FRANCIS, RED LION COURT, FLEET STREET.

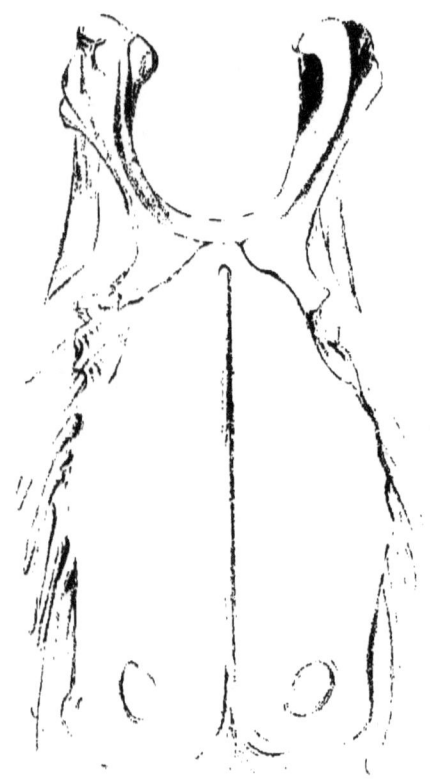

Front View of Sternum of Machæramphus Anderssoni, ♀ nat size.
Drawn from Nature by C. J. Andersson.
vide. p 23

Trachea, Larynx and Tongue of Machæramphus Anderssoni, ♀ nat size
Drawn from Nature by C J Andersson.
vide. p. 23.

SCALE OF SIX INCHES DIVIDED INTO LINES.

www.ingramcontent.com/pod-product-compliance
Lightning Source LLC
Chambersburg PA
CBHW022138300426
44115CB00006B/243